Luminescent Materials in Display and Biomedical Applications

Editors

Vikas Dubey
Department of Physics
Bhilai Institute of Technology
Raipur, India

Sudipta Som
Department of Chemical Engineering
National Taiwan University
Taipei, Taiwan

Vijay Kumar
Department of Physics,
National Institute of Technology
Srinagar, J&K, India

CRC Press
Taylor & Francis Group
Boca Raton London New York

CRC Press is an imprint of the
Taylor & Francis Group, an **informa** business

A SCIENCE PUBLISHERS BOOK

Cover credit:
Figure used on the cover has been taken from Chapter 9. Reproduced by kind courtesy of the authors.

CRC Press
Taylor & Francis Group
6000 Broken Sound Parkway NW, Suite 300
Boca Raton, FL 33487-2742

First issued in paperback 2022

© 2021 by Taylor & Francis Group, LLC
CRC Press is an imprint of Taylor & Francis Group, an Informa business

No claim to original U.S. Government works

Version Date: 20200714

ISBN-13: 978-0-367-54117-0 (pbk)
ISBN-13: 978-0-367-11212-7 (hbk)

DOI: 10.1201/9780429025334

**Visit the Taylor & Francis Web site at
http://www.taylorandfrancis.com**

**and the CRC Press Web site at
http://www.routledge.com**

Library of Congress Cataloging-in-Publication Data

Names: Dubey, Vikas, 1985- editor. | Som, Sudipta, 1987- editor. | Kumar, Vijay, 1983- editor.
Title: Luminescent materials in display and biomedical applications / editors, Vikas Dubey, Department of Physics, Bhilai Institute of Technology, Raipur, India, Sudipta Som, Department of Chemical Engineering, National Taiwan University, Taipei, Taiwan, Vijay Kumar, Department of Physics, Chandigarh University, Punjab, India.
Description: First. | Boca Raton : CRC Press, Taylor & Francis Group, [2021] | Includes bibliographical references and index.
Identifiers: LCCN 2020028380 | ISBN 9780367112127 (hardcover)
Subjects: LCSH: Luminescence. | Inorganic compounds. | Biomedical engineering.
Classification: LCC QC476.4 .L8717 2021 | DDC 620.1/1295--dc23
LC record available at https://lccn.loc.gov/2020028380

Preface

Human civilization relies mainly on mankind's ability to harvest light and therefore the subject on luminescence requires the foremost scientific protagonists. Starting with knowledge of fire, the lighting industry deals with incandescence lamps, fluorescent lamps to light emitting diodes for generating well-illuminated outdoor and indoor spaces. Moreover, the biomedical industry also greatly relies on various powders which can illuminate light after excitation via several excitation sources. The key role for this development is the knowledge of luminescence. The book contains fourteen chapters and each chapter provides the basic knowledge and deep ongoing research in recent days. The present book assesses the delineation of luminescence, the types and related appliances including nanomaterials, organic and inorganic light emitting diode materials and devices in thin film and powder, which will be an asset to beginners. The book will provide the execution of luminescence into practical devices via the knowledge of physics along with materials features. However, our main aim with this book is to inspire and develop young minds towards luminescence research for future prospects. We are grateful to authors of the chapters for their exceptional assistance for the completion of the book. We would also like to acknowledge the admirable control at CRC for their proficiency and commitment to publish the book. We hope the present book will help the readers to comprehend luminescence at their end.

Vikas Dubey
Sudipta Som
Vijay Kumar

Acknowledgement

Editor Dr. Vikas Dubey is very grateful to TEQIP-III CSVTU Bhilai for funding organization of International Conference on Recent Trends in Renewable Energy and Sustainable Development 2020.

Editor (Dr. Vikas Dubey) is very grateful to TEQIP-III CSVTU Bhilai for funding through Collaborative Research Project CSVTU Project Ref. No. CSVTU/CRP/TEQIP-III/16 date 17/08/2019. Also very grateful to Principal Dr. T. Ramarao Bhilai Institute of Technology for kind support. Grateful to Dr. Jagjeet Kaur Saluja Ma'am for lab support and facilities and guidance. Very grateful to Faculty and staff BIT Raipur for moral support. Last but not least my wife Dr. Neha Dubey for her continuous support for editing of books and My daughter Ms. Vaishnavi Dubey.

Contents

Tb³⁺ Activated High-Color-Rendering Green Light Yttrium Oxyorthosilicates Phosphors for Display Device Application

Dhananjay K. Deshmukh[1]*, Jayant Nirmalkar[2] and Mozammel Haque[3]

[1] Chubu University, Kasugai 487-8501, Japan
[2] Korea Research Institute of Standards and Science, Daejeon 305-340, Republic of Korea
[3] Nanjing University of Information Science and Technology, Nanjing 210044, China

Introduction

The industrial phosphors of the last few years are either silicate-based or sulfide-based phosphors. Silicate-based phosphors have attracted much consideration due to their low price, high luminescence efficiency, excellent stability and inertness against chemical and thermal degradations [Naik et al. 2014]. The Rare-Earth (RE) oxyorthosilicates (RE_2SiO_5) that have been doped with Eu^{3+}, Ce^{3+} and Tb^{3+} are well-known luminescent materials. Among the RE_2SiO_5 categories, Y_2SiO_5 (YSO), which was doped with rare-earth ions were studied extensively regarding display applications. Yttrium silicate (Y_2SiO_5) is an important luminescent host material for various RE activators. Y_2SiO_5 has been synthesized by a chemical method since 1964. As a dopant the trivalent terbium ion (Tb^{3+}) displays efficient radiative recombination channels that are mainly observed in the green region of the visible spectral range. It is an important part of the Red (R), Green (G) and Blue (B) display (RGB) panels since the maximum of the human-eye sensibility falls in this region [Song et al. 2014; Penilla et al. 2013; Li et al. 2011; Xie et al. 2010].

The present chapter reports on the synthesis and characterization as well as effect of a variable Tb^{3+} concentration (0.1–2 mol%) in a photoluminescence (PL) and thermoluminescent (TL) analysis of the Y_2SiO_5 phosphor. The

---

*Corresponding author: deshmukh@isc.chubu.ac.jp

samples show well-resolved spectra in the green region for the various concentrations. Our results describe high color purity of Tb^{3+} activated YSO phosphor that can be useful for Light Emitting Diode (LED) application for intense green emission. We also report on the TL glow curve analysis of $YSO:Tb^{3+}$.

Experimental

The Y_2SiO_5 phosphor that had been doped with Tb^{3+} ions with a variable Tb^{3+} molar concentration (0.1–2 mol%) was prepared using a modified solid-state-reaction method. We used the precursors Y_2O_3, SiO_2, Tb_4O_7 and H_3BO_3 for the synthesis of $Y_2SiO_5:Tb^{3+}$. The composition of each chemical was weighed according to a proper stoichiometric ratio and then they were mixed thoroughly using an agate mortar and pestle for 45 minutes. The grinded samples were placed in an alumina crucible and then combusted in a muffle furnace under 1000 °C for 1 hour for calcinations followed by firing at 1250 °C for 3 hours for sintering. Every heating stage was followed by an intermediate grinding. Lastly, the samples were cooled slowly to room temperature in the furnace and grinded into powder for the subsequent characterization.

The observation of morphology of particle was conducted using the JSM-7600F Field Emission Gun Scanning Electron Microscope device of Japan Electron Optics Laboratory. The PL emission and excitation spectra were recorded at room temperature using the Shimadzu spectrofluorophotometer model RF-5301PC. We used Xenon Lamp as a excitation source. The obtained phosphor under the TL examination was subjected to ultraviolet (UV) radiation using a 254 nm UV source. The TL glow curves were recorded at room temperature using TL-dosimeter (TLD) reader model I1009 supplied by Nucleonix Systems Pvt. Ltd. (Dubey et al. 2014; Kaur et al. 2013).

Results and Discussion

X-ray Analysis

The XRD pattern of the sample is shown in Fig. 1.1 where a monoclinic-body-centered structure with the unit cell dimensions of $a = 12.50$, $b = 6.728$ and $c = 10.421$ are displayed. These values match those of the International Centre for Diffraction Data (JCPDS) card No. 36-1476 (Qi et al. 2003). The crystallite size of the phosphor was calculated using the Scherrer equation (Lee et al. 2008; McMurdie et al. 1986). The calculated crystallite sizes for the different glancing angles are shown in Table 1.1. The average crystallite size of the phosphor was found to be 33 nm.

Morphology of the Phosphor

The SEM images of the synthesized phosphor in different resolutions are presented in Fig. 1.2. The $YSO:Tb^{3+}$ (1.5 mol%) phosphor displays a

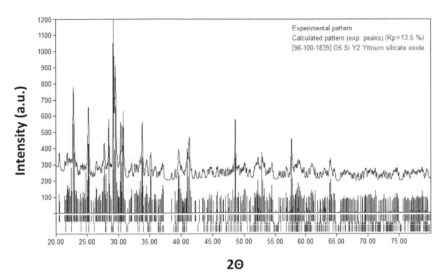

Fig. 1.1. X-ray diffraction (XRD) pattern of the Y_2SiO_5:Tb^{3+} (1.5 mol%) phosphor.

Table 1.1. The structural parameters and crystallite size of the prepared Y_2SiO_5:Tb^{3+} phosphore (1.5 mol%)

2θ	FWHM (Degree)	D-spacing	D Crystallite size (nm)
22.700	0.288	3.914	28
25.062	0.262	3.550	31
28.297	0.275	3.151	29
29.077	0.249	3.068	31
30.237	0.223	2.953	36
33.410	0.183	2.679	45
35.054	0.275	2.557	30
36.908	0.236	2.433	35
40.889	0.236	2.205	35
48.458	0.275	1.877	31
52.136	0.236	1.752	37
52.781	0.288	1.733	30
57.558	0.328	1.600	27
60.851	0.328	1.521	28

sound morphology and particle-size distribution above 200 nm. The high-temperature synthesis method resulted in an appearance comparable to that of the foam-like structure with an irregular shape.

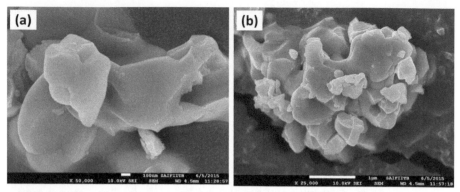

Fig. 1.2. Scanning electron microscope (SEM) images of prepared Y_2SiO_5:Tb^{3+} phosphor at 10 k (a) and 50 k (b) resolutions.

Photoluminescence (PL) Observation

The PL excitation spectra of the YSO:Tb^{3+} (1.5 mol%) phosphor under a 543 nm emission wavelength is displayed in Fig. 1.3. The results presented two intense peaks at 243 nm and 273 nm. The peak at 243 nm corresponds to the spin that is allowed by the $4f^8$–$4f^75d$ transition ($\Delta S = 0$), whereas the peak at 273 nm is from the spin-forbidden component of the $4f^8$–$4f^75d$ transition ($\Delta S = 1$) (Han et al. 2004; Lin and Su 1995).

The PL emission spectra for a different Tb^{3+} concentration at a 243 nm excitation wavelength is shown in Fig. 1.4. When the YSO:Tb^{3+} phosphors (0.1–2 mol%) are excited by the spin-allowed $4f^8 \rightarrow 4f^75d$ band at 243 nm the displayed characteristics are the blue and green emission lines of Tb^{3+}

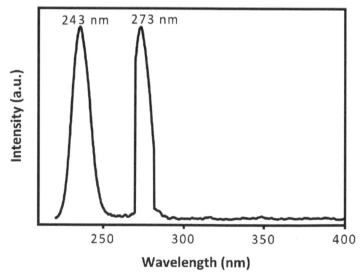

Fig. 1.3. Photoluminescence (PL) excitation spectra of the prepared YSO:Tb^{3+} phosphor at a 543 nm emission wavelength.

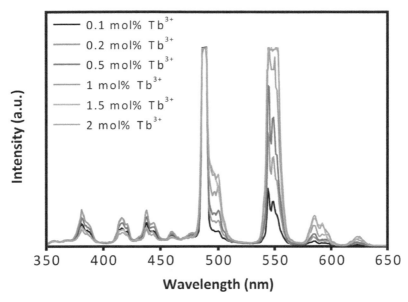

Fig. 1.4. Photoluminescence (PL) emission spectra for a different Tb^{3+} concentration at a 243 nm excitation wavelength.

$(^5D_{3,4} \rightarrow {}^7F_J; J = 3, 4, 5, 6)$. It shows the green emission color due to the strongest peak at 549 nm $(^5D_4 \rightarrow {}^7F_5)$ and the blue emission peak at 485 nm $(^5D_4 \rightarrow {}^7F_6)$. The small peaks at 380, 415 and 438 nm are attributed to the $^5D_3 \rightarrow {}^7F_5$ transition whereas the peak at 460 nm is due to the $^5D_3 \rightarrow {}^7F_4$ transition of the Tb^{3+}. The peaks at both 585 and 592 nm correspond to the $^5D_4 \rightarrow {}^7F_4$ transition whereas the peak at 623 nm is revealed due to the $^5D_4 \rightarrow {}^7F_3$ transition of the Tb^{3+}.

The energy-level diagram of the Tb^{3+} ion is shown in Fig. 1.5. The intensity of the magnetic-dipole-allowed transition $^5D_4 \rightarrow {}^7F_{5,6}$ is much stronger than the electric-dipole-allowed transition $^5D_4 \rightarrow {}^7F_{3,4}$ (Shi et al. 2015). The variation in the PL intensity with a variable Tb^{3+} concentration is presented in Fig. 1.6. The intensity of the emission spectra increased up to the 1.5 mol% Tb^{3+}-doped sample and then the intensity decreased due to the concentration quenching. The concentration quenching phenomenon is due to the ion-ion interaction between rare earth activated phosphors.

Commission International de l'Eclairage (CIE) Coordinates

The CIE chromaticity coordinates of the YSO:Tb^{3+} (1.5 mol%) phosphor was calculated from the corresponding emission spectrum and presented in Fig. 1.7. Their corresponding locations have been marked in the green region of Fig. 1.7 with a cross. Our results clearly show that the Tb^{3+}-doped YSO sample can be used for green-light-emission applications such as solid-state lighting and display devices. Its chromaticity coordinates are $x = 0.252$ and $y = 0.494$.

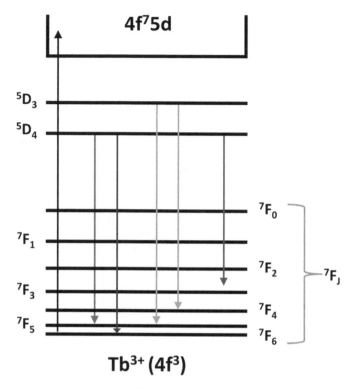

Fig. 1.5. Possible transitions of the Tb^{3+} ion.

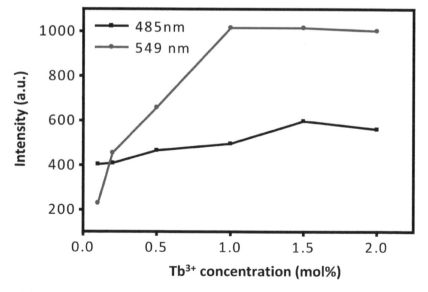

Fig. 1.6. Variation in the PL intensity with a variable Tb^{3+} concentration.

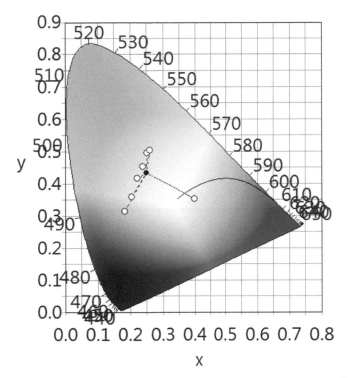

Fig. 1.7. Commission International de l'Eclairage (CIE) chromaticity diagram of the YSO:Tb³⁺ phosphors. X and Y axis are color coordinate axis.

Thermoluminescence (TL) Observation

The effect of the UV dose on the TL intensity for the 1 mol% Tb³⁺-doped Y_2SiO_5 is shown in Fig. 1.8. We found that the TL intensity increased linearly up to the 20 minute UV dose and then decreased for the 25 minute and 30 minute UV dose. It became evident that the TL intensity increased almost linearly with the UV irradiation. Further, there was no appreciable shift in the glow peak position for the higher irradiation doses. We predicted that the release of a greater number of charge carriers would increase the trap density with the increase of the UV exposure, while resulting in an increase of the TL intensity. Nevertheless, the traps would start to destroy the results of the TL-intensity decrease after a specific exposure (Parganiha et al. 2015; Shrivastava et al. 2015; Chen 1969).

The TL glow curve of YSO:Tb³⁺ (1 mol%) for a variable ultraviolet (UV) exposure is presented in Fig. 1.9. Figure 1.10 shows the Computerized Glow Curve Deconvolution (CGCD) curve of the Y_2SiO_5:Tb³⁺ (1 mol%) sample for the 20 minute UV dose at the heating rate of 5 °C s⁻¹. Table 1.2 shows the calculated kinetic parameters using the first-order kinetics for the 20 minute UV exposure for the three deconvoluted peaks.

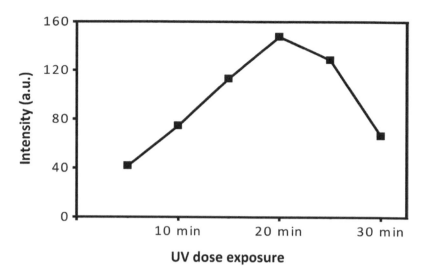

Fig. 1.8. Ultraviolet (UV) dose versus the TL intensity plot for YSO:Tb³⁺ (1 mol%).

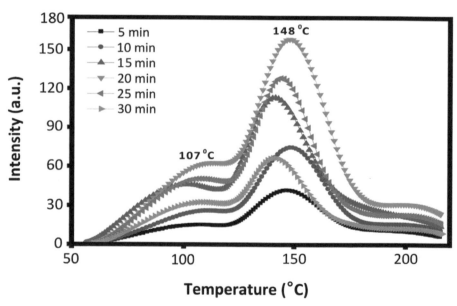

Fig. 1.9. The TL glow curve of YSO:Tb³⁺ (1 mol%) for a variable
ultraviolet (UV) exposure.

    Glowfit is based on the Halperin and Barner equations that describe the flows of the charges between the various energy levels during a trap emptying that is because of thermal heating. The kinetic parameters of the trap levels were determined for each deconvoluted peak using this program (Tiwari et al. 2014; Chung et al. 2005). The theoretical generated glow curves

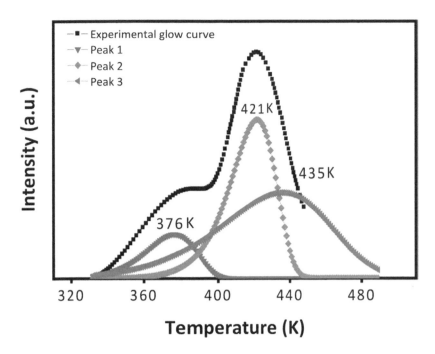

Fig. 1.10. Fitted glow curve of YSO:Tb³⁺ (1 mol%) for a 20 minute ultraviolet (UV) exposure.

were fitted with the experimental glow curves. The quality of the fitting was checked by calculating the Figure Of Merit (FOM) for each fitting. This summation extends across all of the available experiment-data points. The quality of the fitting and the choice of the appropriate number of peaks were refined by repeating the fitting process to obtain the minimum FOM with the minimum number of possible peaks. The fits were considered adequate when the observed FOM values were less than 5%, and most of the actual values were less than 2%. Our results show the FOM is 1.67%, which confirm a very sound agreement between the theoretical generated glow curves and the experimental recorded glow curves.

Table 1.2. Calculation of the kinetic parameters for the deconvoluted glow peaks for a 20 minute UV exposure.

Peak	T_1 (K)	T_m (K)	T_2 (K)	τ	δ	ω	μ (δ/ω)	Activation energy (E)	Frequency factor (s)
Peak 1	358.30	376.60	389.60	18.30	13.00	31.30	0.42	0.91	3×10^8
Peak 2	404.10	421.40	433.50	17.30	12.10	29.40	0.41	1.23	3×10^6
Peak 3	393.90	435.40	467.20	41.50	31.80	73.30	0.43	0.48	1×10^6

Glow Curve Shape Method

The method based on the shape of glow curve proposed by Chen and Mckeever 1997 (Som et al. 2015; Singh et al. 2008; Furetta 2003) was used to verify the trapping parameters calculation.
The following shape parameters were determined.

Total half intensity width $(\omega) = T_2 - T_1$
The high temperature half width $(\delta) = T_2 - T_m$
The low temperature half width $(\tau) = T_m - T_1$

where T_m is the peak temperature and T_1 and T_2 are two temperatures on either side of T_m corresponding to half peak intensity.

Order of Kinetics

The order of kinetics (b) was determined by calculating the symmetry factor (μ) of the glow peak using the known values of the shape parameters

$$\left\{ \mu_g = \frac{T_2 - T_m}{T_2 - T_1} \right\} \tag{1}$$

Activation energy: Activation energy (E) was calculated by using the Chen equations, giving the trap depth in terms of τ, δ, and ω. A general formula for E was given by Chen and Mckeever (1997) as follows:

$$E_\alpha = c_\alpha \left[\frac{kT}{} \right] - b_\alpha \left(kT \right) \tag{2}$$

Conclusion

Y_2SiO_5:Tb^{3+}-doped phosphors were synthesized using a modified solid-state-reaction method. The XRD pattern confirms that the synthesized sample shows a monoclinic-body-centered structure. The average crystallite size was 33 nm. The PL emission was observed in the range of 350–630 nm for the Y_2SiO_5 phosphor doped with the Tb^{3+}. The excitation spectra were found at 243 and 273 nm. In the emission spectra, sharp intense peaks of a high intensity were found at around 485 and 549 nm. The present phosphors can act as a host for the green-light emission in solid-state lighting and display devices. The CIE chromaticity coordinates of the YSO:Tb^{3+}, was $x = 0.252$ and $y = 0.494$, and exhibited the green light. The chromaticity point is in the green region, indicating its high color purity. The TL kinetic parameters were also calculated using the peak-shape method. The results indicate that the TL glow curves of the samples followed the first-order kinetics. The activation energy was found in between 0.48 and 1.23 eV, and the frequency factor was found in between 1×10^6 and 3×10^8. The simple glow-curve structure and the wide linearity range of the samples show that they might be useful in terms of TL dosimetry for short exposure times.

References

Chen, R. 1969. Thermally stimulated current curves with non-constant recombination lifetime, Brit. J. Appl. Phys. 2, 371-375.

Chen, R., McKeever, S.W.S. 1997. Theory of thermoluminescence and related phenomena. World Scientific, Singapore. https://doi.org/10.1142/2781.

Chung, K.S., Choe, H.S., Lee, J.I., Kim, J.L., Chang, S.Y. 2005. A computer program for the deconvolution of the thermoluminescence glow curves. Rad. Prot. Dosim. 115, 345-349.

Dubey, V., Kaur, J., Agrawal, S., Suryanarayana, N.S., Murthy, K.V.R. 2014. Effect of Eu^{3+} concentration on photoluminescence and thermoluminescence behavior of YBO_3:Eu^{3+} phosphor. Superlattices and Microstruct. 67, 156-171.

Furetta, C. 2003. Handbook of Thermoluminescence. World Scientific, Singapore. https://doi.org/10.1142/5167.

Han, X.M., Lin, J., Fu, J., Xing, R.B., Yu, M., Zhou, Y.H., Pang, M.L. 2004. Fabrication, patterning and luminescence properties of X_2-Y_2SiO_5:A (A = Eu^{3+}, Tb^{3+}, Ce^{3+}) phosphor films via sol-gel soft lithography. Solid State Sci. 6(4), 349-355.

Kaur, J., Parganiha, Y., Dubey, V. 2013. Luminescence studies of Eu^{3+}-doped calcium bromofluoride phosphor. Phys. Res. International, https://doi.org/10.1155/2013/494807.

Lee, W.G., Lee, D.H., Kim, Y.K., Kim, J.K., Park, J.W. 2008, Growth and characteristics of Gd_2SiO_5 crystal-doped Ce^{3+}. J. Nucl. Sci. Tech. 5, 572-574.

Li, R., Fu, R., Song, X., He, H., Yu, X., He, B., Shi, Z. 2011. Green emission from Tb-doped $SrSi_2O_2N_2$ phosphors under ultraviolet light irradiation. J. Phys. Chem. Solids 72, 233-235.

Lin, J., Su, Q. 1995. Luminescence and energy transfer of rare-earth-metal ions in $Mg_2Y_8(SiO_4)_6O_2$. J. Mater. Chem. 5(8), 1151-1154.

McMurdie, H.F., Morris, M.C., Evans, E.H., Paretzkin, B., Wong-Ng, W., Hubbard, C.R. 1986. Standard X-ray diffraction powder patterns from the JCPDS research associateship. Powder Diff. 1, 264-275.

Naik, R., Prashantha, S.C., Nagabhushana, H., Sharma, S.C., Nagabhushana, B.M., Nagaswarupa, H.P., Premkumar, H.B. 2014. Low temperature synthesis and photoluminescence properties of red emitting Mg_2SiO_4:Eu^{3+} nanophosphor for near UV light emitting diodes. Sens. Actuat. B: Chemical 195, 140-149.

Parganiha, Y., Kaur, J., Dubey, V., Shrivastava, R. 2015. $YAlO_3$:Ce^{3+} powders: Synthesis, characterization, thermoluminescence and optical studies. Superlattices and Microstruct. 85, 410-417.

Penilla, E.H., Kodera, Y., Garay, J.E. 2013. Blue-green emission in terbium-doped alumina (Tb:Al_2O_3) transparent ceramics. Adv. Funct. Mater. 23, 6036-6043.

Qi, Z., Shi, C., Zhang, G., Han, Z., Hung, H.H. 2003. Preparation and characterization of nanocrystalline Gd_2SiO_5:Ce. Phys. Stat. Sol. (a) 195, 311-316.

Shi, M., Zhang, D., Chang, C. 2015. Tunable emission and concentration quenching of Tb^{3+} in magnesium phosphate lithium. J. Alloy. Compd. 627, 25-30.

Shrivastava, R., Kaur, J., Dubey, V., Jaykumar, B., Loreti, S. 2015. Photoluminescence and thermoluminescence investigation of europium and dysprosium-doped dibarium magnesium silicate phosphor. Spectrosc. Lett. 48(3), 179-183.

Singh, L.R., Ningthoujam, R.S., Sudarsan, V., Srivastava, I., Singh, S.D., Dey, G.K., Kulshreshtha, S.K. 2008. Luminescence study of Eu^{3+} doped Y_2O_3 nanoparticles: Particle size, concentration and core-shell formation effects. Nanotechnol. 19, https://doi.org/10.1088/0957-4484/19/05/055201.

Som, S., Chowdhury, M., Sharma, S.K. 2015. Kinetic parameters of γ-irradiated Y_2O_3 phosphors: Effect of doping/codoping and heating rate. Radiat. Phys. Chem. 110, 51-58.

Song, Y., Liang, S., Li, F., You, C. 2014. Synthesis and photoluminescence properties of $(Y,Gd)SiO_5:Tb^{3+}$ under VUV excitation. Ceramics International, 40, 15985-15990.

Tiwari, N., Kuraria, R.K., Tamrakar, R.K. 2014. Thermoluminescence glow curve for UV induced $ZrO_2:Ti$ phosphor with variable concentration of dopant and various heating rate. J. Radiat. Res. Appl. Sci. 7, 542-549.

Xie, Y., Ma, Z., Liu, L., Su, Y., Zhao, H., Liu, Y., Zhang, Z., Duan, H., Li, J., Xie, E. 2010. Oxygen defects-modulated green photoluminescence of Tb-doped ZrO_2 nanofibers. Appl. Phys. Lett. 97, http://dx.doi.org/10.1063/1.3496471.

Versatile Applications of Rare-Earth Activated Phosphate Phosphors: A Review

Sumedha Tamboli[1], Govind B. Nair[2], S.J. Dhoble[1] and H.C. Swart[2*]

[1] Department of Physics, RTM Nagpur University, Nagpur - 440033, India
[2] Department of Physics, University of the Free State, P.O. Box 339, Bloemfontein 9300, South Africa

Introduction to Phosphate-based Lamp Phosphors

Man has always tried different ways to provide light that both refreshes the mind and emotions while giving a pleasing visualization of the surroundings. Not surprisingly, the sun used to be the ultimate source of heat and light during the stone-age period. Thereafter, advanced homosapiens developed their own ways to create fire as a source of heat and light. With the evolution of mankind, the source of light also reached new milestones. The first breakthrough came in the form of a tungsten filament incandescent lamp that could be powered by electricity. The incandescent bulbs ruled the lighting industry for more than a century until fluorescent lamps entered the scenario with their cool light emission. The efficiency entrusted with less heat emissions led fluorescent lamps to reach a new mark in the lighting industry. Much higher levels of efficiency were achieved with their manifestation in the form of Compact Fluorescent Lamps (CFLs). But the latest entrants in the lighting industry have ruthlessly expelled all the earlier mentioned light sources from the scenario.

More than a couple of decades have passed since the invention of blue LEDs that marked the way to a new revolution in the lighting industry. The emergence of solid state lighting resulted in the hope of developing white emitting LEDs that could be developed either by blending the red, green, blue light from LEDs or by developing a single phase white light emitting phosphor. The quest for energy efficient materials for solid state lighting

*Corresponding author: swarthc@ufs.ac.za

has led to the discovery of numerous compounds with high functionality. Phosphates are well-recognized phosphor materials that offer various advantages over other types of host lattices. Rare-earth activated phosphates are known for their brightness in addition to excellent physical, chemical and thermal stability (Deyneko et al. 2019; Fan et al. 2018; Li et al. 2019; Lin et al. 2009; Xu et al. 2019). An exceptional number of orthophosphate phosphors have shown prodigious potential in this regard. The ability of PO_4^{3-} tetrahedron to bond with other structural units has added more appeal to phosphate phosphors (Ray et al. 2018; Shinde and Dhoble 2014). Besides orthophosphates, the lamp industry has acknowledged the role of pyrophosphates and halophosphates in luminaires, display panels, etc. (Durugkar et al. 2018; Yerpude et al. 2018).

Several orthophosphate phosphors with unmatched efficiencies have been reported over the last couple of decades. The family of $ABPO_4$ (where A and B are monovalent and divalent, respectively) phosphors is the most widely investigated monophosphates for their luminescence properties. These provide ideal charge stabilization with their three rigid dimensional tetrahedral PO_4^{3-} matrix. Almost all the elements in the first and second group of the periodic table have been tried and tested in the $ABPO_4$ family. Although the composition looks similar, the occupancy of A and B atom by different elements can often lead to different crystal structures. This has probably resulted in varying photoluminescence properties in different hosts belonging to this family. $NaMgPO_4$:Eu^{2+} showing excellent red emission under blue excitation with an internal quantum efficiency or more than 80% (Kim et al. 2013). It is quite normal to obtain red emission with Eu^{3+} ions, but in $NaMgPO_4$, the position has something else to deal with. Europium has conveniently chosen to adapt the +2 oxidation state to give out red luminescence. This phosphor gives a broad emission centered at 628 nm when excited by blue light at 450 nm. Another important feature leading to its applicability as red phosphor is the thermal stability that it exhibits during the temperature quenching effect. The PL emission intensity of $NaMgPO_4$:Eu^{2+} phosphor at 150 °C reaches 86% of the value that it shows at 25 °C. Other $ABPO_4$ phosphates have shown the normally seen blue emission when they are activated with Eu^{2+} ions (Lin et al. 2010; Nair et al. 2017; Zhang et al. 2010a, 2010b). Solid solutions of $ABPO_4$ phosphors have also come under widespread investigation and a lot more controlled synthesis techniques were implemented to achieve their composition. One such way is the polymerizable complex method, which was used to synthesize solid solutions of $NaBa_{0.5}Ca_{0.5}PO_4$ and $Na_3Ba_2Ca(PO_4)_3$ (Kim et al. 2015). Doping Eu^{2+} in $NaBaPO_4$ would have yielded a broad band emission centered at 438 nm at an excitation wavelength of 330 nm. On the other hand, Eu^{2+} in $NaBa_{0.5}Ca_{0.5}PO_4$ and $Na_3Ba_2Ca(PO_4)_3$ gave a strong blue emission centered at 460 nm for the same excitation. But this was achieved only after making a compromise with its efficiency. The two solid solutions managed to show quantum efficiency of merely 51% and 7%, respectively, as against the 84% efficiency shown by $NaBaPO_4$:Eu^{2+} phosphor. Figure 2.1 shows

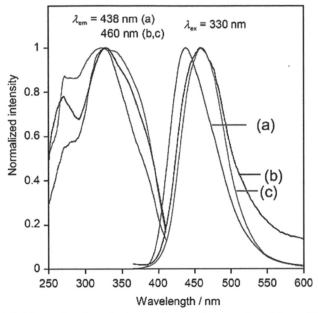

Fig. 2.1. Normalized excitation and emission spectra of 1 mol% Eu^{2+} doped: (a) $NaBaPO_4$, (b) $Na_3Ba_2Ca(PO_4)_3$ and (c) $NaBa_{0.5}Ca_{0.5}PO_4$ (Copyright © 2014, Royal Society of Chemistry, Reproduced with Permission from Ref. (Kim et al. 2015)).

normalized excitation and emission spectra of 1 mol% Eu^{2+} doped $NaBaPO_4$, $Na_3Ba_2Ca(PO_4)_3$ and $NaBa_{0.5}Ca_{0.5}PO_4$ phosphors, respectively.

Eu^{2+} ions are known to produce blue emission in a number of phosphors. There are a vast majority of orthophosphates that emit blue light when suitably doped with Eu^{2+} ions. $Ca_3(PO_4)_2$:Eu^{2+} phosphor has shown a broad band emission peaking at 480 nm under an excitation of 340 nm and 370 nm. This phosphor managed to show a cyan emission rather than the conventional blue emission, although its quantum efficiency seems to be slightly poor (Zhou et al. 2015). Yellow emitting phosphates were achieved in the form of $Ca_{10}Na(PO_4)_7$:Eu^{2+}, $Sr_8MgLa(PO_4)_7$:Eu^{2+} and $Sr_8MgGd(PO_4)_7$:Eu^{2+} phosphors (Huang et al. 2012; Huang and Chen 2011; Yu et al. 2013). The large band gap exhibited by these phosphors is very favorable for luminescence. Additionally, the emission intensity of $Ca_{10}Na(PO_4)_7$:Eu^{2+} phosphors at 200 °C decreases upto 28% of the intensity observed at room temperature. $Ca_9Gd(PO_4)_7$: Eu^{2+}, Mn^{2+} phosphor also showed promising features for a single-phase white light emitting phosphor (Huang et al. 2010). The non-radiative dipole-quadrupole energy transfer mechanism from Eu^{2+} to Mn^{2+} ions is responsible for the color-tunable emission in this host lattice. The color hue can be tuned from blue-green through white to red by varying the Mn^{2+} concentration, but by keeping the concentration of Eu^{2+} constant. $Ca_8MgLu(PO_4)_7$:Eu^{3+} phosphor gives pure red emission under near-UV excitation (Xie et al. 2015). The luminescence intensity of the phosphor goes on increasing with Eu^{2+} concentration and no quenching was observed even at 100% Eu^{2+} concentration. The photoluminescence excitation and emission spectra of $Ca_8MgLu(PO_4)_7:Eu^{3+}$ phosphor are shown in Fig. 2.2. It exhibits an internal quantum efficiency of 69% and has its CIE chromaticity coordinates located at (0.654, 0.346). Color-tunablity from blue to red was achieved with $Sr_9Sc(PO_4)_7:Ce^{3+}$, Mn^{2+} phosphor (Dong et al. 2014). Although it was possible to tune the color from blue to red by altering the Mn^{2+} concentration, it was not possible to produce white light with this composition. The PL excitation and emission spectra of $Sr_9Sc(PO_4)_7:Ce^{3+}$, Mn^{2+} phosphors are shown in Fig. 2.3 and the CIE chromaticity diagram corresponding to different concentrations is shown in Fig. 2.4. Wang et al. successfully fabricated a green emitting phosphor converted LED by combining $Ca_{10}K(PO_4)_7$: Eu^{2+}, Tb^{3+} phosphor with GaInN chip (Wang et al. 2009). Although Ce^{3+} is commonly found in sensitizing the f-f transitions of Tb^{3+} ions for green emission, the energy transfer from Eu^{2+} to Tb^{3+} is rarely implemented. Jia et al. also developed a green emitting phosphor by implementing an energy transfer from Eu^{2+} to Dy^{3+} ions in a whitelock structure of $Ca_9Bi(PO_4)_7$ (Jia et al. 2014). The green emission from $Ca_9Bi(PO_4)_7$ is solely because of Eu^{2+} ions. However, the introduction of Dy^{3+} ions has dramatically increased the decay life time of the phosphor and the phosphorescence was found to last for about 5 hours with a decent intensity level. $Ca_9Bi(PO_4)_7:Eu^{2+}$, Dy^{3+} phosphor turns out to be a promising green emitting long lasting phosphorescent material. Zheng and Wanjun (2016) introduced $Sr_8MgLa(PO_4)_7:Ce^{3+}$, Eu^{2+}, Mn^{2+} phosphor that is itself capable of emitting white light under near-UV excitation. The energy transfer process in this tri-doped phosphor has been explained by

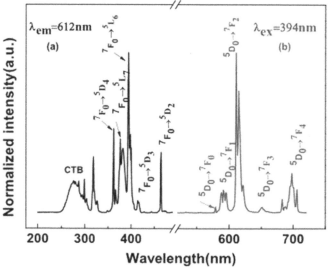

Fig. 2.2. Photoluminescence excitation and emission spectra of $Ca_8MgLu(PO_4)_7:Eu^{3+}$ phosphor (Copyright © 2015 Elsevier Ltd and Techna Group S.r.l., Reproduced with Permission from Ref. (Xie et al. 2015)).

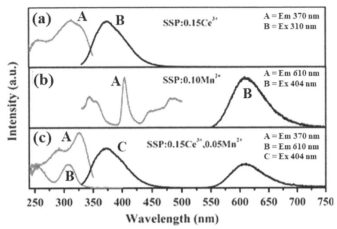

Fig. 2.3. Photoluminescence excitation and emission spectra of (a) $Sr_9Sc(PO_4)_7$: Ce^{3+} phosphor, (b) $Sr_9Sc(PO_4)_7$:Mn^{2+} phosphor and (c) $Sr_9Sc(PO_4)_7$:Ce^{3+}, Mn^{2+} phosphor (Copyright © 2013 Elsevier B.V., Reproduced with Permission from Ref. (Dong et al. 2014)).

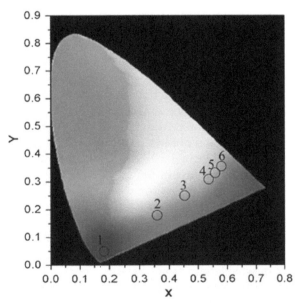

Fig. 2.4. CIE chromaticity diagram of $Sr_9Sc(PO_4)_7$:$0.15Ce^{3+}$, xMn^{2+}. Here (1) x = 0; (2) x = 0.01; (3) x = 0.03; (4) x = 0.05; (5) x = 0.10; and (6) x = 0.20 (Copyright © 2013 Elsevier B.V., Reproduced with Permission from Ref. (Dong et al. 2014)).

a cascade model in which the transfer of energy occurs from $Ce^{3+} \rightarrow Eu^{2+} \rightarrow Mn^{2+}$. $KCaPO_4$:Eu^{2+}, Tb^{3+}, Mn^{2+} phosphor is another example of triply doped phosphors that have been investigated for their white light emitting properties (Fang et al. 2015).

Lately, Dy^{3+} ions have been used to produce white light emission in a number of phosphates. It is possible to tune the color temperature of the white light by simply varying the yellow to blue (Y/B) ratio. With the variation of Dy^{3+} concentration in the host, the Y/B ratio of the emission will also show some variation. But it does not guarantee that the Dy^{3+} ions will show the same intensity for all the concentrations. The yellow emission of the Dy^{3+} ions are due to the electric dipole transition and this transition becomes dominant when the Dy^{3+} ions occupy a site with non-inversion symmetry. $CaZr_4(PO_4)_6:Ce^{3+}$, Dy^{3+} phosphors have shown white light emission close to standard D_{65} illuminant and their CIE chromaticity coordinates on the CIE 1931 diagram were located to be at (0.3114, 0.3501) with a correlated color temperature of 6413 K (Nair and Dhoble 2015). $Ca_3Mg_3(PO_4)_4:Ce^{3+},Li^+$ phosphor exhibited white-light emission with a correlated color temperature of 4739 K (Nair and Dhoble, 2017). $Sr_3Gd(PO_4)_3:Dy^{3+}$ phosphors proved to be thermally stable white light emitting phosphors whose intensity decreased to only 88% from 100 °C to 25 °C. At 150 °C, the PL intensity reached 84% of its initial value with minimum shift in the emission band (Xu et al. 2015).

Among the halophosphates, the apatite type phosphates are well acknowledged for their luminescence efficiency. However, the fascination for apatites may seem to dim after experiencing the critical conditions required in their synthesis. The realization of apatites may often lead to the formation of various intermediates. The synthesis conditions need to be optimized and controlled to obtain a pure apatite structure without developing any trace of impurity-phases. Yet, the synthesis of $Ca_6Y_2Na_2(PO_4)_6F_2:Eu^{2+}$, Mn^{2+} phosphor by solid state reaction successfully yielded a single phase formation (Guo et al. 2014). The emission color of $Ca_6Y_2Na_2(PO_4)_6F_2:Eu^{2+}$, Mn^{2+} phosphor could be tailored from blue to white and ultimately to yellow by varying the concentration of Mn^{2+} ions. Surprisingly, the CIE coordinates for $Ca_6Y_2Na_2(PO_4)_6F_2:0.01Eu^{2+}$, $0.025Mn^{2+}$ phosphor turned out to be at (0.335, 0.337) with a color temperature of 5375 K. This is approximately the same as the standard coordinates for white light, i.e. (0.33, 0.33). Another fluoroapatite $Ba_3LaK(PO_4)_3F:Tb^{3+}$ phosphor showed good thermal stability with color tunability from blue to green (Zeng et al. 2016). A similar version of this phosphor was reported, featuring color tenability and energy transfer from Eu^{2+} to Tb^{3+}. The $Ba_3GdNa(PO_4)_3F:Eu^{2+},Tb^{3+}$ phosphor can absorb the 350 – 400 nm wavelength from a UV LED and give out a strong blue-green emission with color tunability from blue to green depending on the concentration of Eu^{2+} and Tb^{3+} ions (Jiao et al. 2013; Zeng et al. 2015). A greenish blue emitting phosphor without the introduction of Tb^{3+} ions were reported by Xiao et al. wherein Eu^{2+} ions were capable of producing a broad band emission centered at 500 nm in $Na_5Ca_4(PO_4)_4F$ host matrix (Xiao et al. 2012). The phosphor was able to produce 76% quantum efficiency when excited at 365 nm and showed a decline upto 73% of its normalized PL intensity of the initial value as the temperature was raised to 150 °C. Figure 2.5 shows the behavior of $Na_5Ca_4(PO_4)_4F:Eu^{2+}$ phosphor excited at 365 nm in the temperature range 50-250 °C.

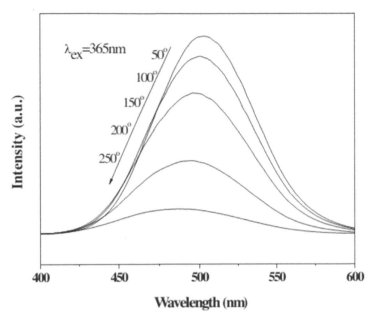

Fig. 2.5. Behavior of $Na_5Ca_4(PO_4)_4F:Eu^{2+}$ phosphor excited at 365 nm in the temperature range 50-250 °C (Copyright © 2012 Elsevier B.V., Reproduced with Permission from Ref. (Xiao et al. 2012)).

Phosphate Phosphors for Phototherapy Lamp Application

Treatment of diseases, especially skin diseases, with the aid of light is called phototherapy. In phototherapy, the affected portion of the skin is exposed to light of a specific wavelength for a particular time period. Interaction of light with the skin tissue gives rise to a biological reaction which results in improvement of the skin. Ultraviolet radiation having wavelengths 200 nm to 400 nm and infrared light with wavelength 700 nm to 1200 nm are used in phototherapy application. Infrared light is mostly absorbed by bones, joints and muscle. It is used in healing and relieving pain. UV light gets absorbed in the skin directly and produces beneficial as well as adverse effects on the skin (Grossweiner et al. 2005). Ultraviolet radiations are again divided into UV-C (220 nm-290 nm), UV-B (290 nm-315 nm) and UV-A (315 nm-400 nm). UV-C light is highly carcinogenic in nature and causes skin cancer. UV-C light is used in germicidal application such as purification of air and water. UV-B radiation results in vitamin-D production, erythema, hyper-pigmentation and skin cancer induction. UV-A is weakly absorbed by the skin and can be used in addition with the psoralens drug for treatment. UV-A light has 1000 times less erythema effect than a shorter wavelength. Phototherapy can be classified into different classes such as broadband (BB) UV-B (290 nm-315 nm), narrowband (NB) UV-B (311 nm-313 nm) and PUVA phototherapy (psoralens drug + UVA). In earlier times, broadband UV-B therapy was in practice but was replaced by narrow band UV-B phototherapy. Broadband therapy consists of shorter wavelength in the UVB region which may induce skin cancer instead of healing it, thus proving fatal at times. It is found that light of 313 nm wavelength is the most effective in curing skin (Dogra and Kanwar 2004). Therefore, NBUV-B phototherapy is the most trusted. It can cure nearly 40 types of skin diseases such as psoriasis, vitiligo, Ofuji's disease, erythropoietic protoporphyria, chronic graft-versus-host disease, atopic dermatitis, scleroderma, cutaneous T-cell lymphoma, morphea and acne vulgaris (Shinde et al. 2015; Thakare et al. 2007).

Phototherapy unit for narrow band UVB phototherapy consist of florescent lamps coated inside with white luminescent material called phosphor. This phosphor is made up of inorganic compound doped with Gadolinium (Gd^{3+}) ions. Gadolinium is the rare earth lanthanide element located at the middle of lanthanides series in the periodic table. Rare earth ions are characterized by incompletely filled 4f orbitals. They possess characteristic intraband 4f-4f transitions. These transitions remain unperturbed even after doping of these ions in different hosts since 4f orbitals are shielded by outermost $5s^2$ and $5p^6$ orbitals. Gd^{3+} ions possess large energy differences between the ground state and excited metastable state, which corresponds to the UV region. Gd^{3+} ions doped phosphors show excitation at about 274 nm corresponding to $^8S_{7/2} \rightarrow {}^6I_{11/2}$ transition and emission at 313 nm corresponding to $^8P_{7/2} \rightarrow {}^8S_{7/2}$ transition. Emission of light at this wavelength is essential for curing 40 different types of skin diseases. Gd^{3+} possesses f-f transition which is parity forbidden and therefore, its emission and excitation intensities seem to be low. However, the electric-dipole transition of Gd^{3+} can be very intense if the ions occupy a non-inversion site symmetry site in the host matrix. The other way to enhance the intensity of Gd^{3+} emission is by introducing a sensitizer capable of absorbing in the UV region and transferring the energy to the activator, i.e., Gd^{3+} ions. Pr^{3+}, Bi^{3+}, Ce^{3+} and Pb^{2+} ions can be selected as the sensitizer depending upon host materials (Tamboli et al. 2018; Tamboli and Dhoble 2017). These sensitizers absorb energies from the excitation sources and then transfer it to Gd^{3+} ions. Gd^{3+} ion absorbs energy from the sensitizer and then emit at 313 nm. LaB_3O_6:Gd, Bi is a phosphor used in commercial narrow UVB phototherapy lamps.

It is well-known that phosphates are excellent host materials for rare earth ion doping and also give bright luminescence. So many rare earth ions doped phosphate compounds are used in lighting applications such as fluorescent lamps, Plasma Display Panel (PDP), White Light Emitting Diode (W-LED) etc. Researchers also tried to develop phosphate phosphor for phototherapy application and found that phosphates are a good candidate for making phototherapy lamp phosphor. Mokoena et al. reported $Ca_5(PO_4)_3OH$:Gd^{3+}, Pr^{3+} (Mokoena et al. 2014a) and $Ca_3(PO_4)_2$:Gd^{3+}, Pr^{3+} (Mokoena et al. 2014b) phosphate phosphors for phototherapy application. $Ca_5(PO_4)_3OH$ belongs to the hydroxyapatite family of compounds. This naturally occurring mineral consists of base apatite calcium and phosphate bonded to hydroxide. In

$Ca_3(PO_4)_5OH$ compound, the calcium site is replaced by Gd^{3+}, Pr^{3+} ions. $Ca_5(PO_4)_3OH:Gd^{3+}$, Pr^{3+} compound was synthesized by the precipitation method and annealed at 900 °C for 2 hours. Optical band gap of this compound was found to be 5 eV with absorption peak at 207 nm. PL emission of $Ca_5(PO_4)_3OH:Gd^{3+}$ compound with 313 nm emission showed characteristic excitation at 245 nm, 254 nm and 275 nm corresponds to $^8S_{7/2}$-^6D_j, $^8S_{7/2}$-^6I_j and $^8S_{7/2}$-^6G_j transitions of Gd^{3+} and host excitation at 207 nm. It is assigned to the interband excitation of the $Ca_5(PO_4)_3(OH)$ host. PL emission was observed at 313 nm with 275 nm excitation. $Ca_5(PO_4)_3OH:Gd^{3+}$ co-doped with Pr^{3+} found energy transfer from Pr^{3+} to Gd^{3+}. Excitation spectrum of $Ca_5(PO_4)_3OH:Gd^{3+}$, Pr^{3+} compound monitored at 313 nm showed broad excitation band at 222 nm which corresponds to 4f-5d transition of Pr^{3+} ions. At 222 nm excitation, intense emission was observed at 313 nm with increased intensity. $Ca_3(PO_4)_2:Gd^{3+}$, Pr^{3+} synthesized by the combustion method was found to exhibit commendable UVB emission. Band gap of the compound was found to be 4.92 eV. PL excitation spectra of $Ca_3(PO_4)_2:Gd^{3+}$, Pr^{3+} compound showed excitation at 220 nm and this excitation corresponds to 4f-5d transition of Pr^{3+} ions. At 220 nm, phosphor shows intense luminescence emission peak at 313 nm. Thus, both the phosphors can be used in narrow band-UV-B phototherapy application. Figure 2.6 shows the photoluminescence excitation and emission spectra of $Ca_3(PO_4)_2:Gd^{3+}$, Pr^{3+} phosphor.

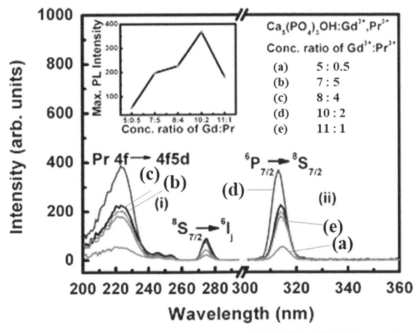

Fig. 2.6. PL excitation and emission spectra of $Ca_3(PO_4)_2:Gd^{3+}$, Pr^{3+} phosphor (Copyright © 2014 Elsevier B.V., Reproduced with Permission from Ref. (Mokoena et al. 2014b)).

Phosphate Phosphors for Thermoluminescence Dosimetry (TLD)

Thermoluminescence dosimetry (TLD) refers to the assessment of dose of ionizing radiations absorbed by an individual or a material and the determination of dose of radiation present in the environment (Bhatt and Kulkarni 2013a). As one is aware, the universe is filled with ionizing and non-ionizing radiations and the ionizing radiations being highly energetic are dangerous for living beings. On earth, decay of radioactive isotope, coal mines, industries dealing with the radioactive materials, nuclear power stations, nuclear weapons, cosmic rays, medical uses of radiations and radiation used for research purpose are natural and manmade sources of the ionizing radiations (Byju et al. 2012; Horowitz 2014; Noh et al. 2001). Ionizing radiations are hazardous for occupational workers as well as the public when its limits exceed. Excess exposure of ionizing radiations can cause cancer, skin diseases, burning, cognitive decline and heart disease. Therefore, proper safety precautions should be taken so as to keep occupational workers and others healthy and free of disorders due to exposure from radiation. International Commission on Radiological Protection (ICRP) recommends Maximum Permissible Dose (MPD) values for occupational workers; these values are expected not to cause any injuries or adverse effects on the health of workers for their lifetime. For non-penetrating radiations (X-rays < 15 KeV, β-rays) MPD value is 5-10 mg/cm^{-2} on the basal layers of epidermis and for penetrating radiations (X-rays > 15 KeV, γ-rays, neutrons) MPD value is 400-1000 mg/cm^{-2} below the surface of the body (Bhatt and Kulkarni 2013b). For personnel dosimetry, i.e., for measuring the dose of radiation absorbed by the workers, TLD badges are used in which TLD phosphor is pelletized or filled directly in the powder form. TLD phosphor having tissue equivalent Z_{eff} value (7.14) is used for personnel dosimetry so as to estimate an accurate amount of radiation absorbed by tissues. When the dose of radiation absorbed by workers seems to exceed the above mentioned limits, they are not allowed to work further in the radiation environment for a specific duration of time (Moscovitch and Horowitz 2007; Nakajima et al. 1978).

Determining the amount of radiation present in the environment is called environmental monitoring and is done by non-tissue equivalent TLD phosphor. Other than these, TLD phosphor finds application in dating, biology, studying defects formed in materials, etc (McKeever et al. 1995). Commercial TLD phosphors used for personnel dosimetry are LiF: Mg, Ti (Moscovitch and Horowitz 2007), Li$_2$B$_4$O$_7$: Mn (Chandra and Bhatt 1981), Li$_2$B$_4$O$_7$: Cu, Ag (Rawat et al. 2012) and BeO (Mukherjee 2015). All these phosphors possess near tissue equivalent Z_{eff} value. Among these LiF:Mg, Ti (TLD-100) is highly sensitive for radiation. For environmental dosimetry CaSO$_4$:Dy (Mathur et al. 1999), CaSO$_4$:Tm (Chagas et al. 2010), CaF$_2$:Mn (Bakshi et al. 2009) and CaF$_2$:Dy (Binder and Cameron 1969) are used as a commercial phosphor. Among these CaSO$_4$:Dy (TLD – 900) is extensively studied and a highly sensitive dosimeter. Radiation assessment using TLD is the three-step process which involves irradiation of TL detector, formation of radicals or trapping of the electron-hole in defect energy levels and finally heating or TL readout. Ionizing radiation possesses energy to detach the electron from its outermost orbital, and therefore, when the material is exposed to ionizing radiations, electrons get free from the valence band and jump to the conduction band. If the material consists of any kind of defect or impurity in it, electrons get trapped in this impurity level formed by the defects. Electrons remain trapped in this trapping level unless and until they are provided with energy equal to or more than that energy which is required by the electron to cross the energy barrier of the defect level called activation energy. When we heat the material, activation energy is gained by the electron in the form of heat and they get detrapped from the trapping energy level, thereby, jumping back to conduction band. These electrons then return to the valence band and combine with holes giving out the energy of recombination in the form of light called thermoluminescence (TL). This process is called TL readout (Kadari and Kadri 2010). Linearity of intensity of emitted light against different doses of radiation, gives the estimate of radiation absorbed by materials. Linearity over a wide range, sensitivity to the lowest dose of radiations, less fading, tissue equivalence, stability and reproducibility are some of the requirements for the materials so that they can be used for radiation dose calculations. Though TL phenomenon is less studied in phosphate materials, recent studies on phosphate materials show that these materials possess good TL properties. Here, we have summarized some phosphate phosphors having appreciable TL characteristics.

In earlier times, less consideration was given for developing phosphate materials for TL dosimetry, as the field of TL dosimetry was mainly dominated by sulfate, borate and fluoride based TL dosimeters. Two decades before, researchers started exploring TL characteristics of phosphate phosphors. Lapraz and Baumer (1983) studied TL properties of natural and synthetic fluorapatite $[Ca_5(PO_4)_3F]$. Fukuda et al. discovered TL in hydroxapatite $[Ca_{10}(PO_4)_6(OH)_2]$ (Fukuda 2008). TL in commercial halophosphate lamp phosphors $[Ca_5(PO_4)_3(Cl, F):Sb^{3+},Mn^{2+}]$ was investigated by Bulur et al. (Bulur et al. 1996). Ohtaki studied TL emission in tricalcium phosphate $[Ca_3(PO_4)_2]$ and gave researchers a way for investigating phosphates phosphors for TL applications (Ohtaki et al. 1993). Their study showed that phosphates possess a good TL emission peak from 200 °C to 300 °C and the mechanism of TL was found to lie in the line emission of rare earth ions. In the series of calcium phosphate materials such as $CaHPO_4$, $Ca_3(PO_4)_2$, $Ca_4(PO_4)_2O$, $Ca_{10}(PO_4)_6(OH)_2$ differ in chemical stoichiometry by Ca/P ratio which lies between 1 and 2. Among these, $Ca_3(PO_4)_2$ was found to be a good candidate for TL.

Phosphate materials are of great interest for TL application because they are the main inorganic components of bones and teeth. Calcium phosphates (especially tricalcium phosphate) are equivalent to bone materials. Hydroxyapatite [$Ca_{10}(PO_4)_6(OH)_2$] is synthesized in the laboratory and used for implants and strengthening bones. Tricalcium phosphates and calcium hydroxyapatite are used as ceramic bone substitute. In all areas of orthopedics and orthodontics, partially or totally, parts of bone tissue are replaced by hydroxyapatite (Alvarez et al. 2014). Again, TL characteristics of calcium phosphate materials have special importance in radiation dosimetric ascribed to use of X-rays in diagnostic and therapeutic application related to the bone. Developing phosphate materials equivalent to the bone tissue will help in dosimetry of X-rays as well as other ionizing radiations. Considering these aspects of phosphate materials, research is ongoing for developing phosphate TL materials. In recent years, many phosphate phosphors have been reported for TL dosimetry, some of which are tissue equivalent and some are non-tissue equivalent. Emphasis is also given for developing nano phosphate materials to detect high dose of radiation. Different synthesis routes are applied for making TL materials as synthesis plays a major role in producing defects in the materials. The main focus is on developing tissue equivalent materials which satisfy all the characteristics needed so as to use this material for TL dosimetry.

$LiCaPO_4$:Eu, $LiMgPO_4$:Tb and $KMgPO_4$:Tb

$LiCaPO_4$, $LiMgPO_4$ and $KMgPO_4$ compounds belong to $ABPO_4$ family of phosphates, in which A refers to monovalent ions (A = Li^+, K^+, Na^+, Rb^+ and Cs^+) and B refers to divalent ions (B = Sr^{2+}, Ba^{2+}, Ca^{2+} and Mg^{2+}). The crystal structure of these compounds varies according to the ionic sizes of the constituent elements and subsequently TL emission varies from host to host. These types of compounds are thermally and chemically very stable and good optical as well as ferroelectric materials too. Researchers are attracted to these compounds for studying and exploring their TL properties as these compounds are highly sensitive radiation dosimeters. Among the researchers, More et al. (More et al. 2011) were the first who synthesized $LiCaPO_4$ compound doped with Eu^{2+} and explored its thermoluminescence properties. $LiCaPO_4$ crystallizes in the hexagonal structure. They synthesized $LiCaPO_4$ via the two-step solid state synthesis method by mixing Li_2CO_3 and freshly precipitated $CaHPO_4$ powder. The mixture was heated at 300 °C, 650 °C and 800 °C. TL glow curve was obtained for the samples irradiated by γ-rays. TL glow curve for doped and undoped $LiCaPO_4$ compound were found to have the same type of glow curve structure with two peaks at 155 °C and 210 °C. $LiCaPO_4$:Eu compound was found to be more sensitive than pristine $LiCaPO_4$. TL and PL emission comparison shows that Eu^{2+} was not a luminescent center in TL mechanism but it results in increasing the number of traps. TL peak at 210 °C shows four times higher sensitivity than LiF-TLD 100 dosimeter. Figure 2.7 shows the comparison between the TL glow curves of $LiCaPO_4$:Eu with LiF-TLD 100. The compound has a linear dose response upto 264Gy. Z_{eff} value of $LiCaPO_4$ is 14.8 and, therefore, it is a high

Z_{eff} compound. This compound has the potential to be used in radiation dosimetry.

LiMgPO$_4$ compound has an orthorhombic crystal structure with space group *Pnma*. LiMgPO$_4$:Tb (Menon et al. 2012) compound was first prepared by solid state synthesis and TL glow curve was obtained after γ-ray irradiation that showed a single glow peak at 170 °C. Intensity of TL glow curve for LiMgPO$_4$:Tb (2 mole%) was found to be 2.5 times higher than CaSO$_4$:Dy (0.5 mole %), as shown in Fig. 2.8. This phosphor has a linear dose response upto 1 kGy. Due to its high dose linearity, it can

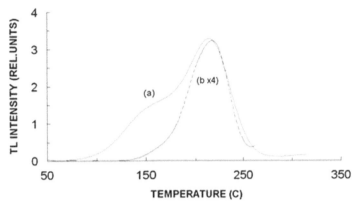

Fig. 2.7. Comparison between the TL glow curves of LiCaPO$_4$:Eu (curve a) with LiF-TLD 100 (curve b) exposed to 8.25×10^{-3}C/kg (32 R). To fit the data in the same figure, curve b is plotted with gain of 4 (Copyright © 2010 Elsevier B.V., Reproduced with Permission from Ref. (More et al. 2011)).

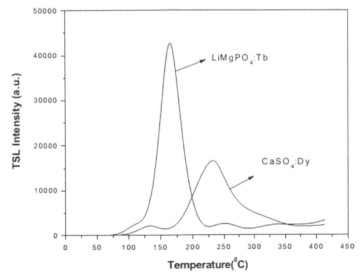

Fig. 2.8. Comparison of TL glow curves of LiMgPO$_4$:Tb (2 mole%) and CaSO$_4$:Dy (0.5 mole %) phosphors (Copyright © 2012 Elsevier B.V., Reproduced with Permission from Ref. (Menon et al. 2012)

be used for high dose radiation measurement. The mechanism of TL was proposed on the basis of TL and PL emission. Both the emission matched well with the PL emission indicating that Tb is a luminescent center in the TL process. $KMgPO_4$ was also synthesized by solid state synthesis (Palan et al. 2016). This phosphor shows two TL peaks at 117 °C and 295 °C. Intensity of 295 °C peak is 10 times that of Al_2O_3:C-TLD 500, as it can be seen in Fig. 2.9. Although, the TL mechanism, TL peaks and their intensities are different for $LiCaPO_4$:Eu, $LiMgPO_4$:Tb and $KMgPO_4$:Tb phosphate phosphors, they have potential to be used for TL dosimetry.

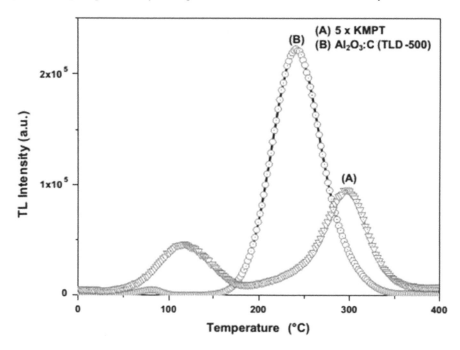

Fig. 2.9. TL glow curve comparison between $KMgPO_4$:Tb and Al_2O_3:C-TLD 500 (Copyright © 2016 Elsevier B.V., Reproduced with Permission from Ref. (Palan et al. 2016)

NaLi₂PO₄:Eu³⁺

The $NaLi_2PO_4$ compound has an orthorhombic crystal structure with space group Pmnb. Its Z_{eff} value is 10.8 which is the tissue equivalent value and therefore TL from $NaLi_2PO_4$ is of great interest amongst researchers. Its Z_{eff} value is low and, therefore, it can be used for personnel dosimetry. Singh et al. (2013). synthesized $NaLi_2(PO_4)_2$:Eu³⁺ compound by solid state synthesis by using LiOH, NaH_2PO_4 and $EuCl_3$ as precursors. LiOH and NaH_2PO_4 were added in a 2:1 proportion and heated at 400 °C, 600 °C and 800 °C. At each stage of heating, the compound was kept for 12 hours in the furnace. Samples of $NaLi_2PO_4$:Eu³⁺ were irradiated with γ-rays of different doses and the TL glow curve was obtained. The TL curve showed a single glow peak at 205 °C. Intensity of this peak when compared with commercial TLD phosphors were found to be 18, 2 and 1.5 times more than TLD-100, TLD-400, and TLD-900, as shown in Fig. 2.10. The compound showed linear dose response up to 1 kGy. Singh et al. (2013) proposed TL mechanism for these phosphors based on electron trapping at Eu^{3+} ions and hole trapping at Na^+ or Li^+ ions after irradiation. During TL readout, hole and electron combines to give TL emission. With high sensitivity, low Z_{eff} value and linearity over a wide region of dose, this phosphor is suitable as a candidate for personnel dosimetry.

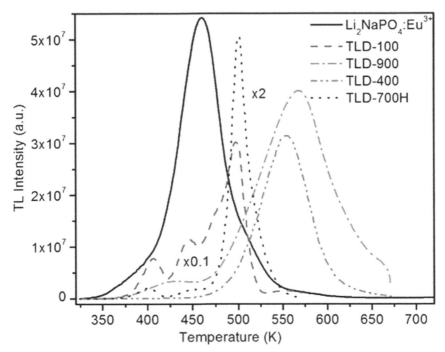

Fig. 2.10. Comparison of the TL sensitivity with some standard commercially available phosphors exposed to 5.0 Gy of γ-rays from a ^{137}Cs source (Copyright © 2013 Elsevier B.V., Reproduced with Permission from Ref. (Singh et al. 2013)).

β-Ca₃(PO₄)₂:Dy

The β-$Ca_3(PO_4)_2$:Dy compound has a rhombohedral crystal structure. Nakashima et al. (2005) synthesized β-$Ca_3(PO_4)_2$ and β-$Ca_3(PO_4)_2$:Dy by rapid mixing of a solution of $Ca(NO_3)_2$, $(NH_4)_2H_2PO_4$ and $Dy(NO_3)_3\ 5H_2O$. Ammonia was added for maintaining pH of the solution. The gel formed by adding the above solution was filtered, washed and heated at 1100 °C to get β-$Ca_3(PO_4)_2$:Dy^{3+}. PL results showed that emission consisting of peaks at 480 nm, 571 nm, 660 nm and 751 nm that correspond to line emissions of Dy^{3+}. TL peak obtained by irradiating compound with X-rays showed a peak at 64 °C.

No ESR signal was found for pristine samples of β-Ca$_3$(PO$_4$)$_2$:Dy. ESR signal for X-ray irradiated sample showed formation of hydrogen atom center by a reaction between ^{1}H$^+$ ion and electron during irradiation. The temperature dependence on ESR signal intensity of β-Ca$_3$(PO$_4$)$_2$:Dy is shown in Fig. 2.11.

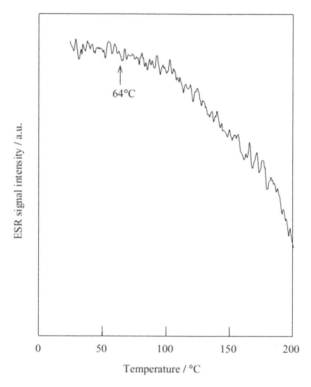

Fig. 2.11. The temperature dependence on ESR signal intensity of β-Ca$_3$(PO$_4$)$_2$:Dy phosphor (Copyright © 2004 Elsevier B.V., Reproduced with Permission from Ref. (Nakashima et al. 2005)).

K$_3$Gd(PO$_4$)$_2$:Tb^{3+} and K$_3$Y(PO$_4$)$_2$:Eu^{3+}

K$_3$Gd(PO$_4$)$_2$:Tb^{3+}(Gupta et al. 2014) and K$_3$Y(PO$_4$)$_2$:Eu^{3+}(Gupta et al. 2016) belong to the orthophosphate family of compounds having the formula M$_3$RE(PO$_4$)$_2$ (M = alkali metal ion and RE = rare earth ion). Both the compounds have a monoclinic crystal structure with the space group P2$_{1/m}$. These compounds were synthesized by combustion synthesis and their thermoluminescence characteristics were analyzed. KNO$_3$, NH$_4$H$_2$PO$_4$, Gd$_2$O$_3$/Y(NO$_3$)$_3$.6H$_2$O and urea were mixed and kept in a muffle furnace at 550 °C. Thereafter, the product was annealed at 800°C for 2 hours and the final product was formed. K$_3$Gd(PO$_4$)$_2$ and K$_3$Gd(PO$_4$)$_2$:Tb^{3+} phosphors were irradiated with 0.1 KGy dose of γ-rays and the TL glow curves were recorded. Intensity and glow curve structure for K$_3$Gd(PO$_4$)$_2$ and

$K_3Gd(PO_4)_2:Tb^{3+}$ phosphors were found to vary significantly from each other. As shown in Fig. 2.12, TL glow curve for $K_3Gd(PO_4)_2$ showed TL peak at 302 °C (575 K) and TL peak of $K_3Gd(PO_4)_2: Tb^{3+}$ was found to be at 200 °C (473 K). Incorporating Tb^{3+} in $K_3Gd(PO_4)_2$ resulted in enhancement of trap density. $K_3Y(PO_4)_2:Eu^{3+}$ (Gupta et al. 2016) compound was irradiated with γ-rays and TL glow curve was obtained with TL peaks at 134 °C and 205 °C. This compound had a linear dose response from 400 to 5000 Gy, as shown in Fig. 2.13.

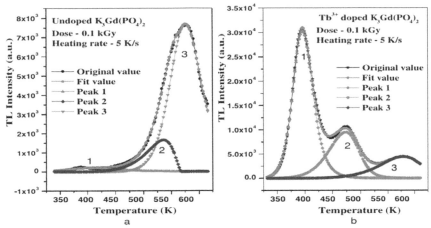

Fig. 2.12. Deconvoluted curves of (a) undoped $K_3Gd(PO_4)_2$ and (b) $K_3Gd(PO_4)_2:Tb^{3+}$ (1.5 mol%) nanophosphors exposed to 0.1 kGy of gamma rays at heating rate of 5 K/s (Copyright © 2014 Elsevier B.V., Reproduced with Permission from Ref. (Gupta et al. 2014)).

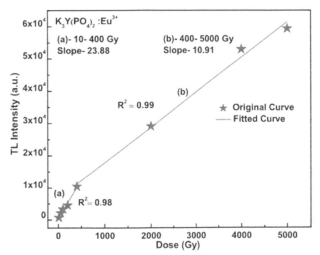

Fig. 2.13. Dose–response of $K_3Y(PO_4)_2:Eu^{3+}$ (2.5 mol%) nanophosphors (Copyright © 2014 Elsevier B.V., Reproduced with Permission from Ref. (Gupta et al. 2016)

$Sr_6Al(PO_4)_5:Dy^{3+}$

Shinde et al. (2011) synthesized $Sr_6Al(PO_4)_5:Dy^{3+}$ compound by combustion synthesis using urea as a fuel. The compound was investigated for photoluminescence as well as thermoluminescence properties. When irradiated with γ-rays of dose 0.048 kGy, the TL glow curve showed a single peak at 194 °C. The highest intensity of the TL glow peak was found when the doping of Dy^{3+} ions was 0.3 mol%. This TL peak was compared with the commercially used $CaSO_4:Dy$ thermoluminescence dosimetric phosphor. Sensitivity of this peak was found to be 33 times less than $CaSO_4:Dy$ phosphor. This phosphor showed linearity upto 4.8 kGy (which is very high dose), as shown in Fig. 2.14.

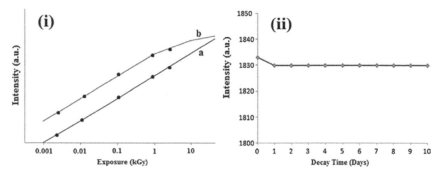

Fig. 2.14. (i) Response curve (a) $Sr_6Al(PO_4)_5:Dy^{3+}$ and (b) $CaSO_4:Dy$ phosphors. **(ii)** The fading characteristic of the $Sr_6Al(PO_4)_5:Dy^{3+}$ phosphor (Copyright © 2011 Elsevier B.V., Reproduced with Permission from Ref. (Shinde et al. 2011)).

$Sr_2P_2O_7:Mn, Pr$ and $Ca_2P_2O_7:Ce^{3+}$

Pyrophosphates are generally good hosts for rare earth ions doping and show intense photoluminescence emission. TL characteristics of pyrophosphate compounds were investigated by several researchers. Some of the pyrophosphates with appropriate TL properties are discussed here. $Sr_2P_2O_7$ belongs to the pyrophosphate family of compounds and has an orthorhombic crystal structure. Natarajan et al. (2004) were the first who studied TL and EPR properties of europium doped $Sr_2P_2O_7$ compound. They recorded EPR spectra of gamma-ray irradiated $Sr_2P_2O_7$ at various temperatures, as it can be seen in Fig. 2.15. They showed that no TL peak is obtained for undoped $Sr_2P_2O_7$ compound, but $Sr_2P_2O_7:Eu$ possess two glow peaks at 192 °C and 292 °C, which was attributed to Eu^{3+} ions. Ilkay et al. (2014) prepared Cu–Ag, Cu–In and Mn–Pr doped $Sr_2P_2O_7$ phosphor by solid state synthesis and exposed their TL properties. For different dopant combination, different types of TL glow structure were obtained. $Sr_2P_2O_7:Cu$, In and $Sr_2P_2O_7:Cu$, Ag phosphor showed TL peaks at 100 °C. TL peaks for $Sr_2P_2O_7:Mn$, Pr were found to be at 100 °C and 175 °C, as shown in Fig. 2.16. The TL glow curve of $Sr_2P_2O_7$:Mn (7 mole%) ,Pr (1 mole%) showed the highest intensity for 175 °C peak. This peak is useful for dosimetric application. TL properties of $Ca_2P_2O_7$:Ce^{3+}phosphor was studied by Lozano et al. (2016). They prepared

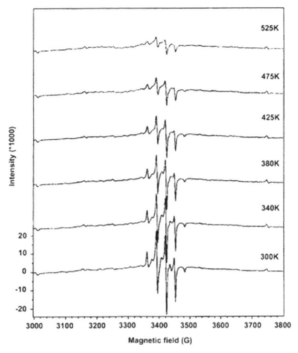

Fig. 2.15. EPR spectra of gamma-ray irradiated $Sr_2P_2O_7$ at various temperatures (Copyright © 2004 Elsevier B.V., Reproduced with Permission from Ref. (Natarajan et al. 2004)).

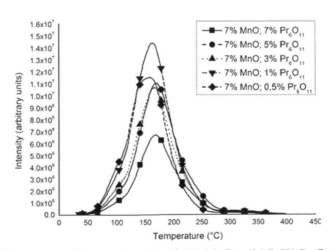

Fig. 2.16. Glow curves of $Sr_2P_2O_7$doped with 7% MnO and 0.5–7% Pr_6O_{11} (Copyright © 2014 Elsevier B.V., Reproduced with Permission from Ref. (Ilkay et al. 2014)).

Ca$_2$P$_2$O$_7$:Ce^{3+} phosphor by the precipitation method and annealed at 800 °C and 900 °C. TL of the prepared compound showed a different type of TL glow curve for different annealing temperature. TL peak for the precipitated and annealed (900 °C) sample was obtained at 175 °C. TL of Ca$_2$P$_2$O$_7$:Ce^{3+} compound annealed at 800 °C was found to have three TL peaks at 130 °C, 170 °C and 225 °C.

Li$_2$BaP$_2$O$_7$:Dy^{3+} and LiNa$_3$P$_2$O$_7$:Tb^{3+}

Li$_2$BaP$_2$O$_7$:Dy^{3+} (Wani et al. 2015) and LiNa$_3$P$_2$O$_7$:Tb^{3+} (Munirathnam et al. 2015) compounds were easily prepared by solid state reaction and analyzed for TL emission. TL emission of γ-irradiated and C^{5+} irradiated Li$_2$BaP$_2$O$_7$:Dy^{3+} compound was compared by Wani et al. (2015). TL glow curve for both the sources of irradiation were found to be nearly the same with the glow peak situated at 137 °C. For γ-ray irradiation two small shoulders were seen, but for C^{5+} beam irradiation only one shoulder was seen. TL intensity of C^{5+} beam irradiated sample was found to be more as compared to γ-ray but less than CaSO$_4$:Dy sample. The effect of dose variation on TL intensity of Li$_2$BaP$_2$O$_7$:Dy^{3+} phosphor under two different excitation sources can be seen in Fig. 2.17. LiNa$_3$P$_2$O$_7$:Tb^{3+} is the green emitting phosphor with appreciable

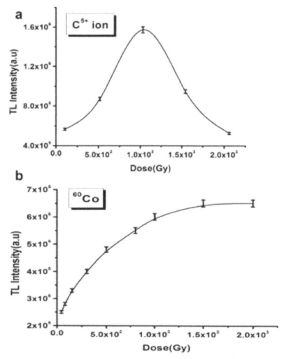

Fig. 2.17. Effect of dose variation on TL intensity of Li$_2$BaP$_2$O$_7$:Dy^{3+} phosphor under two different excitation sources Pr$_6$O$_{11}$ (Copyright © 2015 Elsevier B.V., Reproduced with Permission from Ref. (Wani et al. 2015)).

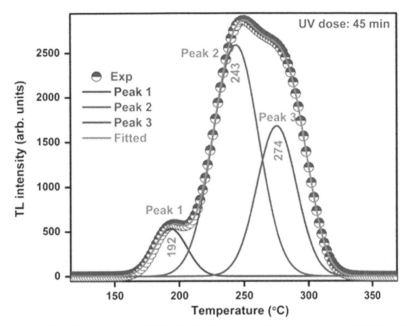

Fig. 2.18. TL glow curve of LiNa$_3$P$_2$O$_7$:Tb^{3+} phosphor exposed to UV for 45 min (Copyright © 2015 Elsevier B.V., Reproduced with Permission from Ref. (Munirathnam et al. 2015)).

TL emission when irradiated with UV light. For UV exposure of 110 minutes, maximum intensity for TL emission was observed with three overlapping glow peaks with maxima at 192 °C, 243 °C and 274 °C, as shown in Fig. 2.18.

Sr$_5$(PO$_4$)$_3$Cl:Eu^{2+}

Sr$_5$(PO$_4$)$_3$Cl:Eu^{2+} is the member of apatite family and is recognized by the general formula M$_5$(PO$_4$)$_3$X, where M = Sr, Ba, Ca and X = Cl, F, Br, OH. This compound has a hexagonal system with the space group P63/m. Sr$_5$(PO$_4$)$_3$Cl:Eu^{2+} compound is used as efficient blue phosphor for tricolor fluorescent lamp. Dhoble et al. (Dhoble, 2000; Dhoble et al. 2007) synthesized Sr$_5$(PO$_4$)$_3$Cl:Eu^{2+} phosphor by solid state synthesis and studied its TL characteristics. TL peak for Sr$_5$(PO$_4$)$_3$Cl:Eu^{2+} compound was found to be situated at 157 °C with a simple glow curve structure. The sensitivity of the phosphor was found to be two times of the CaSO$_4$:Dy phosphor. This phosphor was projected to be useful in ionizing radiation dosimetry. Dhoble et al. (2007) studied the correlation between PL, TL and Electron Spin Resonance (ESR) characteristics of the Sr$_5$(PO$_4$)$_3$Cl:Eu^{2+} phosphor. PL emission was observed at 446 nm with 350 nm excitation which corresponds to 4f-5d transition of Eu^{2+} ions. ESR spectra, obtained after irradiation, showed that irradiation leads to the formation of (PO$_4$)$^{2-}$ radical which is responsible for TL emission.

Conclusion

Phosphates have proven their worth as excellent host materials for doping rare-earth ions and provided opportunities to explore different possibilities in solid-state lighting, phototherapy and thermoluminescence applications. The number of phosphors available for these applications is very vast and hence, few selected phosphors are only described in this article. Some notable phosphate-based phosphors for TLD are listed in the Table 2.1. Phosphates have exhibited excellent materials for phototherapy lamp and solid-state lighting phosphors due to their large band gap, moderate phonon-energies and high absorption in the VUV and UV regions. They are chemically and thermally stable materials as well as demonstrate excellent optical damage threshold.

Table 2.1. List of phosphate TLD materials

Sr. No.	Compound	Crystal structure	Synthesis	TL peak	Sensitivity
1	$LiCaPO_4$:Eu (More et al. 2011)	Hexagonal	Precipitation	155 °C and 210 °C	5 times higher than LiF-TLD 100
2	$LiMgPO_4$:Tb (Menon et al. 2012)	Orthorhombic	Solid state	170 °C	2.5 times more sensitive than $CaSO_4$:Dy
3	$KMgPO_4$:Tb (Palan et al. 2016)	Monoclinic	Solid state	117 °C and 295 °C	10 times less than Al_2O_3:C (TLD 500)
4	$NaLi_2PO_4$:Eu^{3+} (Singh et al. 2013)	Orthorhombic	Solid state	205 °C	18, 2 and 1.5 times more than the TLD-100, TLD-400, TLD-900.
5	β-$Ca_3(PO_4)_2$:Dy (Nakashima et al. 2005)	Rhombohedral	Wet chemical	64 °C	–
6	$K_3Gd(PO_4)_2$:Tb^{3+} (Gupta et al. 2014)	Monoclinic	Combustion	302 °C	–
7	$K_3Y(PO_4)_2$$Eu^{3+}$ (Gupta et al. 2016)	Monoclinic	Combustion	200 °C	–
8	$Sr_6Al(PO_4)_5$:Dy^{3+} (Shinde et al. 2011)	Tetragonal	Combustion	194 °C	33 times less than $CaSO_4$:Dy^{3+}
9	$Sr_2P_2O_7$: Mn, Pr (Ilkay et al. 2014)	Orthorhombic	Solid state	192 °C and 292 °C	–
10	$Ca_2P_2O_7$:Ce^{3+} (Lozano et al. 2016)	Tetragonal	Precipitation	175 °C	–

(Contd.)

Table 2.1. *(Contd)*

Sr. No.	Compound	Crystal structure	Synthesis	TL peak	Sensitivity
11	$Li_2BaP_2O_7$:Dy^{3+} (Wani et al. 2015)	Monoclinic	Solid state	137 °C	–
12	$LiNa_3P_2O_7$:Tb^{3+} (Munirathnam et al. 2015)	Orthorhombic	Solid state	192 °C, 243 °C and 274 °C	–
13	$Sr_5(PO_4)_3Cl$:Eu^{2+} (Dhoble, 2000)	Hexagonal	Solid state	157 °C	2 times more than $CaSO_4$:Dy^{3+}

Acknowledgement

This research is supported by the South African Research Chairs Initiative of the Department of Science and Technology and National Research Foundation of South Africa (Grant 84415). The financial assistance from the University of the Free State is greatly appreciated. One of the authors SJD is thankful to Department of Science and Technology (DST), India (Nano Mission) (Sanction Project Ref. No. DST/NM/NS/2018/38(G), dt. 16/01/2019) for financial assistance.

References

Alvarez, R., Rivera, T., Guzman, J., Piña-barba, M.C., Azorin, J. 2014. Thermoluminescent characteristics of synthetic hydroxyapatite (SHAp). Appl. Radiat. Isot. 83, 192-195. https://doi.org/10.1016/j.apradiso.2013.04.011

Bakshi, A.K., Dhabekar, B., Rawat, N.S., Singh, S.G., Joshi, V.J., Kumar, V. 2009. Study on TL and OSL characteristics of indigenously developed CaF_2:Mn phosphor. Nucl. Instruments Methods Phys. Res. Sect. B Beam Interact. with Mater. Atoms 267, 548-553. https://doi.org/10.1016/j.nimb.2008.12.007

Bhatt, B.C., Kulkarni, M.S. 2013a. Thermoluminescent phosphors for radiation dosimetry. Defect Diffus. Forum 347, 179. https://doi.org/10.4028/www.scientific.net/DDF.347.179

Bhatt, B.C., Kulkarni, M.S. 2013b. Worldwide status of personnel monitoring using Thermoluminescent (TL), Optically Stimulated Luminescent (OSL) and Radiophotoluminescent (RPL) dosimeters. Int. J. Lumin. Appl. 3, 6-10.

Binder, W., Cameron, J.R. 1969. Dosimetric properties of CaF_2:Dy. Health Phys. 17, 613-618.

Bulur, E., Goksu, H.Y., Wieser, A., Figel, M., Ozer, A.M. 1996. Thermoluminescence properties of fluorescent materials used in commercial lamps. Radiat Prot Dosim. 65, 373-379.

Byju, S. Ben, Koya, P.K.M., Sahoo, B.K., Jojo, P.J., Chougaonkar, M.P., Mayya, Y.S. 2012. Inhalation and external doses in coastal villages of high background radiation area in Kollam, India. Radiat. Prot. Dosimetry 152, 154-158. https://doi.org/10.1093/rpd/ncs213

Chagas, M.A.P., Nunes, M.G., Campos, L.L., Souza, D.N. 2010. TL properties of anhydrous $CaSO_4$:Tm improvement. Radiat. Meas. 45, 550-552. https://doi.org/10.1016/j.radmeas.2010.01.025

Chandra, B., Bhatt, R.C. 1981. Isothermal decay characteristics of $Li_2B_4O_7$: Mn phosphor. Nucl. Instruments Methods 184, 557-559. https://doi.org/10.1016/0029-554X(81)90763-1

Deyneko, D.V., Nikiforov, I.V., Lazoryak, B.I., Spassky, D.A., Leonidov, I.I., Stefanovich, S.Y., Petrova, D.A., Aksenov, S.M., Burns, P.C. 2019. $Ca_8MgSm_{1-x}(PO_4)_7$:xEu^{3+}, promising red phosphors for WLED application. J. Alloys Compd. 776, 897-903. https://doi.org/10.1016/j.jallcom.2018.10.317

Dhoble, S.J. 2000. Preparation and characterization of the $Sr_5(PO_4)_3Cl$:Eu^{2+} phosphor. J. Phys. D. Appl. Phys. 33, 158. https://doi.org/10.1088/0022-3727/33/2/310

Dhoble, S.J., Moharil, S.V., Gundu Rao, T.K. 2007. Correlated ESR, PL and TL studies on $Sr_5(PO_4)_3Cl$:Eu thermoluminescence dosimetry phosphor. J. Lumin. 126, 383-386. https://doi.org/10.1016/j.jlumin.2006.08.098

Dogra, S., Kanwar, A.J. 2004. Review Article Narrow band UVB phototherapy in dermatology. Indian Journal of Dermatology, Venereology and Leprology. 70, 205-209.

Dong, X., Zhang, J., Zhang, X., Hao, Z., Luo, Y. 2014. Luminescence and energy transfer mechanism in $Sr_9Sc(PO_4)_7$:Ce^{3+}, Mn^{2+} phosphor. J. Lumin. 148, 60-63. https://doi.org/10.1016/j.jlumin.2013.11.083

Durugkar, A., Tamboli, S., Dhoble, N.S., Dhoble, S.J. 2018. Novel photoluminescence properties of Eu^{3+} doped chlorapatite phosphor synthesized via sol-gel method. Mater. Res. Bull. 97, 466-472. https://doi.org/10.1016/j.materresbull.2017.09.043

Fan, J., Zhang, W., Dai, S., Yan, G., Deng, M., Qiu, K. 2018. Effect of charge compensators A^+ (A = Li, Na and K) on luminescence enhancement of $Ca_3Sr_3(PO_4)_4$:Sm^{3+} orange-red phosphors. Ceram. Int. 44, 20028-20033. https://doi.org/10.1016/j.ceramint.2018.07.276

Fang, H., Huang, S., Wei, X., Duan, C., Yin, M., Chen, Y. 2015. Synthesis and luminescence properties of $KCaPO_4$:Eu^{2+}, Tb^{3+}, Mn^{2+} for white-light-emitting diodes (WLED). J. Rare Earths 33, 825-829. https://doi.org/10.1016/S1002-0721(14)60491-9

Fukuda, Y. 2008. Thermoluminescence in calcium fluoride doped with terbium and gadolinium ions. Radiat. Meas. 43, 455-458. https://doi.org/10.1016/j.radmeas.2007.10.034

Grossweiner, L.I., Grossweiner, J.B., Gerald Rogers, B.H. 2005. The Science of Phototherapy: An Introduction. Springer-Verlag, Berlin/Heidelberg. https://doi.org/10.1007/1-4020-2885-7

Guo, N., You, H., Jia, C., Ouyang, R., Wu, D. 2014. A Eu^{2+} and Mn^{2+} coactivated fluoro-apatite-structure $Ca_6Y_2Na_2(PO_4)_6F_2$ as a standard white-emitting phosphor via energy transfer. Dalt. Trans. 43, 12373. https://doi.org/10.1039/C4DT01021C

Gupta, P., Bedyal, A.K., Kumar, V., Khajuria, Y., Lochab, S.P., Pitale, S.S., Ntwaeaborwa, O.M., Swart, H.C. 2014. Photoluminescence and thermoluminescence properties of Tb^{3+} doped $K_3Gd(PO_4)_2$ nanophosphor. Mater. Res. Bull. 60, 401-411. https://doi.org/10.1016/j.materresbull.2014.09.001

Gupta, P., Bedyal, A.K., Kumar, V., Singh, V.K., Khajuria, Y. 2016. Thermoluminescence and glow curves analysis of γ-exposed Eu^{3+} doped $K_3Y(PO_4)_2$ nanophosphors. Mater. Res. Bull. 73, 111-118. https://doi.org/10.1016/j.materresbull.2015.08.030

Horowitz, Y.S. 2014. Thermoluminescence dosimetry: State-of-the-art and frontiers of future research. Radiat. Meas. 71, 2-7. https://doi.org/10.1016/j.radmeas.2014.01.002

Huang, C.-H., Wang, D.-Y., Chiu, Y.-C., Yeh, Y.-T., Chen, T.-M. 2012. $Sr_8MgGd(PO_4)_7$:Eu^{2+}: yellow-emitting phosphor for application in near-ultraviolet-emitting diode based white-light LEDs. RSC Adv. 2, 9130-9134. https://doi.org/10.1039/c2ra20646c

Huang, C., Liu, W., Chen, T. 2010. Single-phased white-light phosphors $Ca_9Gd(PO_4)_7$:Eu^{2+}, Mn^{2+} under near-ultraviolet excitation. J. Phys. Chem. C 114, 18698-18701. https://doi.org/10.1021/jp106693z

Huang, C.H., Chen, T.M. 2011. Novel yellow-emitting $Sr_8MgLn(PO_4)_7$:Eu^{2+} (Ln = Y, La) phosphors for applications in white LEDs with excellent color rendering index. Inorg. Chem. 50, 5725-5730. https://doi.org/10.1021/ic200515w

Ilkay, L.S., Ozbayoglu, G., Yilmaz, A. 2014. Synthesis, characterizations and investigation of thermoluminescence properties of strontium pyrophosphate doped with metals. Radiat. Phys. Chem. 104, 55-60. https://doi.org/10.1016/j.radphyschem.2014.04.028

Jia, Y., Li, H., Zhao, R., Sun, W., Su, Q., Pang, R., Li, C. 2014. Luminescence properties of a new bluish green long-lasting phosphorescence phosphor $Ca_9Bi(PO_4)_7$:Eu^{2+}, Dy^{3+}. Opt. Mater. 3, 6-11. https://doi.org/10.1016/j.optmat.2014.04.006

Jiao, M., Guo, N., Lü, W., Jia, Y., Lv, W., Zhao, Q., Shao, B., You, H. 2013. Tunable blue-green-emitting $Ba_3LaNa(PO_4)_3$:Eu^{2+}, Tb^{3+} phosphor with energy transfer for near-UV white LEDs. Inorg. Chem. 52, 10340-10346. https://doi.org/10.1021/ic401033u

Kadari, A., Kadri, D. 2010. Numerical model for thermoluminescence of MgO. Phys. B Phys. Condens. Matter 405, 4713-4717. https://doi.org/10.1016/j.physb.2010.08.062

Kim, M., Kobayashi, M., Kato, H., Yamane, H., Sato, Y., Kakihana, M. 2015. Crystal structures and luminescence properties of Eu^{2+}-activated new $NaBa_{0.5}Ca_{0.5}PO_4$ and $Na_3Ba_2Ca(PO_4)_3$. Dalt. Trans. 44, 1900-1904. https://doi.org/10.1039/C4DT03024A

Kim, S.W., Hasegawa, T., Ishigaki, T., Uematsu, K., Toda, K., Sato, M. 2013. Efficient red emission of blue-light excitable new structure type $NaMgPO_4$:Eu^{2+} phosphor. ECS Solid State Lett. 2, R49. https://doi.org/10.1149/2.004312ssl

Lapraz, D., Baumer, A. 1983. Thermoluminescent properties of synthetic and natural fluorapatite, $Ca_5(PO_4)_3F$. Phys. Status Solidi 80, 353-366.

Li, L., Tang, X., Wu, Z., Zheng, Y., Jiang, S., Tang, X., Xiang, G., Zhou, X. 2019. Simultaneously tuning emission color and realizing optical thermometry via efficient $Tb^{3+} \rightarrow Eu^{3+}$ energy transfer in whitlockite-type phosphate multifunctional phosphors. J. Alloys Compd. 780, 266-275. https://doi.org/10.1016/j.jallcom.2018.11.378

Lin, C.C., Tang, Y.S., Hu, S.F., Liu, R.S. 2009. $KBaPO_4$:Ln (Ln=Eu, Tb, Sm) phosphors for UV excitable white light-emitting diodes. J. Lumin. 129, 1682-1684. https://doi.org/10.1016/j.jlumin.2009.03.022

Lin, C.C., Xiao, Z.R., Guo, G.Y., Chan, T.S., Liu, R.S. 2010. Versatile phosphate phosphors $ABPO_4$ in white light-emitting diodes: Collocated characteristic analysis and theoretical calculations. J. Am. Chem. Soc. 132, 3020-3028. https://doi.org/10.1021/ja9092456

Lozano, I.B., Roman-Lopez, J., Sosa, R., Díaz-Góngora, J.A.I., Azorín, J. 2016. Preparation of cerium doped calcium pyrophosphate: Study of luminescent behavior. J. Lumin. 173, 5-10. https://doi.org/10.1016/j.jlumin.2015.12.032

Mathur, V., Lewandowski, A., Guardala, N., Price, J. 1999. High dose measurements using thermoluminescence of $CaSO_4$:Dy. Radiat. Meas. 30, 735-738. https://doi.org/10.1016/S1350-4487(99)00244-9

McKeever, S.W.S., Moscovitch, M., Townsend, P.D. 1995. Thermoluminescence Dosimetric Materials: Properties and Uses. Nuclear Technology Publishing, Ashford, Kent, England.

Menon, S.N., Dhabekar, B., Alagu Raja, E., Chougaonkar, M.P., Raja, E.A., Chougaonkar, M.P. 2012. Preparation and TSL studies in Tb activated $LiMgPO_4$ phosphor. Radiat. Meas. 47, 236-240. https://doi.org/10.1016/j.radmeas.2011.12.013

Mokoena, P.P., Nagpure, I.M., Kumar, V., Kroon, R.E., Olivier, E.J., Neethling, J.H., Swart, H.C., Ntwaeaborwa, O.M. 2014a. Enhanced UVB emission and analysis of chemical states of $Ca_5(PO_4)_3OH:Gd^{3+}$, Pr^{3+} phosphor prepared by co-precipitation. J. Phys. Chem. Solids 75, 998-1003. https://doi.org/10.1016/j.jpcs.2014.04.015

Mokoena, P.P., Gohain, M., Kumar, V., Bezuidenhoudt, B.C.B., Swart, H.C., Ntwaeaborwa, O.M. 2014b. TOF SIMS analysis and enhanced UVB photoluminescence by energy by urea assisted combustion. J. Alloys Compd. 595, 33-38.

More, S.D., Wankhede, S.P., Kumar, M., Chourasiya, G., Moharil, S.V. 2011. Synthesis and dosimetric characterization of $LiCaPO_4$:Eu phosphor. Radiat. Meas. 46, 196-198. https://doi.org/10.1016/j.radmeas.2010.11.001

Moscovitch, M., Horowitz, Y.S. 2007. Thermoluminescent materials for medical applications: LiF : Mg, Ti and LiF : Mg, Cu, P. Radiat. Meas. 41, 71-77. https://doi.org/10.1016/j.radmeas.2007.01.008

Mukherjee, B., 2015. LiBe-14: A novel microdosimeter using LiF and BeO thermoluminescence dosimeter pairs for clinical and aerospace applications. Radiat. Meas. 72, 31-38. https://doi.org/10.1016/j.radmeas.2014.11.003

Munirathnam, K., Dillip, G.R., Ramesh, B., Joo, S.W., Prasad Raju, B.D. 2015. Synthesis, photoluminescence and thermoluminescence properties of $LiNa_3P_2O_7:Tb^{3+}$ green emitting phosphor. J. Phys. Chem. Solids 86, 170-176. https://doi.org/10.1016/j.jpcs.2015.07.011

Nair, G.B., Bhoyar, P.D., Dhoble, S.J. 2017. Exploration of electron-vibrational interaction in the 5d states of Eu^{2+} ions in $ABaPO_4$ (A = Li, Na, K and Rb) phosphors. Luminescence 32, 22-29. https://doi.org/10.1002/bio.3143

Nair, G.B., Dhoble, S.J. 2017. White light emission through efficient energy transfer from Ce^{3+} to Dy^{3+} ions in $Ca_3Mg_3(PO_4)_4$ matrix aided by Li^+ charge compensator. J. Lumin. 192, 1157-1166. https://doi.org/10.1016/j.jlumin.2017.08.047

Nair, G.B., Dhoble, S.J. 2015. Highly enterprising calcium zirconium phosphate $[CaZr_4(PO_4)_6:Dy^{3+}, Ce^{3+}]$ phosphor for white light emission. RSC Adv. 5, 49235-49247. https://doi.org/10.1039/C5RA07306E

Nakajima, T., Murayama, Y., Matsuzawa, T., Koyama, A. 1978. Development of a new highly sensitive LiF thermoluminescence dosimeter and its applications. Nucl. Instrum. Meth. A 157, 155-162.

Nakashima, K., Takami, M., Ohta, M., Yasue, T., Yamauchi, J. 2005. Thermoluminescence mechanism of dysprosium-doped β-tricalcium phosphate phosphor. J. Lumin. 111, 113-120. https://doi.org/10.1016/j.jlumin.2004.07.002

Natarajan, V., Bhide, M.K., Dhobale, A.R., Godbole, S.V., Seshagiri, T.K., Page, A.G., Lu, C.-H. 2004. Photoluminescence, thermally stimulated luminescence and electron paramagnetic resonance of europium-ion doped strontium pyrophosphate. Mater. Res. Bull. 39, 2065-2075. https://doi.org/10.1016/j.materresbull.2004.07.009

Noh, A.M., Amin, Y.M., Mahat, R.H., Bradley, D.A. 2001. Investigation of some commercial TLD chips/discs as UV dosimeters. Radiat. Phys. Chem. 61, 497-499. https://doi.org/10.1016/S0969-806X(01)00313-9

Ohtaki, H., Fukuda, Y., Takeuchi, N. 1993. Thermoluminescence in calcium phosphate doped with Dy_2O_3. Radiat Prot Dosim. 47, 119-122.

Palan, C.B., Bajaj, N.S., Soni, A., Omanwar, S.K. 2016. A novel $KMgPO_4$:Tb^{3+} (KMPT) phosphor for radiation dosimetry. J. Lumin. 176, 106-111. https://doi.org/10.1016/j.jlumin.2016.03.014

Rawat, N.S., Kulkarni, M.S., Tyagi, M., Ratna, P., Mishra, D.R., Singh, S.G., Tiwari, B., Soni, A., Gadkari, S.C., Gupta, S.K. 2012. TL and OSL studies on lithium borate single crystals doped with Cu and Ag. J. Lumin. 132, 1969-1975. https://doi.org/10.1016/j.jlumin.2012.03.008

Ray, S., Nair, G.B., Tadge, P., Malvia, N., Rajput, V., Chopra, V., Dhoble, S.J. 2018. Size and shape-tailored hydrothermal synthesis and characterization of nanocrystalline $LaPO_4$:Eu^{3+} phosphor. J. Lumin. 194, 64-71. https://doi.org/10.1016/j.jlumin.2017.10.015

Shinde, K.N., Dhoble, S.J. 2014. Europium-activated orthophosphate phosphors for energy-efficient solid-state lighting: A review. Crit. Rev. Solid State Mater. Sci. 39, 459-479. https://doi.org/10.1080/10408436.2013.803456

Shinde, K.N., Dhoble, S.J., Brahme, N. 2011. Combustion synthesis of $Sr_6AlP_5O_{20}$:Dy^{3+} submicron phosphor for high dose TL dosimetry. Radiat. Meas. 46, 1886-1889. https://doi.org/10.1016/j.radmeas.2011.01.015

Shinde, V.V., Kunghatkar, R.G., Dhoble, S.J. 2015. UVB-emitting Gd^{3+}-activated M_2O_2S (where M = La , Y) for phototherapy lamp phosphors. Luminescence 30(8), 1257-1262. https://doi.org/10.1002/bio.2889

Singh, M., Sahare, P.D.D., Kumar, P. 2013. Synthesis and dosimetry characteristics of a new high sensitivity TLD phosphor $NaLi_2PO_4$:Eu^{3+}. Radiat. Meas. 59, 8-14. https://doi.org/10.1016/j.radmeas.2013.10.002

Tamboli, S., Dhoble, S.J. 2017. Influence of Li^+ charge compensator ion on the energy transfer from Pr^{3+} to Gd^{3+} ions in $Ca_9Mg(PO_4)_6F_2$:Gd^{3+}, Pr^{3+}, Li^+ phosphor. Spectrochim. Acta Part A Mol. Biomol. Spectrosc. 184, 119-127. https://doi.org/10.1016/j.saa.2017.05.001

Tamboli, S., Nair, G.B., Dhoble, S.J., Burghate, D.K. 2018. Energy transfer from Pr^{3+} to Gd^{3+} ions in BaB_8O_{13} phosphor for phototherapy lamps. Phys. B Condens. Matter 535, 232-236. https://doi.org/10.1016/j.physb.2017.07.042

Thakare, D.S., Omanwar, S.K., Moharil, S.V., Dhopte, S.M., Muthal, P.L., Kondawar, V.K. 2007. Combustion synthesis of borate phosphors. Opt. Mater. 29, 1731-1735. https://doi.org/10.1016/j.optmat.2006.09.016

Wang, J., Zhang, Z., Zhang, M., Zhang, Q., Su, Q., Tang, J. 2009. The energy transfer from Eu^{2+} to Tb^{3+} in $Ca_{10}K(PO_4)_7$ and its application in green light emitting diode. J. Alloys Compd. 488, 582-585. https://doi.org/10.1016/j.jallcom.2008.09.088

Wani, J.A., Dhoble, N.S., Lochab, S.P., Dhoble, S.J. 2015. Luminescence characteristics of C^{5+} ions and $Li_2BaP_2O_7$:Dy^{3+} phosphor. Nucl. Instruments Methods Phys. Res. B 349, 56-63.

Xiao, H., Xia, Z., Liao, L., Zhou, J., Zhuang, J. 2012. Luminescence properties of a new greenish blue emitting phosphor $Na_5Ca_4(PO_4)_4F$:Eu^{2+}. J. Alloys Compd. 534, 97-100. https://doi.org/10.1016/j.jallcom.2012.04.042

Xie, F., Dong, Z., Wen, D., Yan, J., Shi, J.J., Shi, J.J., Wu, M., 2015. A novel pure red phosphor $Ca_8MgLu(PO_4)_7$:Eu^{3+} for near ultraviolet white light-emitting diodes. Ceram. Int. 41, 9610-9614. https://doi.org/10.1016/j.ceramint.2015.04.023

Xu, D.-D., Zhou, W., Zhang, Z., Li, S.-J., Wang, X.-R. 2019. Improved photoluminescence by charge compensation in Dy^{3+} doped $Sr_4Ca(PO_4)_2SiO_4$ phosphor. Opt. Mater. 89, 197-202. https://doi.org/10.1016/j.optmat.2019.01.041

Xu, Q., Sun, J., Cui, D., Di, Q., Zeng, J. 2015. Synthesis and luminescence properties of novel $Sr_3Gd(PO_4)_3:Dy^{3+}$ phosphor. J. Lumin. 158, 301-305. https://doi.org/10.1016/j.jlumin.2014.10.034

Yerpude, V., Nair, G.B., Ghormare, K.B., Dhoble, S.J. 2018. Electron-vibrational interaction in the 3d states of Cu^+ ions activated in $Na_5Ca_4(PO_4)_4F$. Optik 161, 266-269. https://doi.org/10.1016/j.ijleo.2018.02.050

Yu, H., Deng, D., Li, Y., Xu, S., Li, Y., Yu, C., Ding, Y., Lu, H., Yin, H., Nie, Q. 2013. Electronic structure and photoluminescence properties of yellow-emitting $Ca_{10}Na(PO_4)_7:Eu^{2+}$ phosphor for white light-emitting diodes. J. Lumin. 143, 132-136. https://doi.org/10.1016/j.jlumin.2013.04.036

Zeng, C., Huang, H., Hu, Y., Miao, S., Zhou, J. 2016. A novel blue-greenish emitting phosphor $Ba_3LaK(PO_4)_3F:Tb^{3+}$ with high thermal stability. Mater. Res. Bull. 76, 62-66. https://doi.org/10.1016/j.materresbull.2015.12.008

Zeng, C., Liu, H., Hu, Y., Liao, L., Mei, L., 2015. Color-tunable properties and energy transfer in $Ba_3GdNa(PO_4)_3F:Eu^{2+}$, Tb^{3+} phosphor pumped for n-UV w-LEDs. Opt. Laser Technol. 74, 6-10. https://doi.org/10.1016/j.optlastec.2015.05.003

Zhang, S., Huang, Y., Jin, H. 2010a. Luminescence properties and structure of Eu^{2+} doped $KMgPO_4$ phosphor. Opt. Mater. 32(11), 1545-1548. https://doi.org/10.1016/j.optmat.2010.06.020

Zhang, S., Huang, Y., Shi, L., Seo, H.J. 2010b. The luminescence characterization and structure of Eu^{2+} doped $LiMgPO_4$. J. Phys. Condens. Matter 22, 235402. https://doi.org/10.1088/0953-8984/22/23/235402

Zheng, Z., Wanjun, T. 2016. Tunable luminescence and energy transfer of $Ce^{3+}/Eu^{2+}/Mn^{2+}$-tridoped $Sr_8MgLa(PO_4)_7$ phosphor for white light LEDs. J. Alloys Compd. 663, 731-737. https://doi.org/10.1016/j.jallcom.2015.12.184

Zhou, W., Han, J., Zhang, X., Qiu, Z., Xie, Q., Liang, H., Lian, S., Wang, J. 2015. Synthesis and photoluminescence properties of a cyan-emitting phosphor $Ca_3(PO_4)_2:Eu^{2+}$ for white light-emitting diodes. Opt. Mater. 39, 173-177. https://doi.org/10.1016/j.optmat.2014.11.021

Photoluminescence Mechanism and Key Factors to Improve Intensity of Lanthanide Doped Tungstate/ Molybdate Phosphors with Their Applications

Neha Jain[1], Rajan Kumar Singh[1,2], R.A. Singh[1], Sudipta Som[2], Chung-Hsin Lu[2]* and Jai Singh[1]*

[1] Department of Physics, Dr. Harisingh Gour Central University, Sagar, M.P. - 470003, India
[2] Department of Chemical Engineering, National Taiwan University, Taipei, Taiwan, ROC

Introduction

Recently rare earth ion doped materials have attracted great attention for their astonishing optical property. These materials have been studied due to their application in optical transmission, white Light Emitting Diodes (LEDs), photoluminescence device, solar cells, fluorescent lamps, temperature sensors, plasma display panels, X-ray imaging scintillators, Field Emission Displays (FEDs) etc. (Geng et al. 2012). Rare earth doped metal oxides show luminescence due to the trivalent state of their 4f valence shell electron so that f-f and f-d transition occurs (Qin et al. 2017). The f-f transitions are parity forbidden by the Laporte selection rule, however it was permitted if RE is doped in a suitable host so that these transitions were allowed due to high crystal field strength, occupancy at non-inversion symmetry site and mixing of the wave function of host/RE ion. The f-f transitions have a sharp emission (having a lifetime in micro seconds) whereas f-d transitions have a broad absorption and emission wavelength range (having a lifetime in nano seconds). f-d transition depends on the host crystal field strength and polarizability; it also gets affected with the structure and composition of the host.

*Corresponding authors: chlu@ntu.edu.tw; jai.bhu@gmail.com

These days white LEDs have gained more attention for lighting over conventional Compact Fluorescent Lamps (CFL) because of their low power consumption, longer durability, good color rendering index, higher luminescence, reliability and higher energy efficiency. Rare Earth (RE) ion doped materials could be useful in fabrication of white LEDs. The quality of white light emission has been evaluated by the Commission International de I' Eclairage (CIE) chromacity co-ordinate, Correlated Color Temperature (CCT) and Color Rendering Index (CRI). For high quality white LEDs, the desired value of CIE co-ordinate is (0.33, 0.33), CRI > 80 and CCT values falls in the range 2700-4000 K (Shirasaki et al. 2013). The fabrication of white LEDs has been done on blue or UV (ultra-violet) chip by coating of phosphors. White lighting could be obtained by combining trichromatic emitters, red-green-blue or yellow-red-blue. White LEDs were first developed in 1996 by the combination of blue light emitting InGaN and yellow YAG:Ce^{3+} (Nakamura and Fasol 2013), but in these types of white LED, the red component is absent. The red component can be added into LEDs by using Eu^{3+} doped phosphor as Eu^{3+} ion gives prominent red luminescence when occupying a non-centrosymmetric site. Optical temperature sensors based on upconversion emission from rare earth doped materials have an important application such as measuring temperature of a distant object, photonics, sensing, security and medicine. Such types of sensors are used in temperature measurement by evaluating the change in luminescence absorbed from two closely spaced levels of the emitting center of materials (Li et al. 2007). Before discussing applications of RE based phosphor, it is necessary to understand the basic luminescence obtained due to the transition of electrons between energy levels.

Luminescence Mechanism

Luminescence materials are excited by radiation (X-Rays, Gamma radiation, UV-Visible or near-infrared) and emit electromagnetic radiation. This emission occurs due to electronic transition of electrons from higher energy levels to lower energy level. Lanthanides (Ln^{3+}) doped materials emission takes place due to excitation of 4f valence shell electrons, thus energy level splitting takes place as shown in Fig. 3.1 (Bünzli and Piguet 2005). Splitting of 4f levels occurs due to the crystal field, coulomb interaction of the electron and spin orbit coupling. These levels split into several levels with energy difference 10^3 cm^{-1}, denoted by $^{2s+1}L_J$, where s is the total spin, L is the total orbital angular momentum and J is the total angular momentum. When Ln^{3+} is doped in the host, not only the strengthening of f-f transition takes place but also additional splitting of 4f levels is observed which depends upon site symmetry of Ln^{3+} ions. The excited states of lanthanide ions are relaxed by two ways, light emission and phonon emission processes (Bünzli and Piguet 2005; Dexter 1953). Therefore, transition of electrons from its lower energy state to higher energy state depends on the oxidation state and atomic configuration of lanthanide (Ln^{3+}) ion. The photoluminescence emissions due

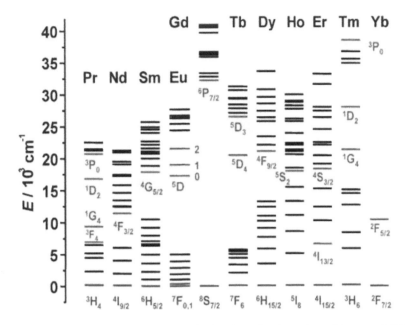

Fig. 3.1. Energy levels diagram for the lanthanides (Bünzli and Piguet, 2005).

to transitions of electrons between various electronic levels in lanthanides are presented in Table 3.1. From Table 3.1, it is clear that lanthanides emission occurs from visible to near-infrared range.

Table 3.1. The various active ions with their transitions

RE	Transition	Emission (nm)
Pr	$^3P_2 \rightarrow ^3H_4$, $^3P_0 \rightarrow ^3H_4$, $^3P_1/^5I_6 \rightarrow ^3H_5$ $^3P_0 \rightarrow ^3H_6$, $^3P_0 \rightarrow ^3F_2$, $^3P_0 \rightarrow ^1G_4$, $^1D_2 \rightarrow ^3F_4$	440, 480, 520, 605, 635, 715, 1037
Nd	$^2P_{3/2} \rightarrow ^4I_{11/2,\ 13/2}$, $^4I_{9/2} \rightarrow ^2K_{15/2} + ^2G_{9/2}$, $^2K_{13/2} + ^4G_{9/2} + ^4G_{7/2}$, $^2G_{5/2} + ^2G_{7/2}$, $^2P_{1/2} \rightarrow ^4F_{3/2,5/2}$, $^2D_{5/2} \rightarrow ^4F_{3/2,\ 5/2}$ $^4F_{3/2} \rightarrow ^4I_{9/2,11/2}$, $^4I_{9/2} \rightarrow ^4F_{9/2}$, $^4F_{7/2} + ^4S_{3/2}$, $^4F_{5/2}$	410, 452, 482, 532, 592, 875, 925, 812, 888, 1064, 684, 738, 823
Sm	$^4G_{5/2} \rightarrow ^6H_{5/2,7/2,9/2}$	562, 601, 647
Eu	$^5D_0 \rightarrow ^7F_{0,1,2,3,4}$	570–720
Tb	$^5D_4 \rightarrow ^7F_{6,5,4,3}$	480–650
Dy	$^4F_{9/2} \rightarrow ^6H_{15/2,\ 13/2,\ 11/2,\ 9/2,\ 7/2}$	486, 575, 654, 750, 840
Ho	$^3K_8 \rightarrow ^5I_8$, $^5F_3 \rightarrow ^5I_8$, $^5S_2$, $^5F_4 \rightarrow ^5I_{8,7,6}$ $^5F_5 \rightarrow ^5I_{8,7}$, $^5I_6 \rightarrow ^5I_8$	465, 490, 540, 759, 1013 644, 978, 1190
Er	$^4S_{3/2} \rightarrow ^4I_{15/2}$, $^4F_{9/2} \rightarrow ^4I_{15/2}$, $^4I_{13/2} \rightarrow ^4I_{15/2}$	545, 665, 1540
Tm	$^1D_2 \rightarrow ^3F_4$, $^3H_4 \rightarrow ^3H_6$	450, 800
Yb	$^2F_{5/2} \rightarrow ^2F_{7/2}$	980

Selection Rule

Luminescence originates from the electronic transition between 4f levels due to electric dipole and magnetic dipole interaction. Electric dipole interaction between free electrons of 4f valence shell is parity-forbidden, but it can be partially allowed by mixing of the orbital with a different parity. Electric dipole transition depends on site symmetry but magnetic dipole is not affected much (Malta et al. 1991). The selection rule for light emission is based on change in the total angular momentum of the electronic state as given below:

Magnetic dipole transition $\Delta J = 0, \pm 1$ (except for $0 \rightarrow 0$)

Electric dipole transition $\Delta J = \pm 2$

Luminescence Quenching

In PL emission if the host is activated by Ln^{3+} then intensity of emission increases up to a certain concentration of Ln^{3+}, beyond optimum concentration, intensity decreases; it is known as luminescence quenching. Luminescence can be quenched by the non-radiative decay of the excited state. This non-radiative decay is brought by different mechanisms involving energy transfer from the excited state of the donor to different types of acceptors like host lattice, organic molecules on the surface, defect levels or nearby ions which may or may not act as an activator. There are different types of non-radiative ways of the excited state such as:

- Multi-phonon emission
- Energy transfer
- Concentration quenching

Luminescence quenching not only depends on the concentration of Ln^{3+} but also on temperature i.e., on heating phosphor PL emission intensity also increases up to a certain temperature, than on further heating at higher temperatures intensity decreases (Anicete-Santos et al. 2007). This also occurs due to non-radiative relaxation (increasing rate of phonon vibration) of excitons.

Lanthanide Doped Materials

Lanthanide does not show the luminescence property by itself. It only shows luminescence when it is doped into a material known as the host. The host should be able to transfer its energy to Ln^{3+} ion so that the Ln^{3+} ions electron can get energy and jump to its higher energy state. It remains in an excited state for few moments, then turns back to the ground state by emitting photons. There are several hosts investigated for doping of lanthanide such as sulfide, fluoride, aluminate, metal tungstate/molybdates, nitrite etc. Among these hosts, metal tungstates/molybdate are garnering attention as hosts for doping of lanthanides because they have several advantages over fluoride and sulfide such as high chemical stability, average refractive index, X-ray absorption coefficient, high light yield, short decay time, excellent light output, higher efficiency, greater stability under electron bombardment and are environmentally friendly (Cavalcante et al. 2012; Singh et al. 2014). In Rare Earth (RE) activated tungstate host, charge transfer takes place in between RE^{3+} ion and $(W/Mo)O_4^{2-}$ group. Metal tungstates shows luminescence property (emission in the blue-green region) due to its $(W/Mo)O_4^{2-}$ group (Cavalcante et al. 2013). Luminescence is observed with doping of rare earth ions in tungstate and molybdates as a number of free charge carriers are generated which traps electrons or holes. These traps are a localized energy state between conduction and valence band due to defects or impurities (Bray 1996). Luminescence is observed due to radiative recombination of electrons and hole traps (Wang et al. 2014). Calcium tungstate is widely used in industrial radiology and medical diagnosis, and it is the host lattice for solid-state LASER action that gives 64% higher efficiency than Nd:YAG laser; it is also used in optical storage applications, sensors and for the detection of γ-rays (Sun et al. 2014). X-ray photography requires a phosphor which is able to x-ray and convert absorbed energy into visible light, $CaWO_4$ suitable for x-ray imaging (Roy and Muller 1974). On the basis of structural property, tungstate and molybdates can be divided in two families known as Scheelite and Wolframite type.

Scheelite Tetragonal Structure

$M(W/Mo)O_4$, where M refers to Ca, Ba, Sr and Pb, belongs to scheelite tetragonal structure family. In this structure metal ions are surrounded by eight oxygen atoms (in polyhedral configuration) and tungsten or molybdenum (W) is surrounded by four oxygen atoms (tetrahedral configuration) (Thongtem et al. 2010), as shown in Fig. 3.2 (a and b). If metal tungstate is doped with a rare earth ion such as Eu^{3+}, it replaces the site of M^{2+} ion (Singh et al. 2014).

Wolframite Monoclinic Structure

$M(W/Mo)O_4$, where M denotes Cd, Mg, Mn and Zn, belongs to the wolframite family and have a monoclinic structure (Sun et al. 2014). Figure 3.3(a and b) shows the wolframite monoclinic crystal structure where metal and tungsten (W) atoms both are co-ordinated to six oxygen atoms, forming distorted octahedral coordination (Su et al. 2008; Sun et al. 2007). Zinc molybdate ($ZnMoO_4$) crystals should be wolframite but exist in two structures, one is α (triclinic) and other is β (monoclinic) depending on synthesis conditions and processing time/temperature (Sun et al. 2007). Triclinic $ZnMoO_4$ contains $[ZnO_6]$ octahedral unit and $[MoO_4]$ tetrahedral unit in crystal as shown in Fig. 3.3(b).

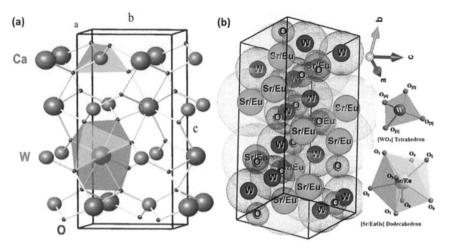

Fig. 3.2. Scheelite tetragonal structure of: (a) $CaWO_4$ (Errandonea and Manjon 2008) (b) $SrWO_4{:}Eu^{3+}$ (Singh et al. 2014).

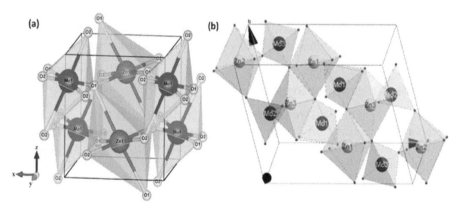

Fig. 3.3. Wolframite (a) monoclinic and (b) triclinic structure of $ZnMoO_4$ (Cavalcante et al. 2013; Spassky et al. 2011).

Synthesis Methods

There are several methods developed for synthesis of Ln^{3+} doped MWO_4 and $MMoO_4$ (M stands for metal like Ca, Sr, Ba, Zn, Pb and Cd) as described.

Ethylene Glycol or Polyol Synthesis Route

MWO_4 and $MMoO_4$ can be prepared by the polyol method. In this method ethylene glycol is used as a capping agent for nano-particles. For synthesis of material, weighed amount of metal ions (M^{2+}) source and Ln^{3+} ion source are dissolved together in de-ionized water. When a transparent mixture is formed, $WO_4{}^{2-}$ solution is added followed by ethylene glycol and urea. The mixture is heated at 150 °C so that the reaction is completed. Crystalline size of particles prepared by this method is 20-50 nm (Singh et al. 2014).

Chemical Precipitation Method

MWO_4 synthesis by this method is suitable at room temperature. In this method aqueous solution of M^{2+} ion source and WO_4^{2-} sources are added together with constant stirring, white precipitate is formed and this precipitate is filtered, dried and calcined at an appropriate temperature. Particle size of material prepared by this method is 5-6 μm (Ruiz-Fuertes et al. 2012).

Hydrothermal Method

In this method, a solution of metal nitrate, lanthanide nitrate and tungstate/ molybdate are mixed together and then a surfactant like oleic acid, ethylene glycol, cetyl trimethyl ammonium bromide (CTAB), EDTA etc. are added. The solution is transferred into a teflon autoclave and heated at 150-180 °C. By this method nanobeans of 20-35 nm diameter are formed (Sun et al. 2014).

Solid State Reaction Method

The starting materials are taken into a stoichiometric ratio, mixed together and ground for about one and half hours by using acetone. The final mixtured powder is kept into an alumina crucible and preheated at 900 °C for 4 hours in an electrical furnace and then further heat treated at 1300 °C for 4 hours in an air medium. Particle size prepared by this method is around 200-300 nm (Spassky et al. 2011).

Supersonic Microwave Co-assisted Synthesis

In this method, a mixture of M^{2+} ion source and WO_4^{2-} sources are put in round bottom flask and connected to the reactor. Set microwave power at 400 W and temperature 70 °C; supersonic power is 300 W for 45 minutes. After completion of the process retrieved by centrifugation, nanoparticles are formed (Arora and Chudasama 2006). PL properties of scheelite tungstate prepared by this method are more than other methods such as co-precipitation and hydrothermal because supersonic power and microwave irradiation accelerate rate of particles.

Molten Salt Method

Stoichiometric amount of starting materials were weighed and dissolved in deionized water with constant stirring. Then these solutions were mixed together with frequent stirring at room temperature. After this, molten salt ($LiNO_3$) was added into the above mixture and then the solution was heated to evaporate the water from the mixture. Then it was put in a crucible and calcined at 300 °C for 6 hours, finally precipitate washed with deionized water and dried. Nanoparticles of nearly 20 nm size were formed (Issler and Torardi 1995).

Reverse Phase Micelle Method

In this method, a mixture of hexadecyl tri-methyl ammonium bromide cyclohexane and 1-pentanol was divided into two parts. Tungstate or molybdates sources were added into one part of the solution and mixture of metal nitrate and lanthanide nitrate were added into the second part. Then these two solutions were stirred for 30 minutes, mixed together and heated into a teflon autoclave at 140 °C for 24 hours. The sample was then washed with distilled water and dried. This method is suitable to prepare nanoparticles of 50-100 nm size (Arora and Chudasama 2006; Basiev et al. 2000).

Factors Affecting Photoluminescence Emission Intensity

Effect of Temperature, Pressure and Synthesis Method on PL Intensity

PL intensity of $SrWO_4:Ln^{3+}$ phosphor depends on the synthesis method. Zheng et al. compared PL intensity of $SrWO_4:Ln^{3+}$ prepared by various methods namely Supersonic Microwave Co-assistance method (SMC), co-precipitation, hydrothermal, sintering and reverse micelle. They noted that PL intensity of $SrWO_4:Ln^{3+}$ prepared by SMC is more than other methods as shown in Fig. 3.4. The reason for their observation was co-assistance in sonochemistry and microwave irradiation that accelerated the rate of particles collision and the distance between the donor (WO_4^{2-}) and the acceptor Tb^{3+} became close, and efficient energy transfer was achieved (Zheng et al. 2012).

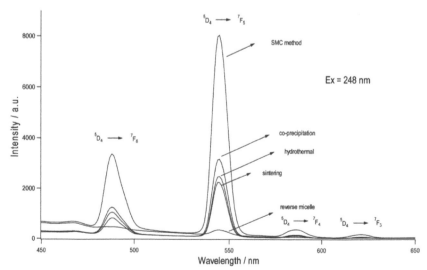

Fig. 3.4. Comparison of PL intensity prepared by various methods (Zheng et al. 2012).

Photoluminescence (PL) emission spectra of $(Ca/Sr/Ba)WO_4$ system occur in the blue region at 420–460 nm under 270 nm excitation (Wang et al. 2014). PL property of $SrWO_4$ depends on morphology of nanoparticles i.e., PL property are different for nanoparticles, nanorods and nanotubes. Sun and co-workers reported PL emission spectra of $SrWO_4$ of nanoparticles, nanopeanuts and nanorods. They prepared these nanocrystals by changing molar ratio of $SrWO_4$ and cetyl-tri-methyl-ammonium bromide (CTAB); as the molar percentage of $SrWO_4$ increases, length of nanorods increases (Sun et al. 2014). From photoluminescence spectra of $SrWO_4$, they observed a blue shift in emission with increasing growth of nanoparticles. This shift is attributed to the difference in their surface roughness. Nanopeanuts have a

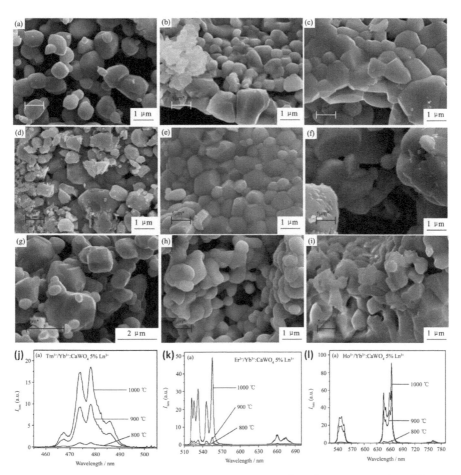

Fig. 3.5. SEM images of $CaWO_4$:Ln^{3+} upconversion phosphors prepared at different temperatures Ln = Tm/Yb: (a) 800 °C, (b) 900 °C, (c) 1000 °C; Ln=Er/Yb: (d) 800 °C, (e) 900 °C, (f) 1000 °C; Ln=Ho/Yb: (g) 800 °C, (h) 900 °C, (i) 1000 °C (j) Upconversion luminescence spectra of Tm^{3+}/Yb^{3+}:$CaWO_4$ (k) Er^{3+}/Yb^{3+}:$CaWO_4$ (l) Ho^{3+}/Yb^{3+}:$CaWO_4$ phosphor (λ_{ex}= 980 nm) (Piskuła et al. 2011).

very rough surface, which can lead to large surface-to-volume ratio and thus there are many dangling bonds and defects on the surface; nanopeanuts can absorb higher energies to emit light than nanoparticles. PL properties also depend on pH value, if pH increases then morphology of sample changes and intensity of PL emission decreases (Liao et al. 2000).

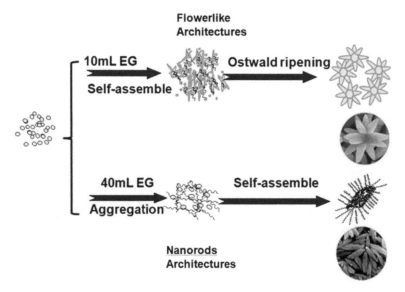

Fig. 3.6. Schematic illustration of the morphology evolution from shuttle-like nanorods to flower-like architectures (Yang et al. 2015).

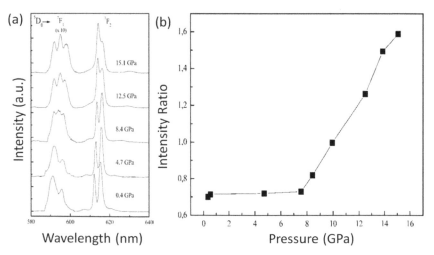

Fig. 3.7. (a) Emission spectra of the $^5D_0 \rightarrow {}^7F_{1,2}$ transitions in a SrWO$_4$ crystal doped with 10% of Eu^{3+} measured at RT at different pressures (0.4-4.7 GPa); (b) Intensity ratio of the two peaks around 615 nm observed in Fig. 3.1 as a function of pressure (Rivera-López et al. 2006).

PL intensity and emission wavelength changes with heating temperature because of structural order-disorder degree in the lattice, charge gradient and the presence of the localized states provide very good conditions for the trapping of electrons and holes (Anicete-Santos et al. 2011). This type of observation has been reported for $BaWO_4$, $PbWO_4$, $SrWO_4$ and $ZnWO_4$. Annealing effect on PL intensity of $CaWO_4:Sm^{3+}$ was reported by Maheshwary et al. (2014). $CaWO_4:Sm^{3+}$ showed excitation intensity improved on annealing due to decrease in non-radiative decay (Singh et al. 2014). PL properties of $SrWO_4$ doped with Ln^{3+} also depend on pressure. The effect of pressure on PL properties of Eu^{3+} doped $SrWO_4$ was reported by Lopez et al. They found that the intensity of PL emission increases with pressure but emission wavelength decreases due to increasing crystal field effect in 4f shell (Rivera-López et al. 2006).

Concentration of Rare Earth Ions

PL properties of tungstate and molybdate depend on doping concentration of rare earth ions and increase with doping concentration until an optimum value (Fig. 3.8). There are several reports based on variation of RE concentration and the optimum amount of rare earth varied with synthesis parameters (Ju et al. 2011; Singh et al. 2014). The concentration quenching occurs due to multiphonon relaxation such as dipole–dipole interaction, dipole quadruple interaction and quadruple -quadruple interaction (Singh et al. 2015).

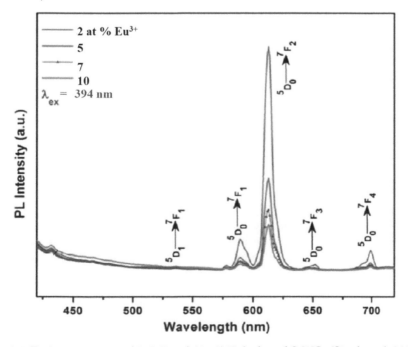

Fig. 3.8. Emission spectra of 2, 5, 7 and 10at% Eu^{3+} doped $SrWO_4$ (Singh et al. 2014).

CaWO$_4$ and CaMoO$_4$ doped with Tb^{3+} exhibit the same luminescence property due to Tb^{3+} excitation; this indicates that the coordination symmetries and Crystal Fields (CF) around the active ions are nearly identical. These two phosphors under 280 nm excitation show maximum intensity at 550 nm due to luminescence overlapping of $^5D_3$ and $^5D_4$ (Cavalli et al. 2010). Luminescence mechanism of CaWO$_4$:Tb^{3+} and CaMoO$_4$:Tb^{3+} are shown in Fig. 3.9. PL properties of Dy^{3+} dope d CaWO$_4$ and CaMoO$_4$ show blue to yellow phosphorescence for CaWO$_4$:Dy^{3+} and blue-green to white for CaMoO$_4$:Dy^{3+}. This multicolor tuning emission could increase their applicability in fluorescent lamps, white LED and display panels. PL emission intensities are maximum for 5 and 7 at% Dy^{3+} in CaWO$_4$ and CaMoO$_4$, respectively; more than this content of Dy^{3+} concentration quenching takes place (Sharma and Singh 2013). Guo et al. reported PL emission by Li$_3$Ba$_2$Ln$_3$(MoO$_4$)$_8$:Eu^{3+} (Ln = La, Gd and Y) and observed that PL intensity varies with concentration and heating temperature (Guo et al. 2009).

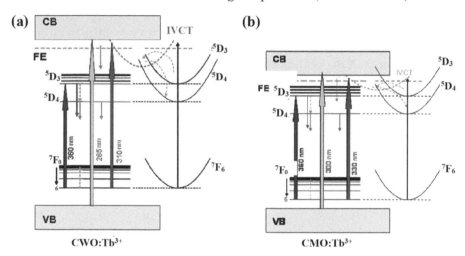

Fig. 3.9. Energy level diagram of (a) CaWO$_4$:Tb^{3+} and (b) CaMoO$_4$:Tb^{3+}. Solid lines represent the absorption (excitation) and radiative emission processes; dotted lines describe the non-radiative processes (Cavalli et al. 2010).

Charge Compensator

RE^{3+} ion doped in M(Mo/W)O$_4$ than the metal site (M) is replaced through rare-earth ion (RE^{3+}). Since oxidation state of both ions are different therefore discrepancy in oxidation number creates defects in crystal. This vacancy in doped material fills with oxygen from the atmosphere and distortion in the PL property increases. To overcome this problem co-doping of alkali metals (Li$^+$, Na$^+$ and K$^+$) works as a charge compensator and play an important role and fill that vacancy as depicted in Fig. 3.10. Several reports are based on enhancement in PL emission and color purity from RE doped tungstate/ molybdate. The co-doping of charge compensation improves crystallinity and also creates oxygen vacancies in crystal so that PL properties can be

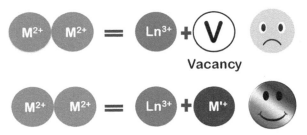

M-Metal, Ln-Lanthanide, M'-Li/Na/K

Fig. 3.10. Schematic representation of vacancies created in crystal by lanthanide ion substitute to metal ion and filling of that vacancy with alkaline metals.

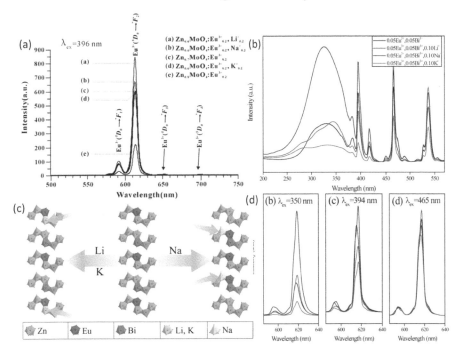

Fig. 3.11. (a) PL emission spectra of $ZnMoO_4$:Eu, Li/Na/K at excitation wavelength 396 nm. (b) PL excitation spectra monitored for 615 nm emission. (c) The schematic diagram of luminescence enhancement mechanisms with remote control device. (d) PL emission spectra at various excitation wavelengths (Ran et al. 2016).

changed. Ju et al. studied variations in emission with co-doping of charge compensator (Li^+, Na^+ and K^+) on $SrWO_4$:Eu^{3+} and found improvement in PL intensity (Ju et al. 2011). The maximum PL intensity obtained for Li^+ co-doped $SrWO_4$:Eu^{3+} was more than the PL intensity of Na^+ and K^+ co-doped samples. Since Li^+ has a small radii as compared to Na^+ and K^+ so that it deftly enters into the crystal (Huang et al. 2011; Shi et al. 2008; Yadav et al. 2016). Moreover, Na^+ and K^+ also give important emission on co-doping due to the creation of oxygen vacancy because of the discrepancy of metal and Na^+/K^+ ion (Li et al. 2009; Ran et al. 2016). The larger ionic radii of K^+ contributes to softer surroundings around the luminescence center and results in larger deviation in metal-ligand distance so that a higher stoke shift was found (Liu et al. 2007). It also reduces the possibility of hypersensitive local crystal symmetry in potassium co-doped tungstates so that a broad PL emission band was found. Upconversion emission also improves with co-doping of charge compensator. Luo and Cao (2008) reported Li^+ co-doping effect on $ZnWO_4$:Er,Yb sample and found remarkable enhancement and red shift in UC emission. Self-stimulated Raman Scattering (SRS) property of doped $SrWO_4$ enhance with co-doping of charge compensator (Na^+ or Nb^{5+}). Nd^{3+} and Er^{3+} doped $SrWO_4$ phosphor give emission in infrared region 1 µm and 1.5 µm, respectively. Na^+ co-doped in this phosphor also improves the emission rate as well as absorption efficiency. Therefore, it can be used in laser applications because it gives stimulated infrared emission from a narrow band pump source (Lupei et al. 2009).

Co-doping of Metal Ion

Metal co-doping with RE^{3+} doped phosphors play a crucial role to improve the radiative emission rate. From literature, it was found that metal co-doping creates oxygen vacancies in crystal which manifest the emission rate. It also showed positive impact on PL intensity till a certain concentration of PL intensity reduced due to crystal lattice collapse. Singh et al. reported Zn^{2+} co-doped $CaMoO_4$:Eu^{3+} with different concentration (Zn^{2+} concentration – 0, 2, 5, 7 and 10 at%0) and found optimum intensity for 2 at% Zn^{2+} (Singh et al. 2015). However, annealing at 900 °C increases PL intensity as well as the value of optimum concentration of Zn^{2+} (10 at%). PL emission intensity enhancement by Bi^{3+} doping from $ZnWO_4$: Eu^{3+} is shown in Fig. 3.12 (Wang et al. 2013). Eu^{3+} and Bi^{3+} ions co-doped in $ZnMoO_4$ when these ions occupy

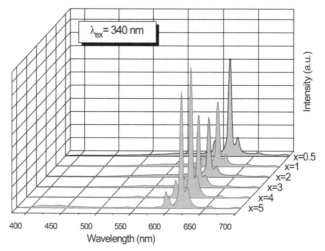

Fig. 3.12. PL spectra of $ZnWO_4$:x mol% Bi^{3+}, 3 mol% Eu^{3+}
(λ_{ex} = 340 nm) (Wang et al. 2013).

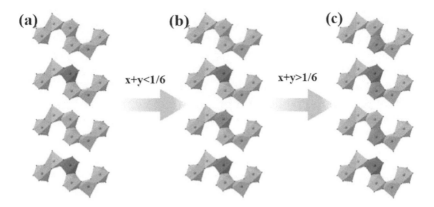

Fig. 3.13. Evolution of occupied Zn sites with the variation of Eu^{3+} (x) and Bi^{3+} (y) concentrations: (a) x < 1/6 and y = 0; (b) x+y < 1/6 and y > 0; (c) x+y > 1/6 and y > 0 (Ran et al. 2015).

two adjacent zinc (Zn^{2+}) sites as depicted in Fig. 3.13, therefore energy transfer from Bi^{3+} to Eu^{3+} is obtained due to the short distance between them.

Co-doping of RE Sensitizer

Rare earth doped with another RE ion improves luminescence efficiency and gives a multicolor emission. In this combination energy is accepted from another so one is known as the accepter and the other as sensitizer. Both are able to absorb incident photons; however the sensitizer has a good photon absorption rate so that it absorbs a higher number of photons as compared to the activator and the absorbed energy is transferred to the activator by energy transfer mechanism. There are three types of energy conversion processes which are shown by the RE co-doped host; these are Down-Conversion (DC) emission, Up Conversion (UC) emission and Quantum Cutting (QC). In DC, ultra-violet photons are absorbed from RE and emit visible photons (Stoke shift). The UC process is that in which visible photon emission is observed via near-infrared excitation (Anti-stoke shift). Moreover, in QC one UV-Visible photon is absorbed by the RE ion and emits two NIR photons.

There are several combinations that have been reported by researchers namely Eu-Tb, Eu-Dy, Eu-Gd, Eu-Sm, Tb-Sm, Tb-Gd, Ho-Yb, Er-Yb, Eu-Yb, Nd-Yb, Tm-Yb etc. (Deng et al. 2014; Dey et al. 2015; Guan et al. 2012; Jain et al. 2017; Jin et al. 2008; Kaczmarek and Van Deun 2013; Lim et al. 2016; Qiao et al. 2016; Ramezani et al. 2015; Sun et al. 2013; Van Deun et al. 2015; Wang et al. 2011; Xu et al. 2014; Yang et al. 2015). These groups give emission from visible to the infrared region. Xu et al. reported Tb^{3+} and Gd^{3+} co-doped $SrWO_4:Eu^{3+}$ (Xu et al. 2014). Emission spectra of $SrWO_4: Tb^{3+}, Eu^{3+}$ shows that as the content of Eu^{3+} increases, intensity of transitions of Eu^{3+} ($^5D_0$–$^7F_2$) increases compared to transitions of Tb^{3+} ($^5D_4$–$^7F_4$). Similarly, emission spectra of $SrWO_4:Gd^{3+}, Eu^{3+}$ phosphor show a red shift if the concentration of Eu^{3+} increases (Xu et al. 2014).

Yang et al. reported PL emission from Eu-Tb co-doped NaLa(MoO$_4$)$_2$ and found multicolor emission with varying Eu and Tb ions concentration (Fig. 3.14) (Yang et al. 2015). Our study based on Eu and Tb co-doped ZnMoO$_4$ showed improved emission of Eu^{3+} ion and good color purity due to energy transfer from Tb^{3+} (sensitizer) to Eu^{3+} (activator) (Jain et al. 2017). Deng and co-workers reported white light emission from Dy^{3+}/Eu^{3+} co-doped BaLa$_2$WO$_7$ phosphors. White light in the CIE diagram found by mixing of Dy^{3+} emission blue (484 nm, $^4F_{9/2} \rightarrow {}^6H_{15/2}$) and yellow (572 nm, $^4F_{9/2} \rightarrow {}^6H_{13/2}$) along with Eu^{3+} red emission (616 nm, $^5D_0 \rightarrow {}^7F_2$) (Fig. 3.15) (Deng et al. 2014).

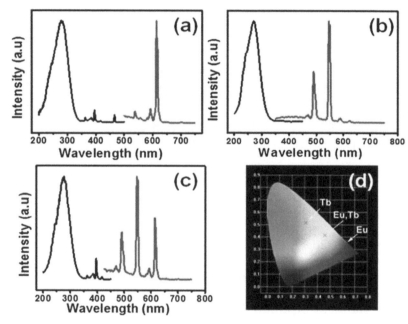

Fig. 3.14. PL excitation and emission spectra of (a) NaLa(MoO$_4$)$_2$:Eu^{3+}, (b) NaLa(MoO$_4$)$_2$:Tb^{3+}, (c) NaLa(MoO$_4$)$_2$:2% Eu^{3+}/3% Tb^{3+} and (d) CIE chromaticity diagram for the emission spectra of the above three samples (Yang et al. 2015).

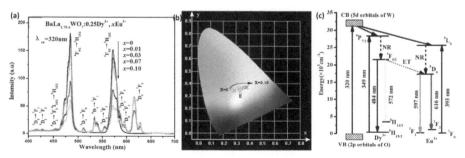

Fig. 3.15. (a) PL emission spectra of BaLaWO$_4$:Dy, Eu under excitation wavelength 320 nm, (b) CIE chromaticity diagram and (c) Energy level diagram for energy transfer between Dy^{3+} and Eu^{3+} ions (Deng et al. 2014).

Singh et al. reported the luminescence property of $CaMoO_4$:Eu^{3+} co-doped with Gd^{3+} (Gd^{3+}= 0, 2, 5, 7 and 10 at%) under 613 nm excitation; strong emission peaks are observed in 250-320 nm range due to overlapping of Eu-O and Mo-O Charge Transfer (CT), sharp peak observed due to f-f transition of Gd^{3+} ion. Intensity of $CaMoO_4$:Eu^{3+} co-doped with Gd^{3+} sample enhances as compared to the $CaMoO_4$:Eu^{3+} sample; this increased intensity indicates there is energy transfer from Gd^{3+} to Eu^{3+}. On annealing at 600 °C and 900 °C PL emission shifts towards the lower wavelength side, related to an increase in the covalent character of Eu-O and Mo-O. The emission spectra indicated under 266 nm excitation for Gd^{3+}, strong emission peaks at 590 and 613 nm. For preparing the sample PL intensity should be maximum for 2 at% Gd^{3+} but on annealing at 600 °C and 900 °C intensity is maximum for 7 and 10 at% Gd^{3+}, respectively. This change may arise due to surface defect or capping ligands on the surface bonds, water molecules adsorbed surface and removal of –NO, –CH group (Singh et al. 2014).

Upconversion emission has also been reported well for tungstate and molybdate hosts. Er-Yb co-doped in hosts $BaMoO_4$, $SrWO_4$, $ZnWO_4$ etc. gives upconversion green emission (Pandey et al. 2015; Rai and Pandey 2016; Soni et al. 2016). Moreover Ho-Yb co-doped into the hosts $CaMoO_4$, $BaMoO_4$, $CaLaMoO_4$, $ZnWO_4$, $NaYb(MoO_4)_2$ emit red and green photons (Lim et al. 2015; Rai 2007; Wei et al. 2016; Xu et al. 2013; Zhang et al. 2017). Moreover, Tm-Yb co-doped hosts are $GdWO_4$, $NaY(WO_4)_2$, $KY(WO_4)_2$, $PbWO_4$, $LiLa(MoO_4)_2$ etc. and give emission in the region of blue and near-infrared (Babu et al. 2011; Demidovich et al. 2002; Hu et al. 2016; Song et al. 2007; Zhang et al. 2017). The upconversion energy transfer mechanism for Er/Ho/Tm and Yb are given in Fig. 3.16.

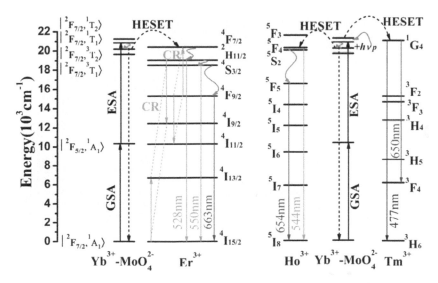

Fig. 3.16. Energy level diagrams and possible upconversion processes of $NaYb(MoO_4)_2$ doped with Er^{3+}, Tm^{3+} and Ho^{3+} by 980 nm laser excitation (Zhang et al. 2017).

Applications of Rare Earth Doped Tungstates and Molybdates

White LED Fabrication

There are several reports based on formation of red LED coated on NUV InGaN chip with red/yellow phosphors namely $AgGdMo_2O_8:Eu^{3+}$, $Gd_2(MoO_4)_3:Eu^{3+}$, $GdEu(MoO_4)_{0.5}(WO_4)_{2.5}$, $CaMoO_4:Dy^{3+},Li^+$, $Gd_2(MoO_4)_3$: Sm^{3+}, $NaEu_{0.96}Sm_{0.04}(MoO_4)_2$, $BaGd_{1.2}Eu_{0.8}(WO_4)_{0.4}$ $(MoO_4)_{3.6}$, $Y_2WO_6:Pr^{3+}$ etc. (Chen et al. 2018; He et al. 2010; Li et al. 2011; Wang et al. 2007; Wang et al. 2008; Xiong et al. 2015; Zeng et al. 2009). White LED fabricate on near-ultraviolet (NUV) InGaN chip or blue InGaN chip with coating of green/yellow and red phosphor. The operating temperature of LED is about 373 K so the first important point is to measure the thermal stability of red phosphor. Dong et al. reported that PL has a red emission intensity of $Gd_{2.20}BW_{0.80}O_9:0.20Mo^{6+}+0.80Eu^{3+}$ which are nearly constant up to 523 and 423 K at excitation wavelengths 385 and 465 nm, respectively (Dong et al. 2014). W-LED with the mixed coating of red phosphor $Gd_3B(W,Mo)O_9:Eu^{3+}$, green phosphor $BaMgAl_{10}O_{17}:Eu^{2+}$, Mn^{2+} and blue phosphor alkaline earth chloroborate coated on an NUV LED chip (excitation wavelength 385 nm). The W-LED lamp glows with 20 mA driven direct current as shown in Fig. 3.17. The CIE co-ordinate (0.320, 0.349) found near to standard white and CCT value is 6031 K (close to day light). In addition, Zeng et al. gives a comparable

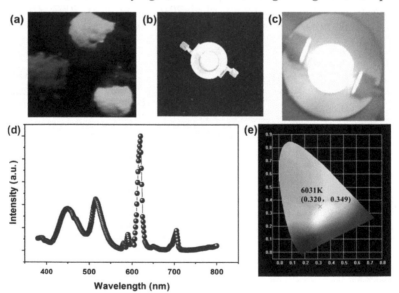

Fig. 3.17. (a) The photos of red/blue/green phosphors under a 365 nm UV lamp. (b) Images of fabricated W-LEDs. (c) Images of fabricated W-LEDs driven by a 20 mA DC current. (d) The emission spectrum of the fabricated W-LEDs lamp operated at 20 mA. (e) The CIE chromaticity coordinates for the lamp operated at 20 mA (Dong et al. 2014).

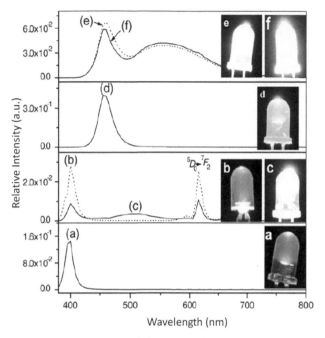

Fig. 3.18. The emission spectra of the original near-UV InGaN chip-LED (a); the red LED by combining a near-UV InGaN chip with $Sr_9Y_{0.4}Eu_{1.6}W_4O_{24}$ (b); W-LED by combining a near-UV InGaN chip with $Sr_9Y_{0.4}Eu_{1.6}W_4O_{24}$ and blue and green phosphors (c); the original blue InGaN chip-LED (d); W-LED by combining a blue InGaN chip with YAG:Ce³⁺ (e); and W-LED by combining a blue InGaN chip with YAG:Ce³⁺ and $Sr_9Y_{0.4}Eu_{1.6}W_4O_{24}$ (f); under excitation of 20 mA forward bias (Zeng et al. 2009).

study of W-LED fabrication on NUV and blue InGaN chip shown in Fig. 3.18 (Zeng et al. 2009). They first prepared W-LED on NUV InGaN chip (395 nm) by coating red phosphor $Sr_9Y_{0.4}Eu_{1.6}W_4O_{24}$ with blue and green phosphor. They found CIE co-ordinate in the white region (0.3461, 0.3517), luminous efficiency 2.38 lm/W, color temperature 4962 K and low CRI value 62.9. Furthermore, they prepared W-LED with red phosphor $Sr_9Y_{0.4}Eu_{1.6}W_4O_{24}$ and YAG:Ce³⁺ on blue InGaN chip (460 nm). For this W-LED, CIE co-ordinates are (0.3278, 0.3632), luminous efficiency is 45.04 lm/W, color temperature is 6710 K and good CRI value 78.2. Therefore, $Sr_9Y_{0.4}Eu_{1.6}W_4O_{24}$ red phosphor gives better performance on blue chip with YAG:Ce³⁺.

Temperature Sensing

Temperature sensing performance is measured by taking Fluorescent Intensity Ratio (FIR) of two thermally coupled levels of lanthanides. If the separation between electronic levels lies in the range 200-2000 cm⁻¹ then it follows the Boltzmann distribution function FIR = $A \exp\left(\dfrac{-\Delta E}{kT}\right)$ where ΔE

is the energy difference between thermally coupled levels, k is Boltzmann constant and T is absolute temperature (Soni et al. 2016). Absolute sensitivity is calculated by using the equation $S = FIR \times \left(\dfrac{\Delta E}{kT^2} \right)$. Temperature dependent photoluminescence emission spectra manifest on increasing temperature, PL intensity gradually decreased with the increase in non-radiative relaxation. Generally, Er-Yb, Ho-Yb, Tm-Yb, Sm and Eu rare earths are used for optical thermometry (Li et al. 2018; Meert et al. 2014; Rai 2007; Soni et al. 2016; Zhang et al. 2017). The Er-Yb combination is mostly used for temperature sensing measurements, its thermally coupled levels are $^2H_{11/2}$ and $^4S_{3/2}$ (Fig. 3.19). FIR are calculated by taking the ratio of emission peaks $^2H_{11/2} \rightarrow {}^4I_{15/2}$ (526 nm) and $^4S_{3/2} \rightarrow {}^4I_{15/2}$ (551 nm) (Van Deun et al. 2015). Table 3.2 summarizes a number of tungstate and molybdate hosts used for Er-Yb doping with absolute sensitivity and temperature. Tm^{3+} doped $NaYb(MoO_4)_2$ phosphors FIR calculated by taking the ratio of thermally coupled levels $^1G_4(1)$, $^1G_4(2)$ $\rightarrow {}^3H_6$ and its sensitivity was 0.0025 K^{-1} at 323 K (Zhang et al. 2017). Along with Sm^{3+} ions temperature sensitivity performance is measured for the transition $^4G_{5/2} \rightarrow {}^6H_{9/2}$ and O_2—Mo^{6+} band. Li et al. reported relative sensitivity for

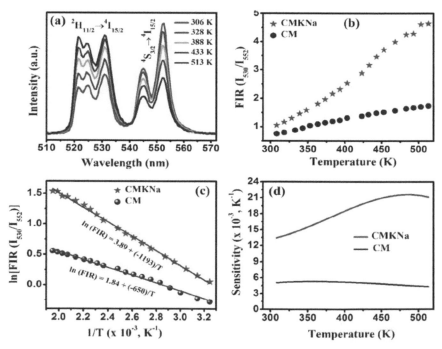

Fig. 3.19. (a) Temperature dependent upconversion emission spectra for 530 nm ($^2H_{11/2}$ $\rightarrow {}^4I_{15/2}$) and 552 nm ($^4S_{3/2} \rightarrow {}^4I_{15/2}$) of $CaMoO_4$: $Er^{3+}/Yb^{3+}/K^+/Na^+$ phosphor excited by 980 nm excitation; (b) variation of FIR with temperature; (c) the monolog plot of the FIR (I_{530}/I_{552}) as a function of inverse absolute temperature; and (d) variation of absolute sensitivity (S) with temperature (Sinha et al. 2017).

Sm^{3+} doped Lu_2MoO_6 at excitation wavelength 402 nm and found 4.9% (423 K) (Li et al. 2018). Meert et al. reported sensitivity for $CaEu_2(WO_4)_4$ in the temperature range 300-500 K. which took FIR of $^5D_0 \rightarrow ^5F_1$ and $^5D_1 \rightarrow ^5F_1$ and its highest sensitivity was 0.014 K^{-1} at room temperature (Meert et al. 2014).

Table 3.2. Temperature sensitivity of lanthanide doped host in temperature range

Lanthanide	Host	Temperature range (K)	Sensitivity (K^{-1})	Reference
Er-Yb	$SrWO_4$	299-518	14.98×10^{-3} (403 K)	(Pandey et al. 2015)
	$ZnWO_4$	299-798	14.63×10^{-3} (798 K)	(Rai and Pandey, 2016)
	$BaMoO_4$	300-575	1.3×10^{-2} (300 K)	(Soni et al. 2016)
	$NaY(MoO_4)_2$	303-523	9.6×10^{-3} (523 K)	(Yang et al. 2015)
	$YbMoO_4$	300-650	1.06×10^{-2} (451 K)	(Wu et al. 2016)
	$NaY(WO_4)_2$	133–773	1.12×10^{-2} (515 K)	(Du et al. 2016)
	$NaYb(MoO_4)_2$	323-573	1.22×10^{-2} (548 K)	(Zhang et al. 2017)
Ho-Yb	$CaMoO_4$	303-543	6×10^{-3} (353K)	(Dey et al. 2015)
	$CaWO_4$	303-923	5×10^{-3} (923K)	(Xu et al. 2013)
	$ZnWO_4$	83-503	6.4×10^{-3} (83K)	(Chai et al. 2017)
	$NaYb(MoO_4)_2$	323-573	3.5×10^{-4} (323 K)	(Zhang et al. 2017)
Tm-Yb	$NaYb(MoO_4)_2$	323-573	2.5×10^{-3} (323 K)	(Zhang et al. 2017)
Eu	$CaEu_2(WO_4)_4$	300-500	1.4×10^{-2} (300 K)	(Meert et al. 2014)
Sm	Lu_2MoO_6	298-473	4.9% (423 K)	(Li et al. 2018)

Furthermore, Ho-Yb doped host have thermally coupled levels $^3K_8$, $^5F_3 \rightarrow ^5I_8$ and $^5F_1 / ^5G_6 \rightarrow ^5I_8$ (460 nm/487 nm) and follow the aforementioned Boltzmann distribution (Dey et al. 2015). Moreover, there are four other levels $^5F_4$, $^5S_2$, $^5I_8$ and $^5I_7$ of Ho^{3+} ion used to measure temperature sensitivity and taken as four level systems as depicted in Fig. 3.20. Additionally, energy difference between $^5F_4$ and $^5S_2$ is about 120 cm^{-1} so the temperature sensing FIR for such type of coupled levels are estimated with the equation given by Haro-Gonzalez et al. (Haro-González et al. 2011; Lojpur et al. 2013)

$$FIR = \frac{I_{547}}{I_{759}} = \frac{I_1 + I_1'}{I_2 + I_2'} = \frac{N_4 \omega_{41} g_4 h \nu_{41} + N_3 \omega_{31} g_3 h \nu_{31}}{N_4 \omega_{42} g_4 h \nu_{42} + N_3 \omega_{32} g_3 h \nu_{32}}$$

where, $N_4 = N_3 \exp \left(\dfrac{-\Delta E}{kT} \right)$

Here N_4 and N_3 are populations of the excited state 4 and 3, ω_{41}/ω_{42} and ω_{31}/ω_{32} define the spontaneous emission rate from $^5F_4 / ^5S_2$ levels to the $^5I_8$ and $^5I_7$, respectively. Moreover, g_3 and g_4 are degeneracy of $^5S_2$ and $^5F_4$ levels,

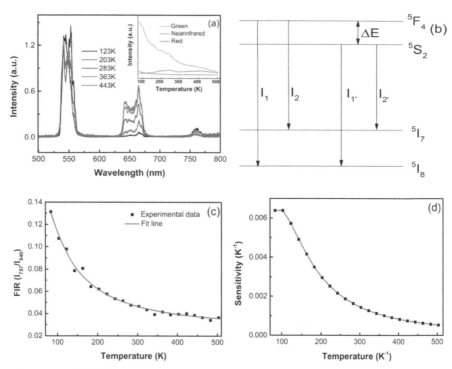

Fig. 3.20. (a) UC emission spectra of $ZnWO_4{:}0.01Ho^{3+}/0.15Yb^{3+}$ at different temperatures; inset, plot of upconversion intensity vs. temperature. (b) A simplified four-level system diagram. (c) FIR of I_{757}/I_{540} at 83–503 K. (d) Sensitivity graph for the I_{757}/I_{540} ratio (Chai et al. 2017).

hv is the energy of the respective transitions. DE depicts the energy gap between $^5F_4$ and $^5S_2$. Therefore, FIR of 547 and 759 nm emission peaks (Ho^{3+} ion) fitted by using this equation.

Biomedical Field

Lanthanide activated nanophosphors are gaining a lot of attention for bioimaging due to their astonishing properties like low photo-bleaching during exposure to light, higher sensitivity and good penetration depth of light in the body tissue. There are some limitations in bioimaging by the visible imaging range 400–750 nm such as absorption and scattering of photons induced by the biological tissue and water molecules that lead to the attenuation of the signal which passes through the feature of interest. This problem could be resolved by using phosphor whose emission falls in the biological transparency window (650–1450 nm) so that scattering and absorption of light by the biological tissue would be less and lead to deeper penetration than the visible light. This observation was reported for $ZnMoO_4{:}Tm$, Yb, K which gives emission around 820 nm at excitation

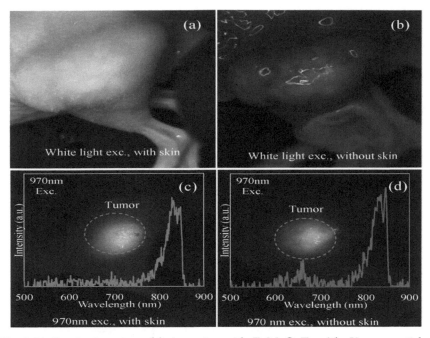

Fig. 3.21. *In vivo* imaging of living mice with ZnMoO$_4$:Tm, Yb, K nanoparticles directly injected into the thigh muscles below the skin. (a and c) Photographs taken before skin removal, and (b and d) photographs taken after removing the skin. The insets (in c and d) show upconverted luminescence when excited by a 970 nm laser (Luitel et al. 2016).

wavelength 980 nm (Luitel et al. 2016). It has good biocompatibility and strong NIR emission under NIR excitation. Figure 3.21 shows vivo imaging of living mice by injecting ZnMoO$_4$:Tm, Yb, K nanoparticles, and the image taken under NIR light by 970 nm laser excitation, as this penetration depth is good to detect a tumor. UCNPs are an effective way to non-invasively visualize biological distribution and real time imaging of the biological process with deep tissue and light penetration. Another study based on confocal imaging of HePG2 cancer cell by Eu^{3+} and Li$^+$ co-doped sodium zinc molybdate illustrates that oxide phosphor showed bright red fluorescence with cells (Jain et al. 2019).

Moreover, biocompatibility of NaLa(MoO$_4$)$_2$:Eu,Tb was determined on ARPE – 19 cells by CCK – 8 tests within the concentration range 50–150 µg/mL (Yang et al. 2015). Cytotoxicity assay were done for microrodes, microflower and nanorods for 24, 48, 72 hours. They observed that particle size at microscale and the cell viability remained constant. However, when size of the particles belongs in nanoscale then significant increment in cell viability with increasing concentration of the nanomaterial was observed. This might be because the surface area of nanomaterial is enlarged to provide a suitable environment for the cell adhesion and growth. Guo et al. reported multimodal imaging by NaGd(WO$_4$)$_2$:Eu^{3+} and in this the phosphor host absorbed the X-rays and transferred energy to Eu^{3+} ion which shows strong red florescence (Fig. 3.22) (Guo et al. 2018). It dispersed in water by modifying the surface of the nano-phosphor (nanorodes) modified by PEG. Here biocompatibility was measured by cytotoxicity assay on HeLa, HePG to and MCF–7 cells. It was measured under 5 minutes of X-ray irradiation (40 kV, 70 µA) with different concentration (50–500 µg/mL). The viability was nearly constant until 500 µg/mL. Furthermore, confocal fluorescence image and overlay image were taken for HeLa cells with excitation. This phosphor is a good signal to the noise ratio for x ray luminescence as compared to UV-excited fluorescence (Fig. 3.23). Therefore, it can be used in optical imaging, MRI and CT imaging.

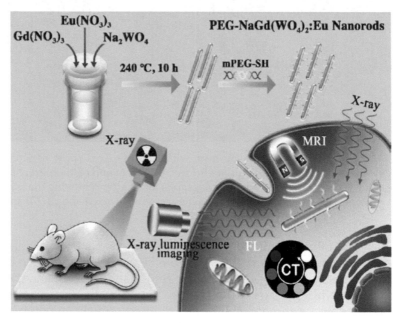

Fig. 3.22. Schematic illustration of the synthesis of PEG-NGW:Eu nanorods and their applications in X-ray luminescence imaging, MRI and CT imaging (Guo et al. 2018).

Additionally, hyperthermia treatment is used to kill cancer cells by raising the temperature of cancer affecting the body part. When AC current is supplied to the heating setup, temperature of MNPs increases and reach 42 °C in a few moments; cancer cells get killed at this temperature. The magnetic nanoparticles were chosen on the basis of less heating time and high specific absorption rate. The theory behind increment in temperature by providing AC magnetic can be understood as magnetic nanoparticles (MNPs) dispersed in a medium and nanoparticles are free to move (because of Brownian motion). There are two types of relaxation in MNPs, one is Brownian relaxation due to particle rotation and other is Neel's relaxation due to magnetic moment rotation. However, Neel's relaxation is applied for superparamagnetic NP (spin relaxation is the order of 10^{-9} s). When AC

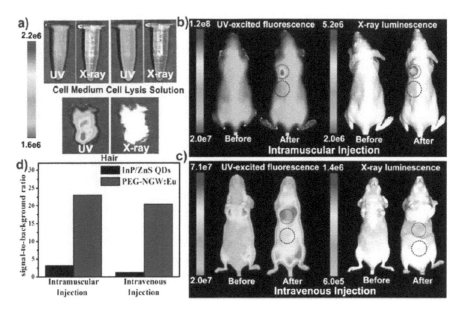

Fig. 3.23. (a) Luminescence imaging of cell medium, cell lysis, mice hair under UV or X-ray excitation. (b) Luminescence imaging of mice before and after intramuscular injection of InP/ZnS QDs (left) and PEG-NGW:Eu nanorods (right) under UV and X-ray excitation, respectively. The red and black dashed circles represent target sites and background sites, respectively. (c) Luminescence imaging of mice before and after (30 minutes) intravenous injection of InP/ZnS QDs (left) and PEG-NGW:Eu nanorods (right) under UV and X-ray excitation, respectively. The red and black dashed circles represent target sites and background sites, respectively. (d) Comparison of signal-to-background ratio in mice under UV or X-ray excitation with InP/ZnS QDs or PEG-NGW:Eu nanorods, respectively (Guo et al. 2018).

magnetic field is applied, then the direction of current changes with time, so that the magnetic spins direction changes with time. Therefore, MNPs solution exhibits heat loss and power dissipation to hysteresis loss, Brownian motion, Neel's spins relaxation and eddy current. The magnetic moment of MNPs is aligned in the direction of applied AC magnetic field. During this alignment, particles collide with the surrounding medium which causes heating (Prasad et al. 2013). The performance of MNPs was detected by calculating Specific Absorption Rate (SAR). SAR is the rate of absorption of energy by the material when the AC field is applied on the material. Parchur et al. reported $CaMoO_4$:Eu@Fe_3O_4 Hybrid MNPS for hyperthermia (Parchur et al. 2013). They observed heating 200, 300 and 400 mA current and their respective SAR values are 10, 19 and 26 W/g. Moreover, vitro cytotoxicity has been performed by MTT assay on HePG2 cells. Hybrid MNPs were taken in concentration 5–640 µg/mL and found IC_{50} was 193.26 µg/mL and MNPS treated hTERT cells 60-80% alive (viable).

Conclusion

In this chapter, luminescence properties for lanthanide doped metal molybdate and metal tungstate have been discussed. Luminescence mechanism for electronic transition of lanthanide has been described. With all these mechanisms, the reason behind luminescence quenching has been elaborated here. It can also be concluded that several factors such as, temperature, pressure, pH, surfactant, concentration etc., affect the luminescence intensity. PL intensity affect with surface or crystal defect which can improve with doping of the charge compensator i.e. alkali metals. Furthermore, metal and lanthanide co-doped molybdates/tungstates showed ramifications on PL emission. The concise mechanism for improving PL emission with co-doping of two lanthanides has been discussed. Applications of lanthanide doped materials like LED, temperature sensor, and bioimaging are explored briefly. However, some of the disadvantages of metal tungstate and molybdate hosts are that its phonon vibration energy is high (400-1000 cm^{-1}); thereby, its upconversion emission property gets affected with phonon vibration. Again, tungstate and molybdate cannot dissolve in water so that it is hard to use such hosts for bio-medical application. Hence, in future it would be of interest to prepare water dispersible tungstate/molybdate by its surface functionalization for biological applications.

Acknowledgements

One of the authors, Neha acknowledges the Maulana Azad National Fellowship (MANF) provided by University Grants Commission (UGC), Govt. of India. Jai Singh would like to acknowledge UGC-India and DST for providing project under UGC Start-up Grant FT30-56/2014 (BSR)3(A)a and DST Fast Track Grant no. SR/FTP/PS-144/2012.

References

Anicete-Santos, M., Orhan, E., De Maurera, M., Simoes, L., Souza, A., Pizani, P., . . ., Beltrán, A. 2007. Contribution of structural order-disorder to the green photoluminescence of PbWO$_4$. Physical Review B 75(16), 165105.

Anicete-Santos, M., Picon, F.C., Alves, C.N., Pizani, P.S., Varela, J.A., Longo, E. 2011. The role of short-range disorder in BaWO$_4$ crystals in the intense green photoluminescence. The Journal of Physical Chemistry C 115(24), 12180-12186.

Arora, S., Chudasama, B. 2006. Crystallization and optical properties of CaWO$_4$ and SrWO$_4$. Crystal Research and Technology: Journal of Experimental and Industrial Crystallography 41(11), 1089-1095.

Babu, A.M., Jamalaiah, B., Chengaiah, T., Reddy, G.L., Moorthy, L.R. 2011. Upconversion luminescence in Tm^{3+}/Yb^{3+} co-doped lead tungstate tellurite glasses. Physica B: Condensed Matter 406(15-16), 3074-3078.

Basiev, T., Sobol, A., Voronko, Y.K., Zverev, P. 2000. Spontaneous Raman spectroscopy of tungstate and molybdate crystals for Raman lasers. Optical Materials 15(3), 205-216.

Bray, K.L. 1996. Luminescent Materials by G. Blasse (University of Utrecht, The Netherlands) and BC Grabmaier (Siemens Research Laboratories). Springer: New York. 1994. ISBN 0-387-58109-0. In: ACS Publications.

Bünzli, J.-C.G., Piguet, C. 2005. Taking advantage of luminescent lanthanide ions. Chemical Society Reviews 34(12), 1048-1077.

Cavalcante, L., Batista, F., Almeida, M., Rabelo, A., Nogueira, I., Batista, N., . . . Li, M.S. 2012. Structural refinement, growth process, photoluminescence and photocatalytic properties of $(Ba_{1-x} Pr_{2x/3}) WO_4$ crystals synthesized by the coprecipitation method. RSC Advances 2(16), 6438-6454.

Cavalcante, L.S., Moraes, E., Almeida, M., Dalmaschio, C., Batista, N., Varela, J.A., . . ., Beltrán, A. 2013. A combined theoretical and experimental study of electronic structure and optical properties of β-ZnMoO$_4$ microcrystals. Polyhedron 54, 13-25.

Cavalli, E., Boutinaud, P., Mahiou, R., Bettinelli, M., Dorenbos, P. 2010. Luminescence dynamics in Tb^{3+}-doped CaWO$_4$ and CaMoO$_4$ crystals. Inorganic Chemistry 49(11), 4916-4921.

Chai, X., Li, J., Wang, X., Li, Y., Yao, X. 2017. Upconversion luminescence and temperature-sensing properties of Ho^{3+}/Yb^{3+}-codoped ZnWO$_4$ phosphors based on fluorescence intensity ratios. RSC Advances 7(64), 40046-40052.

Chen, Y., Liang, Y., Cai, M., Ke, T., Zhang, M., He, X., Zeng, Q. 2018. Luminescent property and application research of red molybdate phosphors for W-LEDs. Journal of Materials Science: Materials in Electronics 29(14), 11930-11935.

Demidovich, A., Kuzmin, A., Nikeenko, N., Titov, A., Mond, M., Kueck, S. 2002. Optical characterization of Yb, Tm: KYW crystal concerning laser application. Journal of Alloys and Compounds 341(1-2), 124-129.

Deng, Y., Yi, S., Huang, J., Xian, J., Zhao, W. 2014. White light emission and energy transfer in Dy^{3+}/Eu^{3+} co-doped BaLa$_2$WO$_7$ phosphors. Materials Research Bulletin 57, 85-90.

Dexter, D.L. 1953. A theory of sensitized luminescence in solids. The Journal of Chemical Physics 21(5), 836-850.

Dey, R., Kumari, A., Soni, A.K., Rai, V.K. 2015. CaMoO$_4$:Ho^{3+}–Yb^{3+}–Mg^{2+} upconverting phosphor for application in lighting devices and optical temperature sensing. Sensors and Actuators B: Chemical 210, 581-588.

Dong, S., Ye, S., Wang, L., Chen, X., Yang, S., Zhao, Y., . . . Zhang, Q. 2014. Gd$_3$B (W, Mo) O$_9$: Eu^{3+} red phosphor: From structure design to photoluminescence behavior and near-UV white-LEDs performance. Journal of Alloys and Compounds 610, 402-408.

Du, P., Luo, L., Yu, J.S. 2016. Upconversion emission, cathodoluminescence and temperature sensing behaviors of Yb^{3+} ions sensitized NaY (WO$_4$) 2: Er^{3+} phosphors. Ceramics International 42(5), 5635-5641.

Errandonea, D., Manjon, F.J. 2008. Pressure effects on the structural and electronic properties of ABX4 scintillating crystals. Progress in Materials Science 53(4), 711-773.

Geng, D., Li, G., Shang, M., Yang, D., Zhang, Y., Cheng, Z., Lin, J. 2012. Color tuning via energy transfer in Sr$_3$In (PO$_4$)3:Ce^{3+}/Tb^{3+}/Mn^{2+} phosphors. Journal of Materials Chemistry 22(28), 14262-14271.

Guan, L., Wei, W., Guo, S., Su, H., Li, X., Shang, Y., . . ., Guangsheng, F. 2012. Fabrication and luminescent properties of Tb^{3+} doped double molybdate phosphors. Journal of the Electrochemical Society 159(4), D200-D203.

Guo, C., Gao, F., Liang, L., Choi, B.C., Jeong, J.-H. 2009. Synthesis, characterization and luminescent properties of novel red emitting phosphor $Li_3Ba_2Ln_3$ (MoO_4) 8: Eu^{3+} (Ln = La, Gd and Y) for white light-emitting diodes. Journal of Alloys and Compounds 479(1-2), 607-612.

Guo, T., Lin, Y., Zhang, W.-J., Hong, J.-S., Lin, R.-H., Wu, X.-P., . . . Yang, H.-H. 2018. High-efficiency X-ray luminescence in Eu^{3+}-activated tungstate nanoprobes for optical imaging through energy transfer sensitization. Nanoscale 10(4), 1607-1612.

Haro-González, P., León-Luis, S., González-Pérez, S., Martín, I. 2011. Analysis of Er^{3+} and Ho^{3+} codoped fluoroindate glasses as wide range temperature sensor. Materials Research Bulletin 46(7), 1051-1054.

He, X., Zhou, J., Lian, N., Sun, J., Guan, M. 2010. Sm^{3+}-activated gadolinium molybdate: An intense red-emitting phosphor for solid-state lighting based on InGaN LEDs. Journal of Luminescence 130(5), 743-747.

Hu, W., Hu, F., Li, X., Fang, H., Zhao, L., Chen, Y., . . . Yin, M. 2016. Optical thermometry of a Tm^{3+}/Yb^{3+} co-doped LiLa $(MoO_4)_2$ upconversion phosphor with a high sensitivity. RSC Advances 6(88), 84610-84615.

Huang, J., Xu, J., Luo, H., Yu, X., Li, Y. 2011. Effect of alkali-metal ions on the local structure and luminescence for double tungstate compounds AEu $(WO_4)_2$ (A = Li, Na, K). Inorganic Chemistry 50(22), 11487-11492.

Issler, S.L., Torardi, C.C. 1995. Solid state chemistry and luminescence of X-ray phosphors. Journal of Alloys and Compounds 229(1), 54-65.

Jain, N., Paroha, R., Singh, R.K., Mishra, S.K., Chaurasiya, S.K., Singh, R., Singh, J. 2019. Synthesis and Rational design of europium and lithium doped sodium zinc molybdate with red emission for optical imaging. Scientific Reports 9(1), 2472.

Jain, N., Singh, B.P., Singh, R.K., Singh, J., Singh, R. 2017. Enhanced photoluminescence behaviour of Eu^{3+} activated $ZnMoO_4$ nanophosphors via Tb^{3+} co-doping for light emitting diode. Journal of Luminescence 188, 504-513.

Jin, Y., Zhang, J., Lü, S., Zhao, H., Zhang, X., Wang, X.-j. 2008. Fabrication of Eu^{3+} and Sm^{3+} codoped micro/nanosized $MMoO_4$ (M = Ca, Ba, and Sr) via facile hydrothermal method and their photoluminescence properties through energy transfer. The Journal of Physical Chemistry C 112(15), 5860-5864.

Ju, Z., Wei, R., Gao, X., Liu, W., Pang, C. 2011. Red phosphor $SrWO_4:Eu^{3+}$ for potential application in white LED. Optical Materials 33(6), 909-913.

Kaczmarek, A.M., Van Deun, R. 2013. Rare earth tungstate and molybdate compounds – from 0D to 3D architectures. Chemical Society Reviews 42(23), 8835-8848.

Li, G., Guoqi, J., Baozhu, Y., Xu, L., Litao, J., Zhiping, Y., Guangsheng, F. 2011. Synthesis and optical properties of Dy^{3+}, Li^+ doped $CaMoO_4$ phosphor. Journal of Rare Earths 29(6), 540-543.

Li, J., Jia, G., Zhu, Z., You, Z., Wang, Y., Wu, B., Tu, C. 2007. Optical spectroscopy of Ho^{3+}-doped $SrWO_4$ scheelite crystal. Journal of Physics D: Applied Physics 40(7), 1902.

Li, L., Fu, S., Zheng, Y., Li, C., Chen, P., Xiang, G., . . . Zhou, X. 2018. Near-ultraviolet and blue light excited Sm^{3+} doped Lu_2MoO_6 phosphor for potential solid state lighting and temperature sensing. Journal of Alloys and Compounds 738, 473-483.

Li, X., Yang, Z., Guan, L., Guo, J., Wang, Y., Guo, Q. 2009. Synthesis and luminescent properties of $CaMoO_4:Tb^{3+}$, R^+ (Li^+, Na^+, K^+. Journal of Alloys and Compounds 478(1-2), 684-686.

Liao, H.-W., Wang, Y.-F., Liu, X.-M., Li, Y.-D., Qian, Y.-T. 2000. Hydrothermal preparation and characterization of luminescent $CdWO_4$ nanorods. Chemistry of Materials 12(10), 2819-2821.

Lim, C.S., Aleksandrovsky, A., Molokeev, M., Oreshonkov, A., Atuchin, V. 2015. The modulated structure and frequency upconversion properties of $CaLa_2$ $(MoO_4)_4$:Ho^{3+}/Yb^{3+} phosphors prepared by microwave synthesis. Physical Chemistry Chemical Physics 17(29), 19278-19287.

Lim, C.S., Aleksandrovsky, A.S., Molokeev, M.S., Oreshonkov, A.S., Ikonnikov, D.A., Atuchin, V.V. 2016. Triple molybdate scheelite-type upconversion phosphor NaCaLa $(MoO_4)_3$: Er^{3+}/Yb^{3+}: structural and spectroscopic properties. Dalton Transactions 45(39), 15541-15551.

Liu, J., Lian, H., Shi, C. 2007. Improved optical photoluminescence by charge compensation in the phosphor system $CaMoO_4$:Eu^{3+}. Optical Materials 29(12), 1591-1594.

Lojpur, V., Nikolic, M., Mancic, L., Milosevic, O., Dramicanin, M. 2013. Y_2O_3:Yb, Tm and Y_2O_3:Yb, Ho powders for low-temperature thermometry based on upconversion fluorescence. Ceramics International 39(2), 1129-1134.

Luitel, H.N., Chand, R., Hamajima, H., Gaihre, Y.R., Shingae, T., Yanagita, T., Watari, T. 2016. Highly efficient NIR to NIR upconversion of $ZnMoO_4$:Tm^{3+}, Yb^{3+} phosphors and their application in biological imaging of deep tumors. Journal of Materials Chemistry B 4(37), 6192-6199.

Luo, X.-x., Cao, W.-h. 2008. Upconversion luminescence properties of Li^+-doped $ZnWO_4$:Yb, Er. Journal of Materials Research 23(8), 2078-2083.

Lupei, A., Achim, A., Lupei, V., Gheorghe, C., Gheorghe, L., Hau, S. 2009. RE^{3+} doped $SrWO_4$ as laser and nonlinear active crystals. Rom. J. Phys 54(9-10), 919-928.

Malta, O., Ribeiro, S., Faucher, M., Porcher, P. 1991. Theoretical intensities of 4f-4f transitions between stark levels of the Eu^{3+} ion in crystals. Journal of Physics and Chemistry of Solids 52(4), 587-593.

Meert, K.W., Morozov, V.A., Abakumov, A.M., Hadermann, J., Poelman, D., Smet, P.F. 2014. Energy transfer in Eu^{3+} doped scheelites: Use as thermographic phosphor. Optics Express, 22(103), A961-A972.

Nakamura, S., Fasol, G. 2013. The Blue Laser Diode: GaN Based Light Emitters and Lasers. Springer Science & Business Media.

Pandey, A., Rai, V.K., Kumar, V., Kumar, V., Swart, H. 2015. Upconversion based temperature sensing ability of Er^{3+}–Yb^{3+} codoped $SrWO_4$: An optical heating phosphor. Sensors and Actuators B: Chemical 209, 352-358.

Parchur, A., Ansari, A., Singh, B., Hasan, T., Syed, N., Rai, S., Ningthoujam, R. 2013. Enhanced luminescence of $CaMoO_4$:Eu by core@ shell formation and its hyperthermia study after hybrid formation with Fe_3O_4: Cytotoxicity assessment on human liver cancer cells and mesenchymal stem cells. Integrative Biology 6(1), 53-64.

Piskuła, Z., Staninski, K., Lis, S. 2011. Luminescence properties of Tm^{3+}/Yb^{3+}, Er^{3+}/Yb^{3+} and Ho^{3+}/Yb^{3+} activated calcium tungstate. Journal of Rare Earths 29(12), 1166-1169.

Prasad, A., Parchur, A., Juluri, R., Jadhav, N., Pandey, B., Ningthoujam, R., Vatsa, R. 2013. Bi-functional properties of Fe_3O_4@YPO_4:Eu hybrid nanoparticles: hyperthermia application. Dalton Transactions 42(14), 4885-4896.

Qiao, X., Tsuboi, T., Seo, H.J. 2016. Correlation among the cooperative luminescence, cooperative energy transferred Eu^{3+}-emission, and near-infrared Yb^{3+} emission of Eu^{3+}-doped LiYb $(MoO_4)_2$. Journal of Alloys and Compounds 687, 179-187.

Qin, X., Liu, X., Huang, W., Bettinelli, M., Liu, X. 2017. Lanthanide-activated phosphors based on 4f-5d optical transitions: Theoretical and experimental aspects. Chemical Reviews 117(5), 4488-4527.

Rai, V.K. 2007. Temperature sensors and optical sensors. Applied Physics B, 88(2), 297-303.

Rai, V.K., Pandey, A. 2016. Efficient color tunable $ZnWO_4:Er^{3+}-Yb^{3+}$ phosphor for high temperature sensing. Journal of Display Technology 12(11), 1472-1477.

Ramezani, M., Hosseinpour-Mashkani, S.M., Sobhani-Nasab, A., Estarki, H.G. 2015. Synthesis, characterization, and morphological control of $ZnMoO_4$ nanostructures through precipitation method and its photocatalyst application. Journal of Materials Science: Materials in Electronics 26(10), 7588-7594.

Ran, W., Wang, L., Tan, L., Qu, D., Shi, J. 2016. Remote control effect of Li^+, Na^+, K^+ ions on the super energy transfer process in $ZnMoO_4:Eu^{3+}$, Bi^{3+} phosphors. Scientific Reports 6, 27657.

Ran, W., Wang, L., Zhang, W., Li, F., Jiang, H., Li, W., . . . Shi, J. 2015. A super energy transfer process based S-shaped cluster in $ZnMoO_4$ phosphors: Theoretical and experimental investigation. Journal of Materials Chemistry C 3(32), 8344-8350.

Rivera-López, F., Martin, I., Da Silva, I., Gonzalez-Silgo, C., Rodriguez-Mendoza, U., Lavin, V., . . . Mestres, L. 2006. Analysis of the Eu^{3+} emission in a $SrWO_4$ laser matrix under pressure. High Pressure Research 26(4), 355-359.

Roy, R., Muller, O. 1974. The Major Ternary Structural Families. Springer-Verlag, Berlin.

Ruiz-Fuertes, J., López-Moreno, S., López-Solano, J., Errandonea, D., Segura, A., Lacomba-Perales, R., . . . Gospodinov, M. 2012. Pressure effects on the electronic and optical properties of AWO_4 wolframites (A = Cd, Mg, Mn, and Zn): The distinctive behavior of multiferroic $MnWO_4$. Physical Review B 86(12), 125202.

Sharma, K.G., Singh, N.R. 2013. Synthesis and luminescence properties of $CaMO_4:Dy^{3+}$(M = W, Mo) nanoparticles prepared via an ethylene glycol route. New Journal of Chemistry 37(9), 2784-2791.

Shi, S., Gao, J., Zhou, J. 2008. Effects of charge compensation on the luminescence behavior of Eu^{3+} activated $CaWO_4$ phosphor. Optical Materials 30(10), 1616-1620.

Shirasaki, Y., Supran, G.J., Bawendi, M.G., Bulović, V. 2013. Emergence of colloidal quantum-dot light-emitting technologies. Nature Photonics 7(1), 13.

Singh, B., Parchur, A., Ningthoujam, R., Ansari, A., Singh, P., Rai, S. 2014. Enhanced photoluminescence in $CaMoO_4:Eu^{3+}$ by Gd^{3+} co-doping. Dalton Transactions 43(12), 4779-4789.

Singh, B., Ramakrishna, P., Singh, S., Sonu, V., Singh, S., Singh, P., . . . Rai, S. 2015. Improved photo-luminescence behaviour of Eu^{3+} activated $CaMoO_4$ nanoparticles via Zn^{2+} incorporation. RSC Advances 5(69), 55977-55985.

Singh, B., Singh, J., Singh, R. 2014. Luminescence properties of Eu^{3+}-activated $SrWO_4$ nanophosphors-concentration and annealing effect. RSC Advances 4(62), 32605-32621.

Sinha, S., Mahata, M.K., Swart, H., Kumar, A., Kumar, K. 2017. Enhancement of upconversion, temperature sensing and cathodoluminescence in the K^+/Na^+ compensated $CaMoO_4$: Er^{3+}/Yb^{3+} nanophosphor. New Journal of Chemistry 41(13), 5362-5372.

Song, F., Han, L., Zou, C., Su, J., Zhang, K., Yan, L., Tian, J. 2007. Upconversion blue emission dependence on the pump mechanism for Tm^{3+}-heavy-doped $NaY (WO_4)_2$ crystal. Applied Physics B 86(4), 653-660.

Soni, A.K., Rai, V.K., Kumar, S. 2016. Cooling in Er^{3+}:$BaMoO_4$ phosphor on codoping with Yb^{3+} for elevated temperature sensing. Sensors and Actuators B: Chemical 229, 476-482.

Spassky, D., Vasil'Ev, A., Kamenskikh, I., Mikhailin, V., Savon, A., Hizhnyi, Y.A., ..., Lykov, P. 2011. Electronic structure and luminescence mechanisms in $ZnMoO_4$ crystals. Journal of Physics: Condensed Matter 23(36), 365501.

Su, Y., Li, L., Li, G. 2008. Self-assembly and multicolor emission of core/shell structured $CaWO_4$: Na^+/Ln^{3+} spheres. Chemical Communications 34, 4004-4006.

Sun, J., Sun, Y., Cao, C., Xia, Z., Du, H. 2013. Near-infrared luminescence and quantum cutting mechanism in $CaWO_4$:Nd^{3+}, Yb^{3+}. Applied Physics B 111(3), 367-371.

Sun, L., Guo, Q., Wu, X., Luo, S., Pan, W., Huang, K., ... Hu, C. 2007. Synthesis and photoluminescent properties of strontium tungstate nanostructures. The Journal of Physical Chemistry C 111(2), 532-537.

Sun, X., Sun, X., Li, X., He, J., Wang, B. 2014. Molten salt synthesis, characterization, and luminescence of $SrWO_4$, $SrWO_4$:Tb^{3+} and $SrWO_4$:Eu^{3+} powders. Journal of Materials Science: Materials in Electronics 25(5), 2320-2324.

Thongtem, T., Phuruangrat, A., Thongtem, S. 2010. Microwave-assisted synthesis and characterization of $SrMoO_4$ and $SrWO_4$ nanocrystals. Journal of Nanoparticle Research 12(6), 2287-2294.

Van Deun, R., Ndagsi, D., Liu, J., Van Driessche, I., Van Hecke, K., Kaczmarek, A.M. 2015. Dopant and excitation wavelength dependent color-tunable white light-emitting Ln^{3+}:Y_2WO_6 materials (Ln^{3+} = Sm, Eu, Tb, Dy). Dalton Transactions 44(33), 15022-15030.

Wang, L.-L., Wang, Q.-L., Xu, X.-Y., Li, J.-Z., Gao, L.-B., Kang, W.-K., ... Wang, J. 2013. Energy transfer from Bi^{3+} to Eu^{3+} triggers exceptional long-wavelength excitation band in $ZnWO_4$:Bi^{3+}, Eu^{3+} phosphors. Journal of Materials Chemistry C 1(48), 8033-8040.

Wang, L., Ma, Y., Jiang, H., Wang, Q., Ren, C., Kong, X., ... Wang, J. 2014. Luminescence properties of nano and bulk $ZnWO_4$ and their charge transfer transitions. Journal of Materials Chemistry C 2(23), 4651-4658.

Wang, X.-X., Xian, Y.-L., Shi, J.-X., Su, Q., Gong, M.-L. 2007. The potential red emitting Gd_2-yEuy $(WO_4)_3-x$ (MoO_4) x phosphors for UV InGaN-based light-emitting diode. Materials Science and Engineering: B 140(1-2), 69-72.

Wang, Z., Liang, H., Zhou, L., Wang, J., Gong, M., Su, Q. 2008. $NaEu_{0.96}$ $Sm_{0.04}$ $(MoO_4)_2$ as a promising red-emitting phosphor for LED solid-state lighting prepared by the Pechini process. Journal of Luminescence 128(1), 147-154.

Wang, Z., Wang, Y., Li, Y., Zhang, H. 2011. Near-infrared quantum cutting in Tb^{3+}, Yb^{3+} co-doped calcium tungstate via second-order downconversion. Journal of Materials Research 26(5), 693-696.

Wei, T., Dong, Z., Zhao, C., Ma, Y., Zhang, T., Xie, Y., ... Li, Z. 2016. Upconversion luminescence and temperature sensing properties in Er-doped ferroelectric $Sr_2Bi_4Ti_5O_{18}$. Ceramics International 42(4), 5537-5545.

Wu, J., Cao, B., Lin, F., Chen, B., Sun, J., Dong, B. 2016. A new molybdate host material: Synthesis, upconversion, temperature quenching and sensing properties. Ceramics International 42(16), 18666-18673.

Xiong, F., Guo, D., Lin, H., Wang, L., Shen, H., Zhu, W. 2015. High-color-purity red-emitting phosphors RE_2WO_6:Pr^{3+} (RE = Y, Gd) for blue LED. Journal of Alloys and Compounds 647, 1121-1127.

Xu, B., Cao, X., Wang, G., Li, Y., Wang, Y., Su, J. 2014. Controlled synthesis and novel luminescence properties of string $SrWO_4$:Eu^{3+} nanobeans. Dalton Transactions 43(30), 11493-11501.

Xu, W., Zhao, H., Li, Y., Zheng, L., Zhang, Z., Cao, W. 2013. Optical temperature sensing through the upconversion luminescence from Ho^{3+}/Yb^{3+} codoped $CaWO_4$. Sensors and Actuators B: Chemical 188, 1096-1100.

Yadav, R., Yadav, R., Bahadur, A., Rai, S. 2016. Enhanced white light emission from a $Tm^{3+}/Yb^{3+}/Ho^{3+}$ co-doped $Na_4ZnW_3O_{12}$ nano-crystalline phosphor via Li^+ doping. RSC Advances 6(57), 51768-51776.

Yang, M., Liang, Y., Gui, Q., Zhao, B., Jin, D., Lin, M., . . . Liu, Y. 2015. Multifunctional luminescent nanomaterials from NaLa $(MoO_4)_2$:Eu^{3+}/Tb^{3+} with tunable decay lifetimes, emission colors, and enhanced cell viability. Scientific Reports 5, 11844.

Yang, X., Fu, Z., Yang, Y., Zhang, C., Wu, Z., Sheng, T. 2015. Optical temperature sensing behavior of high-efficiency upconversion: $Er^{3+}-Yb^{3+}$ co-doped NaY $(MoO_4)_2$ phosphor. Journal of the American Ceramic Society 98(8), 2595-2600.

Zeng, Q., He, P., Liang, H., Gong, M., Su, Q. 2009. Luminescence of Eu^{3+}-activated tetra-molybdate red phosphors and their application in near-UV InGaN-based LEDs. 118 Materials Chemistry and Physics (1), 76-80.

Zeng, Q., He, P., Pang, M., Liang, H., Gong, M., Su, Q. 2009. Sr_9R_2- xEuxW_4O_{24}$ (R = Gd and Y) red phosphor for near-UV and blue InGaN-based white LEDs. Solid State Communications 149(21-22), 880-883.

Zhang, A., Sun, Z., Liu, G., Fu, Z., Hao, Z., Zhang, J., Wei, Y. 2017. Ln^{3+} (Er^{3+}, Tm^{3+} and Ho^{3+})-doped NaYb $(MoO_4)_2$ upconversion phosphors as wide range temperature sensors with high sensitivity. Journal of Alloys and Compounds 728, 476-483.

Zhang, Y., Xu, S., Li, X., Sun, J., Zhang, J., Zheng, H., . . . Chen, B. 2017. Concentration quenching of blue upconversion luminescence in Tm^{3+}/Yb^{3+} co-doped $Gd_2 (WO_4)_3$ phosphors under 980 and 808 nm excitation. Journal of Alloys and Compounds 709, 147-157.

Zheng, Y., Lin, J., Wang, Q. 2012. Emissions and photocatalytic selectivity of $SrWO_4$:Ln^{3+}(Eu^{3+}, Tb^{3+}, Sm^{3+} and Dy^{3+}) prepared by a supersonic microwave co-assistance method. Photochemical & Photobiological Sciences 11(10), 1567-1574.

Down Converted Photoluminescence of Trivalent Rare-Earth Activated Glasses for Lighting Applications

Sathravada Balaji[1], Amarnath R. Allu[1], Mukesh Kumar Pandey[2],
Puja Kumari[3] and Subrata Das[4]*

[1] Glass Science and Technology Section, CSIR – Central Glass and Ceramic
Research Institute, Kolkata - 700032, India
[2] Department of Physics, National Taiwan University, Taipei - 10617, Taiwan
[3] Department of Physics, Darbhanga College of Engineering,
Darbhanga - 846005, India
[4] Materials Science and Technology Division, CSIR – National Institute for
Interdisciplinary Science and Technology, Thiruvananthapuram - 695019, India

Introduction

In the present era, Solid-State Lighting (SSL) technology is generating a large market because of their high efficiency in comparison with conventional incandescent lighting (Almeida et al. 2014). The form and function of various SSL products used in homes or offices are significantly different in function. At present, the Light-Emitting Diodes (LEDs) and Organic LEDs (OLEDs) type SSL sources have been highly successful for lighting applications. Despite their promising applications they do have serious drawbacks that need to be resolved (Almeida et al. 2014; Northwest 2011). SSL has many advantages over conventional light sources which include small size, easiness of control, uni-directional distribution, cool beam, good Color Rendering Index (CRI), less energy consumption, high performance at low temperature conditions, long life and new form factors. Meanwhile, major challenges with SSLs are mainly the cost, compatibility, heating management, power eminence, the failure issues, color steadiness and glare issues (Schubert and Kim 2005).

*Corresponding author: subratadas@niist.res.in

The existing research focuses on the making of full-color phosphors based on trivalent rare-earth (RE^{3+}) ions nowadays. The widely used SSL technology has an enormous magnitude to minimize the worldwide electricity expenditure along with the utilization of fossil fuels significantly. LEDs, as an efficient SSL source, have been associated with demanding interest because of their long lifetime, small energy expenditure, and enormous flexiblility in shapes and sizes, elevated dependability and eco-friendly characteristic. These interesting features make LEDs as special candidates to substitute the orthodox lighting sources (incandescent and fluorescent). White colored Light Emitting Diodes (white-LEDs) have very potential uses for many applications including lasers, indicators, backlight, automobile headlights and general illumination. Usually, white-LEDs are constructed by a mixture of the blue-LED chip along with luminescent phosphors. Nevertheless, the blue-LED chip and phosphors degrade individually, which results in chromatic aberration and subsequent reduction of white luminescence. Additionally, the emission from phosphor materials depends highly on ambient temperature, and therefore phosphor materials should also possess high thermal and chemical steadiness to avoid thermal quenching. The important weakness with phosphor based LED materials is their reduced color rendering index. Currently, a blue diode chip combined with yellow colored light emitting $Y_3Al_5O_{12}:Ce^{3+}$ (YAG) is the furthermost recognized technique for formulating profitable white-LEDs. Nonetheless, owing to the absence of thermal stability at higher temperatures throughout the white-LED operation, low color purity when pumped by a blue-LED chip, lack a red-light component, and therefore it is challenging to make YAG based white-LEDs with a high CRI and low correlated color temperature, both of which are important necessities for indoor lighting applications. Moreover, for making the white-LEDs, epoxy resin is utilized to synthesis phosphor powders and enables its coating on the chip. This process in general cuts down the lifetime and function of the LED owing to the unavoidable aging of the coating layer under long-term UV irradiation. The inhomogeneous coating process may cause the halo effect. Therefore, an alternative is vital to make fresh materials that could emit bright white light by ultraviolet (UV) irradiation.

To surmount the above problems, glass and glass-ceramic based materials were introduced as an alternative and nonstop efforts are ongoing to make glass-based LED materials (Gao and Wondraczek 2014; Allu et al. 2017; Reddy et al. 2011). Rare-earth ions-activated glass materials are well known for different fundamental research and for technical applications such as in optical fibers, laser systems, sensing materials, scintillators, display and lighting, optoelectronic communication, waveguides, and several others. The foremost advantages of the glass-based material are the chemical compositions that can be tailored to optimize the required properties for a specific application (Reddy et al. 2015). The additional benefit for glass-based materials is that it is easier to make products with the required shape and size. Usually, LED devices are developed by uniting several LEDs or by combining one or many phosphors with (In)GaN-type LEDs as the main pumping source. In disparity, luminescent glasses are more favorable alternatives for substituting phosphors for LEDs. An alternative method to unite the luminescent glass system along with a LED is to adjust the shape and character of the glass material into a lens and substitute the usual plastic lens in the existing LEDs (Zhu et al. 2014). Such a system may not require additional encapsulation since the glass lens can act like a protective sheet for the LED chips. In comparison with the obtainable phosphor materials used for LED systems, luminescent glasses are preferred owing to distinctive benefits that include consistent light emission, cheaper manufacture price, simpler fabrication processes, improved thermal steadiness, and free of any epoxy based assembly process, thus enhancing curiosity on luminescent glasses to substitute phosphors for LED systems. The inspiration for this chapter is to highlight glasses for solid state lighting. Glassy materials have numerous benefits over crystals such as easy to make, scaling-up capability and low-cost production. Herein, the luminescence features of efficient glasses with various RE ions including their exclusive property of emitting different colored light under the UV sources irradiation are described .

Definition of Glass and Glass Formers

Glass is known to be a physical state rather than a particular composition, and is characterized by its rigid characteristics of a solid. However, its intrinsic behaviors are very similar to that of a liquid. The usual crystal pattern of typical solids is unseen in the glassy state, and the atoms in glass are arranged randomly similar to that in a liquid. Due to this, glasses are sometimes referred to as "supercooled liquid". Several definitions exist for glass which are as follows: (i) A product inorganic in nature that is produced via fusion followed by cooling to a rigid condition without crystallizing. (ii) This is an amorphous solid that displays a glass transition. (iii) A non-crystalline solid produced by melt-cooling and displaying a glass transition phenomenon. (iv) Every isotropic material (inorganic or organic) which is not exhibiting three dimensional periodicities and the viscosity of which is greater than about 10^{14} poise. Similar to crystals, glasses also possess building blocks as cation polyhedral, which are arranged in different patterns such as glasses have broader distributions of bond angles. The atoms of a glass system are arranged in a random manner similar to a liquid, as revealed in Fig. 4.1. The atoms are jumbled together sloppily but cannot move to form an orderly arrangement. This unusual amorphous arrangement forms once the hot liquid glass is cooled to room temperature via a rapid crystallization.

Among the various types of oxides only a few oxides have the empirical formula of A_mO_n, where m and n are numerical and form glasses. There are several theories for glass formation oxides of which Zachariasen suggested a set of four rules and are as follows:

i. Zero oxygen atom are connected to more than two A atoms.

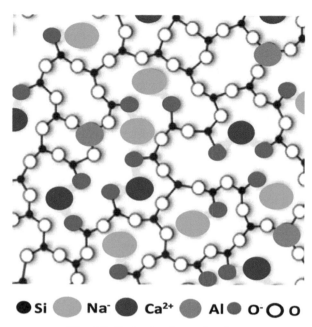

Fig. 4.1. Basic structure of a glass.

ii. The number of oxygen atoms surrounding A must be small (probably 3 or 4).

iii. The oxygen polyhedra are the part of corners but not of any faces or edges.

iv. The corners of a minimum three in numbers of each polyhedron should be shared.

Based on these rules, the following oxides are ready to form glasses B^{3+}, Si^{4+}, P^{3+}, P^{5+}, As^{3+}, Ge^{4+} and also V^{5+}, Sb^{6+}, Cb^{5+}, Ta^{6+} and Al^{3+} which are proficient of substituting Si^{+4} isomorphous. A crucial fact to note are that Al_2O_3 alone cannot form glass. Based on the coordination number, the amount of ions or atoms that straightaway surrounds an atom or ion of attention, and as well field strength, $F = Z/a^2$ where Z is valency of cation and a= r_c (radii of cation) + r_a (radii of anion), of cations, which participate in glass formations, can be classified into three groups:

i. **Network formers:** Cations that have predominantly the coordination numbers 3 and 4 and field strength (F) varies in the range of 1.4 - 2. Ex: Si^{4+}, B^{3+}, P^{5+}, Ge^{4+}, As^{3+} etc.

ii. **Network modifiers:** Cations that have predominantly the coordination number 6 to 8 and F varies in the range of 0.1-0.4. Ex: Na^+, K^+, Ca^{2+}, Ba^{2+}, etc.

iii. **Intermediate oxides:** Cations that have predominantly the coordination numbers 4 to 6 and F varies in the range of 0.5 to 1.0. Ex: Al^{3+}, Mg^{2+}, Zn^{2+}, etc. The intermediate oxides occupy a place between the network formers and the network modifiers. In a multilinked glass they can, according to the composition, reinforce the glass structure as tetrahedra-forming units (with the coordination number 4) or further loosen up the basic structure as network modifiers (with coordination numbers 6 to 8).

Important Glass Host Systems for Lighting

Rare-earth elements can emit photons in the visible region under suitable irradiation conditions. The emission properties of a RE element are weakly depended on the host matrix; however, the intensity and emission wavelength range of the transitions is highly dependent on the crystal field along with the phonon-related energy of the host materials (Zhu et al. 2014). Therefore, the phonon-based energy of the host material for obtaining maximum outcome from the activated rare-earth elements plays a crucial role. Low phonon energy glass host materials reduce the multiphonon transitions and therefore enhances the radiative transitions in the intermediate states or the emitting states, thereby increasing the quantum efficiency. Though different types of glass materials are available, the phonon energy of the materials varies in the following order: borate > phosphate > silicate > tellurite (Table 4.1). Another advantage of the glass-based materials is that it accepts a high concentration of rare-earth ions and can also simultaneously accommodate different types of active elements. Due to this reason luminescent glass-based materials are now becoming alternative materials for phosphor-based LED applications (Ghosh et al. 2017). This brief review has been added for the various types of glass used for LED applications.

Table 4.1. Average phonon frequencies ($\hbar\omega$) of some efficient glasses suitable for luminescent applications

Glass host	$\hbar\omega$ (cm^{-1})
Borate	1400
Phosphate	1200
Silicate	1100
Tellurite	700

Silicate glasses: The phonon-based energy of the silicate glass systems is around 1100 cm^{-1} which is less than that of the (14285 to 25000 cm^{-1}) of visible light. This allows the use of silicate glasses for lighting applications. Moreover, the thermal and chemical steadiness of silicate glasses is high. To accomplish the white colored light emission from rare-earth cations activated silicate glasses, chemical compositions of the glasses can be modified by introducing various type of cations. For example, (i) the addition of a small concentration of B_2O_3 to the glass matrix improves the symmetry of the microenvironments around RE^{3+} ions, thereby increasing the emission intensity (Zhu et al. 2013);

(ii) addition of low phonon network modifying fluoride molecules helps to tune the emission intensity and creates a low phonon energy environment around the active rare-earth cation (Zhang et al. 2013) and (iii) addition of Al_2O_3 increases the rare-earth ion solubility (Miniscalco 1991).

Liu et al. studied the applicability of aluminoborosilicate glass for white-LED applications and reported that Tm^{3+}-Dy^{3+} (Liu et al. 2008b) activated glasses over Eu^{3+}-Dy^{3+} (Liu et al. 2008a) activated glasses showed high potentiality. Figure 4.2 (Fig. 3 in Liu et al. 2008) shows CIE color coordinates of various concentrations of Eu_2O_3 and Dy_2O_3 doped ZABS glasses at the excitation of 360 nm. The luminescence emission color of the Eu^{3+}-Dy^{3+} ions co-activated glasses are influenced by several factors: (i) the reductive reaction that affects the comparative concentration of Eu^{2+} and Eu^{3+} ions; (ii) optical basicity; and (iii) type of modifier cation in glass network (Liu et al. 2008). On the other hand, Eu^{3+}/Tb^{3+} coactivated silicate glasses were also considered for generating white light emission (Zhu et al. 2013). To generate superior white colored light emission additional emission bands covering a broad visible zone are mainly needed. To overcome this problem Eu^{3+}/Tb^{3+}/Dy^{3+} tri-activated oxyfluoride silicate glasses had been considered and obtained white light emission by the synchronized generation of numerous emission bands of Eu^{3+}, Tb^{3+}, and Dy^{3+} ions under the near-UV irradiation (Zhu et al. 2014). The energy transfer mechanism is illustrated in Fig. 4.3. The luminescence properties and color tunability are highly dependent on the chemical composition of glass and are demonstrated in Ce^{3+}, Dy^{3+}, Eu^{3+}

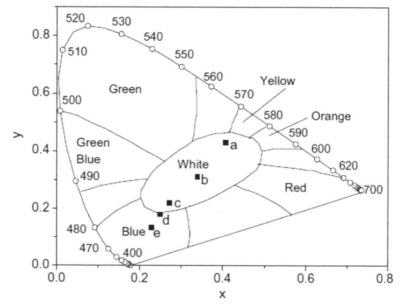

Fig. 4.2. Chromaticity of the photoluminescent emission of xEu_2O_3 - $(0.25 - x)$ - Dy_2O_3 - 99.75ZABS glasses at 360 nm excitation: (a) x = 0, (b) x = 0.0625, (c) x = 0.125, (d) x = 0.1875, and (e) x = 0.25.

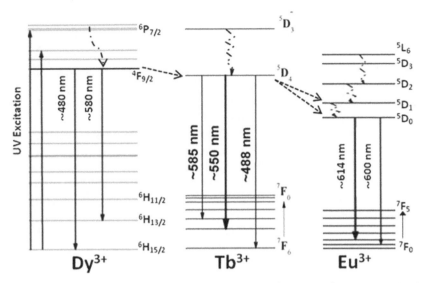

Fig. 4.3. Energy Transfer process (ET) and Cross-Relaxation phenomenon (CR) of trivalent Eu^{3+}, Tb^{3+}, and Dy^{3+} rare earth ions.

activated oxyfluoride aluminoborosilicate glasses (Zhu et al. 2017). With the presence of Ca^{2+} cation, the activated two Eu^{3+} cations could occupy the three Ca^{2+} cation sites and create vacancy defects at the Ca^{2+} sites. The as-formed defect gives the electrons to Eu^{3+} cations via the phonon-assisted process and converts it into Eu^{2+} ions. Nevertheless, the presence of AlO_4 and BO_4 units could prevent the oxidation of Eu^{2+} ions and stabilize the Eu^{2+} ions and therefore influence the luminescence properties of glass. Therefore, the coexistence and presence of Eu^{3+} and Eu^{2+} ions solely depend on the chemical composition of glasses.

Owing to the high efficient broadband luminescence of Ce^{3+} cation can be used as either a sensitizer or an activator in many glass hosts. With this as an advantage, white light emission is attained by doping more than one activator in numerous hosts, such as Ce^{3+} and Dy^{3+} activated oxyfluoride glasses (Luo et al. 2009), Ce^{3+} and Sm^{3+} activated sodium silicate glasses (Ma et al. 2014), Ce^{3+} and Tb^{3+} activated oxyfluoride glasses (Luo et al. 2009), and $Ce^{3+}/Ho^{3+}/Sm^{3+}$ activated oxyfluoride glass (Zhou et al. 2017). In all the cases mentioned above the energy transfer procedure is found to be from Ce^{3+} to RE^{3+} ions. In general, Ce^{3+} cations emit characteristic broad emission band (300-500 nm) centered at around 380-410 nm reliant on the chemical composition of glasses which is due to the allowed electric dipole-based transition of 4f-5d parity. Absorption spectra analysis reveals that strong intense transitions occur in Sm^{3+} at 401 nm (due to $^6H_{5/2} \rightarrow {}^4F_{7/2}$ transition), in Dy^{3+} at 392 nm (due to $^6H_{15/2} \rightarrow {}^4G_{11/2}$), Ho^{3+} ions at 416 nm (due to $^5I_8 \rightarrow {}^5G_5$) and in Tb^{3+} ions at 380 nm (due to $^7F_6 \circledR {}^5D_3$ transition) are well overlapped with the Ce^{3+} broad emission band and therefore result in the energy transfer phenomenon from Ce^{3+} to RE^{3+} cations. By using the appropriate concentration of $Ce^{3+}/Tb^{3+}/$

Sm^{3+} cations the energy transfer process among Ce^{3+}-Tb^{3+}, Ce^{3+}-Sm^{3+} and Tb^{3+}-Sm^{3+} were aroused and white light emission from aluminoborosilicate glasses were reported (Chen et al. 2013).

Phosphate glasses: Phosphate-based glasses exhibit numerous good properties related to low softening as well as melting temperature, low product cost, extraordinary transparency for visible light, intense emission and absorption cross-sections, high dispersion, relatively high refractive indices (compared with silicate-based optical glass) and especially decent solubility of rare-earth ions, which permits higher amounts of rare-earth ions within the glass-based matrix (Reddy et al. 2011). Nevertheless, the applications of these types of glasses are limited because of their hygroscopic nature and poor chemical robustness. Alkali (K, Li, Na) and alkaline-earth elements (Ba, Sr, Ca, Mg, Sr) are frequently included to adjust and increase the strength as well as the chemical robustness of the glass structure. For elevating the thermal, mechanical and chemical steadiness, heavy metallic (M) oxides (Al_2O_3, PbO, Al_2O_3 etc.) having the formation of P-O-M bonds rather than P-O-P bonds can be incorporated in the phosphate glasses. However, phosphate glasses have high phonon energy ranges between 1200 and 1350 cm^{-1}. The addition of fluoride causes the formation of HF after the reaction with OH group and reduces the OH absorption peak and makes the oxyfluoride glasses have low phonon energy (Luewarasirikul et al. 2017). Likewise, Al^{3+}, Zn^{2+} ions are also able to form either ZnO_4 tetrahedron or Zn-O-P bridges via phosphate chain linkages and improve the thermal steadiness of phosphate glasses.

Dy^{3+} activated phosphate glasses were identified as suitable candidates to simulate white colored light emission. In general, Dy^{3+} cations can activate the yellow and blue color emission through $^4F_{9/2} \rightarrow\ ^6H_{15/2}$ (Y) and $^4F_{9/2} \rightarrow\ ^6H_{13/2}$ (B) energy transitions, respectively. White colored light emission can be realized via monitoring the Y/B intensity ratio by altering the chemical composition of glass and also the irradiation wavelengths. Better white light emission has been achieved for 1 mol% Dy^{3+} activated zinc-lead phosphate glass for which the Y/B intensity ratio (Amjad et al. 2013) value lies in the proximity of unity (Amjad et al. 2013). Dy^{3+} cations were also activated in oxyfluoroborophosphate glasses and lanthanum calcium phosphate glasses (Luewarasirikul et al. 2017) and achieved white light emission. On the other hand, color coordinate values are far from the ideal white light color coordinate (x = 0.333, y = 0.333) for those glasses having the yellow vs blue, Y/B, intensity ratio values are higher than unity (Vijayakumar and Marimuthu 2015). Indicating that Y/B ratio demonstrates the characteristics of white light emission, the more the unity of Y/B ratio, the greater likely it is to obtain white light emission. Studies were also carried out to gain white light emission by doping Tm^{3+} cation along with Dy^{3+} ions in phosphate glasses (Chen et al. 2016).

Both Eu^{3+} and Sm^{3+} ions are considered as highly emitting red luminescent centers (Langar et al. 2017). On the other hand, Tm^{3+} (Li et al. 2014) and Tb^{3+}

(Zhu et al. 2013) are acknowledged as efficient blue and green-emitting active cations, respectively. Therefore, the combination of Tb^{3+}, Tm^{3+} and Eu^{3+}/Sm^{3+} active cations can help to understand the highly efficient white light emission. However, singly activated Eu^{3+} and Sm^{3+} cations are not effectively excitable by the near-UV. In order to overcome this, the concentration of Eu^{3+}/Sm^{3+} cations should be increased and need to be optimized to achieve white light emission. Chen et al. (2017) studied the photoluminescence properties $Tm^{3+}/Tb^{3+}/Eu^{3+}$ activated borophosphate glasses, with increasing the concentration of Eu^{3+} cations. Earlier Zhong et al. (2015) studied the white colored light emitting properties of $Tm^{3+}/Tb^{3+}/Sm^{3+}$ activated phosphate glasses. In both the cases the concentration of red emitting Eu^{3+} and Sm^{3+} activator cations were varied and optimized to achieve the white light emission. In addition, it is also interesting to highlight that the while colored light emission is obtained via monitoring the intensity ratio of red (Sm^{3+}/Eu^{3+}), green (Tb^{3+}) and blue (Tm^{3+}) by the energy transfer method from Tm^{3+} to Eu^{3+}/Sm^{3+} and Tb^{3+} to Eu^{3+}/Sm^{3+} (Chen et al. 2017).

Borate glasses: Borate matrix constitutes with the well-defined assembly of BO_4 tetrahedra and BO_3 triangles which build stable borate groups such as di-, tri-, and tetra-borate, etc. These groups usually create the randomized three dimensional network. The additional advantage of borate glasses compared to the other oxide-based glasses is good rare-earth ion solubility. However, high phonon-based energy ranging between 1200 and 1600 cm^{-1} is mainly responsible to decrease the rare-earth emission intensity because of the dominant non-radiative decay, which subsequently reduces the quantum efficiency of the RE emission. Borate glasses, synthesized using boron oxide, are very useful for making optical lenses owing to their high refractive index. The addition of ZnO to virgin borate glass network breakdowns the boroxal rings and creates the BO_3 triangles and BO_4 tetrahedra leading to enhance thermal steadiness, mechanical strength and chemical durability (Kumari and Manam 2016). The addition of various alkali or alkaline earth cations to the borate glass composition mechanical steadiness can be improved and the phonon energy can be reduced (Shamshad et al. 2017). As with silicate and phosphate glass systems, the addition of fluoride can create a low phonon energy environment to the rare-earth cation and enhances the luminescence intensity.

Binary borate glass composition systems were utilized to dope the rare-earth cations and to gain the white light emission. For example, $Li_2B_4O_7$ glass system was utilized to dope various concentrations of Dy^{3+} cations ranges between 0 and 3 mol% (Sun et al. 2013). Photoluminescence studies reveal that the optimal concentration was 0.5 mol% Dy_2O_3 and the strongest white light emission with chromaticity coordinates (0.342, 0.372) was observed by the naked eyes for 0.25 mol% activated glasses. More importantly, the chromaticity coordinates of Dy^{3+}-activated $Li_2B_4O_7$ glass are insensitive to the concentration and selected irradiation wavelength of Dy^{3+} ions. Very recently, Naresh et al. 2017 reported that zinc borate glasses were activated with

$Tm^{3+}/Tb^{3+}/Eu^{3+}$ for white-LED applications. The obtained color coordinates (x = 0.343, y = 0.322) for $1.0Tm^{3+}/0.5Tb^{3+}/0.5Eu^{3+}$ activated ZnB glass are close to the ideal white point (Kumari and Manam 2016; S. Balaji pers. comm.). Various alkaline-earth cations such as Mg, Ca, Sr and Ba were added independently to the lithium borate glass system and their influence on Dy^{3+} spectroscopic properties for white-LED emission were studied. White light emission has been obtained with the color coordinates (x = 0.37, y = 0.40) and lies in the white light region. It is interesting to note that color coordinates are independent of the type of alkaline-earth cation present in the lithium borate glass system though the asymmetric ratio that is the yellow to blue ratio increases with increasing ionic radius of alkaline-earth cation. White light emission was observed for Ce^{3+}-Sm^{3+} activated lead alumina borate glasses under irradiation at 410 nm. However, with the combination of Tb^{3+}, Ce^{3+} and Sm^{3+} cations in oxyfluoroborate glasses the irradiation wavelength was reduced to 310 nm and the obtained color coordinates (x = 0.347, y = 0.343), which is in near estimation along with the ideal white colored light coordinates (x = 0.333, y = 0.333) (Babu et al. 2014).

Tellurite glass: Apart from silicate, phosphate and borate glasses, tellurite glass is also considered as a host material for RE cations to realize the white colored light emission for LEDs. The benefit with the TeO_2 glasses is low melting temperature and the relatively low phonon energy. However, as other glass systems tellurite was not considered as promising candidates for white-LED application; very few studies were conducted. Dy^{3+} activated led germanium tellurite glasses were prepared for white light emission but suggested that different blue emitting luminescent centers are needed to be activated (Shisina et al. 2019). The white light emission is achievable via combining broad blue colored light emission from Ce^{3+}, yellowish green colored light emission from Tb^{3+} and red-orange colored light emission from Sm^{3+} in $Ce^{3+}/Tb^{3+}/Sm^{3+}$ tri-activated TBZN glasses under 350 nm irradiation. Several results showed that telluride glasses coactivated with Ce^{3+}, Tb^{3+}, Sm^{3+} can generate a new podium to propose and construct fresh luminescent systems for white-LEDs (Hong et al. 2015).

The above-mentioned glasses are found to be a potential for LEDs because of the high transparency in a wide range from visible to near-IR regions. However, for the visible emitting electronic transitions, the optically active ions must be introduced inside the glass hosts. In this connection, the optical natures of rare-earth ion activated inorganic glasses suitable for making LEDs have thus attracted noteworthy research attention. The most widely used rare-earth ions in glass for lighting devices are discussed below.

Trivalent Rare-earth Ions for Activating Glasses for Lighting Application

Rare-earth ions have a characteristic $4f^n$ energy level structure and are not affected much by the external electric field because of the shielding properties of electrons situated in outer $5s^2 5p^6$. This causes the energy level position of rare-earth ions to remain invariant irrespective of the host material, but exhibit only minor deviations due to the 'Nephelauxetic effect' caused by the RE - ligand binding typically by a few hundred cm^{-1} (Fuxi 1991). So, such a well-resolved and distinct energy structure of lanthanide ions leads to sharp absorption and emission transitions ranging from the vacuum ultraviolet to the infrared region. Usually, luminescence is mainly originated from electronic transitions among $4f^n$ levels which are either electric and/or magnetic dipole in nature. The transitions based on electric dipoles are parity-forbidden in free 4f ion; however, they become forced or induced transitions due to orbital mixing having dissimilar parity posed by an odd crystal field component. The magnetic dipole transitions (parity allowed) are normally unaffected by the host local symmetry (Wybourne and Smentek 2007). The strength of the oscillator of these forced transitions of the electric dipole is of the order of around 10^{-5} to 10^{-8}, and 10^{-8} for the transitions having the nature of magnetic dipole (Kaminskii 1980). The strength and the probability of these electric dipole transitions can be predicted by the well-known Judd-Ofelt (J-O) theory (Judd 1962; Ofelt 1962). This theory provides useful explanations to understand the character of the fluorescence spectra, as well as the theoretical radiative decay of emission transition and its probability. The quantum efficiency of the emission transition can be estimated by taking the ratio of the experimental decay value to that of theoretical (calculated) lifetime values. Because of the forbidden nature of the $4f^n$ emission transitions, the lifetimes are mostly in the range of few milliseconds (Wybourne and Smentek 2007). The emission usually observed between $4f^n$ levels relaxes either via radiative emission or by a non-radiative way i.e., the host phonon energy to bridge the energy level separation. However, broad irradiation transitions can be observed in certain host materials due to interactions of 4f and 5d orbitals and/or due to the Charge-Transfer process (CT) due to defect centers. The energies of the $4f^{n-1} 5d^1$ and Charge Transfer Band (CTB) are extra reliant on their ligand field than the energies of 4f states. The f-d absorption transitions of Ce^{3+}, Pr^{3+} and Tb^{3+} ions and CTB transition for Eu^{3+} and Yb^{3+} ions are found to be less than 40×10^3 cm^{-1}. However, for certain lanthanide ions like Ce^{3+}, Pr^{3+}, and Eu^{2+} broad luminescence from 5d → 4f transitions can be observed since the energy levels of such states are less than that of 4f states. Such emission transitions vary according to the crystal field environment of the host material. Emission from CTB is already reported for Yb^{3+} (Pieterson et al. 2000). Typically, rare-earth ions, which can be oxidized easily to the tetravalent state, will show lower energetic transitions of 4f → 5d. But those rare-earth ions which can be reduced to their divalent state contain lesser CTB energies. Among all lanthanides, Ce^{3+}, Pr^{3+}, Sm^{3+}, Eu^{3+}, Tb^{3+}, and Dy^{3+} are efficiently visible luminescent centers (Ronda 2008). However, Nd^{3+}, Ho^{3+}, Tm^{3+} and Er^{3+} also exhibit visible emissions; they are more popular for their emissions in the near-infrared region (Fuxi 1995).

Cerium: Trivalent cerium (Ce^{3+}) shows an absorption band of 4f → 5d in the wide wavelength region from 200 to 500 nm and the intensity, as well as the location of the band strongly relies on the environment of the host. Ce^{3+} processes the lowest energy among all lanthanide ions. The larger energy gap between 5d state to 4f: $^2F_{7/2}$ aids 5d state as an effective state for the light emission. The emissive wavelength of Ce^{3+} ions ranges from the ultraviolet region in fluorides-based host materials to the red region in sulfides-based host material (Kojima et al. 1971) because of the emissive transitions from the 5d state to the 4f: $^2F_{7/2}$ and $^2F_{5/2}$ ground states. The strong shift towards the red side in the emission band is due to the Ce^{3+}:5d states' centroid shift because of the crystal field splitting caused by the surrounding anion that further affects the stokes' shift (Andriessen et al. 2005) as shown in Fig. 4.4. The origin of the centroid shift is closely connected with the polarizability as well as the covalency of the anion in the compound. While the crystal field splitting is linked to the form and size of the anion neighboring the Ce^{3+} ion. Generally, the shift varies in the range F > Cl > Br > I > O > S > Se > Te (Dorenbos 2000). Unlike the f-f transition of other lanthanides, Ce^{3+} has high absorption and emission corrections. The Ce^{3+} emission decay times are in the order of 10^{-7} to 10^{-8} s which is shortest compared to other lanthanides because of parity and spin allowed d → f transitions (since $5d^1$ and $4f^1$ states are spin doublets) (Kumari and Manam 2015).

The well-known YAG: Ce phosphor exhibiting broad yellow emission in the region 500-650 nm peaking at around 530 nm is being commercially used as phosphor for white light-emission pumped by (In, Ga)N based blue (Bachmann et al. 2009). The common encapsulation problem of these phosphors allowed the researchers to search for more stable, encapsulate free and cost-effective glass-based luminescent materials. However, the luminescence spectra of some Ce^{3+} activated common glasses like silicates, borates and phosphates are observed in the region 300-500 nm due to $5d^1$ → $4f^1$ transition upon UV irradiation in the region 200-350 nm (Murata et al. 2005). As mentioned earlier the Ce^{3+} ion surrounding ligand field environment dominates the emission band position in the spectrum. Besides, Ce^{3+} ions in oxide glasses generally undergo a redox reaction i.e., conversion of Ce^{3+} → Ce^{4+} (Duffy 1996) state. The luminescence properties of Ce^{3+} ions greatly quench with the increase of this redox reaction due to a strong overlap of Ce^{3+} f → d transitions and Charge Transferring State (CTS) of Ce^{4+} ions facilitate the Ce^{3+} → Ce^{4+} non-radiative energy transfer as well as the electron transfer phenomenon (Sontakke et al. 2016). The redox reaction parameters are affected by melting conditions as well as oxygen fractional pressure of oxygen inside the furnace and the concentration of Ce ions (Murata et al. 2005). The higher Ce^{4+} redox reaction can be observed in the higher basicity of the glass (Vedda 2006). Also, Ce^{3+} ions are often used for the sensitization of other lanthanide ions whose irradiation falls in the emission region of Ce^{3+} ions for white light luminescence applications.

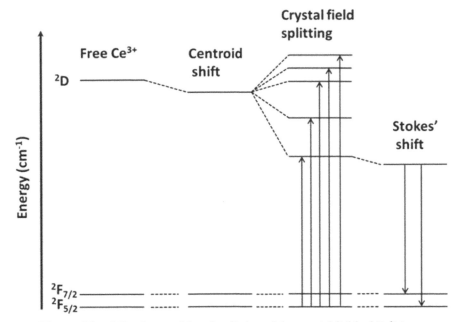

Fig. 4.4. The shift of centroid and splitting of the crystal field of Ce^{3+} ion (shifting and splitting immensely depends on the host environment).

Praseodymium: Trivalent praseodymium (Pr^{3+}) ions are known for a wide quantity of absorption transitions ranging from UV to near-IR and strong emission bands in the visible region. The visible absorption bands spanning from 420-500 nm can be recognized as the ground state transitions from Pr^{3+}: $^3H_4$ to excited states $^3P_{0,1,2}$. The main emission transitions at ~490, ~546, ~617, ~650, ~711 and ~734 nm are due to $^3P_0 \rightarrow {}^3H_4$, $^3H_5$, $^3H_6$, $^3F_2$, $^3F_3$ and $^3F_4$, respectively (Kaminskii 1980). The emission intensities of these transitions are intensely influenced by the host environment. The radiative decay times of these spin allowed $4f^n$ emission transitions are the shortest (in the order ~10^{-5} s) compared to other $4f^n$ transitions of other rare-earth ions (trivalent). The UV emissions originating due to 5d-4f transitions are reported in fluoride-based systems (Piper et al. 1974).

The emissive properties greatly depend on the concentration of Pr^{3+} ions as well as host glass phonon energy. Since many of the absorption and emission transitions of Pr^{3+} ions are overlie in the same wavelength region, the Pr-Pr cross-relaxation mechanism may be intensified at higher concentrations. Figure 4.5 depicts the irradiation, emission spectrum and energy level structure of Pr^{3+} ion in low phonon tellurite glass system and the influence of Pr^{3+} ion emission intensity on the concentration. Usually, the $^3P_0 \rightarrow {}^3H_4$ emission at ~490 nm cannot be observed in high phonon energy glasses like silicates, borates and phosphates owing to the fast relaxation (non-radiative) from $^3P_0$ to $^1D_2$ level.

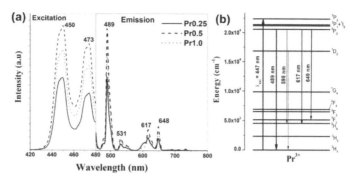

Fig. 4.5. (a) Excitation and emission spectra of Pr³⁺ ions in tellurite glass.
(b) Schematic energy level diagram depicting the transition levels of Pr³⁺

Samarium: Trivalent samarium (Sm³⁺) ions exhibit strong orange-red emission transitions at ~600 and ~650 nm corresponding to $^4G_{5/2} \rightarrow {}^6H_{7/2}$ and $^6H_{9/2}$, respectively. The luminescence decay of these transitions are usually in the order of few milliseconds. Although Sm³⁺ ions have strong emission characteristics, they have very low luminescence efficiency (compared to red-emitting Eu³⁺ ions) due to closely packed energy levels promoting the multi phonon relaxations. Quenching of emission transition can be observed even at moderately high Sm³⁺ ion concentrations. The host material's phonon energy also strongly affects the emission transitions in the higher energy (or) lower wavelength side. Figure 4.6 shows the excited and emission spectrum of Sm³⁺ ions in low phonon tellurite glass systems.

Sm³⁺ activated material are rigorously studied for optical data storage application because of their oxidation state change to Sm²⁺ on electron capture (Keller and Cheroff 1958). The *5d → 4f* emission transition of Sm²⁺ ion has been reported in fluoride-based materials with an emission band in the region 550-800 nm (lifetime in ~ μs).

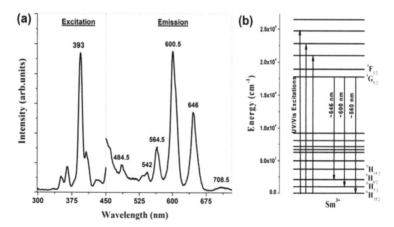

Fig. 4.6. (a) Excited and emission spectrum of Sm³⁺ ion in low phonon tellurite glass, and (b) schematic energy level diagram.

Europium: The europium ion, Eu^{3+}, is the favorable choice to use as the red-emitting optically active element in many commercial display phosphors because of their near-monochromatic red light emission and longer decay times (few milliseconds) with higher luminescence quantum efficiency. Multiple luminescence bands can be observed in the visible region from 400 - 750 nm assigned to $^5D_J \rightarrow {}^7F_{J'}$ emission transitions. The emission in the higher energy side < ~550 nm from $^5D_2$ and $^5D_1$ excited states cannot be observed at higher Eu^{3+} ion concentration in low phonon energy host matrix due to Eu-Eu cross relaxations $[(^5D_J \rightarrow {}^5D_0) \rightarrow (^7F_0 \rightarrow {}^7F_{J'})]$ and at even low concentrations in high phonon energy conventional glasses due to relatively high multi phonon relaxations. Figure 4.7 depicts the excited and emissive transitions of Eu^{3+} ions in a low phonon tellurite glass system where emission transitions at a lower wavelength region can be detected due to less multi phonon effects of the host glass.

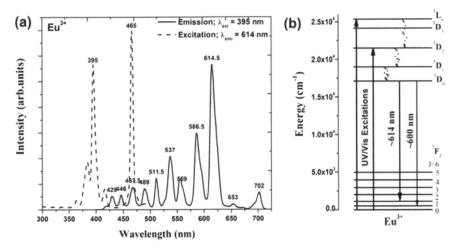

Fig. 4.7. (a) Excited and emission spectrum of Eu^{3+} ions in a low phonon tellurite glass and (b) energy level diagram.

The $^5D_0 \rightarrow {}^7F_1$ (~ 600 nm) is recognized as the host insensitive magnetic dipole transition. The stronger $^5D_0 \rightarrow {}^7F_2$ emission transition (610-630 nm) is ascribed to an electric dipole in nature, which is usually generated because of the non-existence of inversion symmetric sites of Eu^{3+}. The CTB of Eu^{3+} lies well below 40×10^3 cm^{-1}, and in most of the oxide-based glasses, it can be observed in the region 200-300 nm. The broad CTB can be utilized for irradiation which will then transfer the excited Eu^{3+} ions via non-radiative relaxation process to ^5D_J states for its subsequent visible emissions. The CTB position largely varies with the host material, higher the CTB energy, lower the probability of electric dipole transition of Eu^{3+} ions (Blasse and Bril 1969). On the other hand, the $4f^7$ Eu^{2+} ions usually give broader emission due to d-f transition, the band position changes from near-UV to red owing to the effects of crystal field of 5d level in a variety of host materials. The decay time of the Eu^{2+} spin-forbidden emission is in the order of few microseconds at the longer side for an allowed transition (Kumari and Manam 2015).

Terbium: Trivalent terbium (Tb^{3+}) ion shows several sharp emission lines at blue, green and red regions upon UV or blue irradiation. Due to this reason, Tb^{3+} doped materials are explored at length for white light emissions. However, the main drawback of this ion is that even at moderately high concentrations it strongly quenches the emission transition from $^5D_3$ level due to Tb^{3+}-Tb^{3+} cross-relaxations. Among all the emission transitions, $^5D_4{\rightarrow}^7F_5$ is the most prominent with a peak position at ~550 nm in the green region owing to the higher probability of electric-dipole as well as magnetic-dipole nature (Yen et al. 2006). The energy level separation of $^5D_3$ to $^5D_4$ is less than 5000 cm^{-1}, the emission intensity ratio of $^5D_3$ to that of $^5D_4$ strongly relays on the host material phonon energy apart from the Tb^{3+} ion concentration. The $^5D_3$ emissions are very weak in borate glass hosts compared to other low phonon energy glass hosts (Zu et al. 2011). The $4f^8 \rightarrow 4f^75d^1$ transition Tb^{3+} ion shows wide irradiation in the region 200 to 300 nm in a variety of host material that can be potentially used for broadband UV irradiation. The partial energy structure in Fig. 4.8 depicts the most prominent emission transition of Tb^{3+} ions alongside the possible Tb^{3+}-Tb^{3+} cross-relaxation at higher concentrations.

Dysprosium: Trivalent Dy^{3+} ions possess a unique character that the strong emission characteristics in blue (~483 nm, $^4F_{9/2} \rightarrow {}^6H_{15/2}$) and in yellow (~574 nm, $^4F_{9/2} \rightarrow {}^6H_{13/2}$) region (Azeem et al. 2008) can potentially be used for a warm white colored light emission from a single ion. The intensity ratio i.e.,

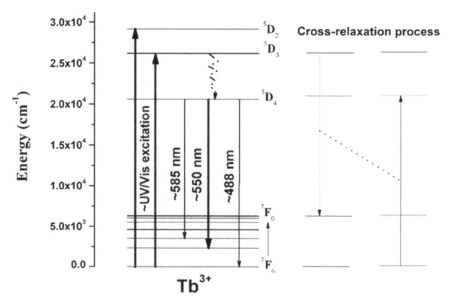

Fig. 4.8. Tb^{3+} ion energy structure depicting the prominent fluorescence transitions and main cross-relaxation process.

Y/B ratio can be tuned to produce white light emission in a suitable host material for an optimized concentration of Dy^{3+} ions (Kaewnuam et al. 2017). The $^4F_{9/2} \rightarrow {}^6H_{13/2}$ blue emission is hypersensitive to the host environment and is electric dipole in nature. Figure 4.9 presents the irradiation and radiation transition of Dy^{3+} ion in tellurite glass alongside the energy level structure depicting the prominent blue and yellow emissions in a tellurite glass host.

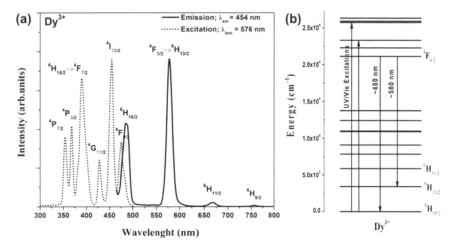

Fig. 4.9. a. Excited and emission spectra of Dy^{3+} in tellurite glass and b. energy level structure depicting the prominent emission transitions.

The CTB energy is well above 50×10^3 cm^{-1} and its influence on the f-f transitions is negligible. There are reports on Dy^{2+} and Dy^{4+} emission transition in the region ~2.3 - 2.7 μm and 500-630 nm respectively in fluoride-based single crystals (Kiss 1965; Watts and Richter 1972), no such reports exist in any glass host materials.

In Table 4.2, some RE electronic structures and transitions have been mentioned which are suitable for lighting and display applications.

Descriptions of Some Important Factors in Glass Photoluminescence

Energy transfer: Energy transfer-based luminescence gained much importance especially for white light generation. White light composing the red, green and blue color components which are very difficult to produce for a single activator rare-earth ion to emit in all three regions equally. Generating pure white light can be possible by choosing the proper combination and concentration of the selected portion emitting rare-earth ions. The energy transfer process among the activator ions can severely affect the luminescence properties as a whole. The donor-acceptor exchange interaction mainly depends on the activator concentration and the interionic separation where

Table 4.2. Some of the important transitions of the RE^{3+} ions and their electronic structures for lighting applications

Ion	Atomic number	Configuration RE^{3+}	Ground state	Transition	Wavelength (nm)	Applications
Ce^{3+}	58	$[Xe]4f^1$	$^2F_{5/2}$	$5d \rightarrow 4f$: $^2F_{7/2}$, $^2F_{5/2}$	~ 400-650	Displays, lighting
Sm^{3+}	62	$[Xe]4f^5$	$^6H_{5/2}$	$^4G_{5/2} \rightarrow {}^6H_{7/2}$	~ 600	Displays, lighting
Pr^{3+}	59	$[Xe]4f^2$	$^3H_4$	$^3P_0 \rightarrow {}^3H_6$	~ 617	Displays, lighting
Eu^{3+}	63	$[Xe]4f^6$	$^7F_0$	$^5D_0 \rightarrow {}^7F_2$	~ 615	Displays, lighting
Tb^{3+}	65	$[Xe]4f^8$	$^7F_6$	$^5D_4 \rightarrow {}^7F_5$	~ 545	Displays, lighting
Dy^{3+}	66	$[Xe]4f^9$	$^6H_{15/2}$	$^4F_{9/2} \rightarrow {}^6H_{13/2}$	~ 575	Displays, lighting

the luminescence and energy transfer probabilities become analogous in the vicinity of some nm. Energy transfer predominantly takes either resonant or non-resonant paths. In the resonant case, the donor and acceptor ion energy levels involved in the process have similar energy level separation gaps whereas the separation gap is dissimilar for the non-resonant energy transfer case. At higher activator ion concentration due to reduced ion-ion interaction length, energy migration among donor ions before reaching the acceptor ion increases the probability that the optical irradiation is trapped at defects or impurity sites, enhancing the non-radiative relaxation process, which leads to concentration quenching. An increase in the activator concentration inspires such non-radiative transition processes. Meanwhile, reducing the activator concentration reduces the energy stored by the ions. Accordingly, the optimum activator concentration, classically between 1 and 5 mol% for trivalent rare-earth ions, ensuing from the above two factors.

The energy transfer efficiency, which can be denoted as η_{ET}, from a donor to an acceptor can be given as, $\eta_{ET} = 1 - I_s/I_0$, where I_s stands for the luminescence intensity of the donor in presence of the activator and I_0 that in the non-appearance of the activator. The energy transfer efficiency immensely relays on the spectral overlapping behavior of the donor's emission band along with the absorption band of the acceptor's. The energy transferring process results either in the enhancement or in the quenching of emission. In glass samples, energy transfer among the 4f levels has usually initiated from interactions which are electric dipole-electric quadrupole in nature (Nakazawa and Shionoya 1967). The close relay of the absorption and emission transition of lanthanide ions, cross-relaxations were predominant at higher concentrations as described for example in the case of Pr^{3+}, Tb^{3+}, and Eu^{3+} ions. In addition to the ion-ion interactions, activator ions also interact with the host matrix to facilitate the energy exchange between the host and ions, like in the case of host sensitized luminescence and are more abundantly seen in phosphors (Yen et al. 2006). The energy transfer analysis and estimation of different energy transfer micro-parameters are well described in the thesis work of Sontakke et al. 2016.

Among various kinds of rare-earth ion combinations of luminescence applications especially for visible lighting, Ce^{3+} ions are commonly used as donor ions because of their strong and broad absorption in the UV-VIS region and broad luminescence in the visible region 440-650 nm (host-dependent) due to f-d and d-f transitions respectively. The emission band of Ce^{3+} ion in common glasses (blue emission in the region 350-450 nm) well matches with the strong f-f absorption transitions of many efficient luminescent trivalent lanthanides such as Tb^{3+} (green), Eu^{3+} (red), Dy^{3+} (blue-yellow), Sm^{3+} (orange-red), Pr^{3+} (red), Ho^{3+} (green), Er^{3+} (green) which can potentially enhance the luminescence properties. Some of the common combinations of rare-earths with Ce^{3+} and other RE^{3+} as efficient donors for white light or other color generation are Ce^{3+}-Tb^{3+}, Ce^{3+}-Eu^{3+}, Ce^{3+}-Sm^{3+}, Ce^{3+}-Tb^{3+}-Eu^{3+}, Dy^{3+}-Eu^{3+}, Ce^{3+}-Dy^{3+}, Dy^{3+}-Tb^{3+}, Dy^{3+}-Tb^{3+}-Eu^{3+}, Tb^{3+}-Sm^{3+} and Ce^{3+}-Tb^{3+}-Sm^{3+}.

Because of the exchange type of interactions, radiation re-absorption or a multipole-multipole interaction the non-radiative energy transfer among rare-earth ions can commonly occur. Normally, the PL intensity is enhanced with the rising amount in rare-earth concentration. Once two adjacent ions come very close so that the probability of interactions between the two ions is very high, the initiation of the non-radiative relaxation can happen which decreases the radiative emission intensity. This phenomenon is known as concentration quenching. Therefore, for obtaining the highest emission intensity or to realize the starting of the non-radiative energy transfer process, it is crucial to estimate the critical distance (R_c) between the two rare-earth ions. The value of R_c can be predicted using a formula proposed by Blasse (1986). If it is assumed that an activator is solely incorporated in the M ionic sites of a host, C is the critical concentration of the activator, N is the number of M ions in the unit cell and V is the volume of the unit cell, then there is on the average one activator ion per {V/(C.N)}volume. Then as per this formula, R_c is nearly equal to double the radius of a sphere with this volume:

$$R_c = 2 \times \left[\frac{3V}{4\pi CN} \right]^{1/3} \tag{1}$$

If the value of R_c is bigger than 5Å, the exchange type of interactions cannot take place. In this situation, the multipolar interaction is mainly responsible for concentration quenching of rare-earth activators. The interactions namely dipole-dipole (d-d), dipole-quadrupole (d-q), quadrupole-quadrupole (q-q), etc. are known as the multipolar type interactions. Meanwhile, Dexter and Schulman (1954) proposed a theory which is very useful to predict the nature of any multipolar interactions. As per this theory, the emission intensity (I) for each activator ion follows the formula given below:

$$\frac{I}{C} = \frac{k_1}{\beta C^{\frac{s}{3}}} \tag{2}$$

where C stands for the activator concentration involved in self concentration quenching, k_1 and β are constants for each interaction in the identical excitation conditions, and s is the sequence of the electric multipolar (for d-d, d-q, and q-q, s values are 6, 8, and 10, respectively). It is worth mentioning that for the donor accepter system, the maximum energy transfer can take place if the distance between the donor and acceptor is equal to R_c.

Direct and indirect excitations: It needs to be mentioned that the optical amplification is not achievable unless supplying energy for bringing RE^{3+} ions in their respective excited states. Such a mechanism can be termed as pumping. One way of excitation is the direct pumping of RE^{3+} ions. Whereas the indirect excitation via a sensitizer is also possible, and the transfer of excitation energy from one ion to another is associated with the electronic transitions. So, the corresponding transitions participating in the energy transfer process needs to be in resonance with each other for a direct donor to an acceptor so that the energy in interaction could be conserved. And thus, despite the type of coupling interactions, the spectral overlap of donor's emission with acceptor's absorption becomes a primary condition for energy transfer to take place in a certain donor-acceptor pair. The energy transfer micro parameter, C_{DA} is an intrinsic property of the material and gives a quantitative measure of the energy transfer process that depends on the spectral overlap function of the donor's emission and the acceptor's absorption using Forster's spectral overlap model (Caird et al. 1991, Braud et al. 1998).

$$C_{DA} = \frac{3c}{8\pi^4 n^2} \int \sigma_{em}^D (\lambda) \sigma_{abs}^A (\lambda) d\lambda \tag{3}$$

where c is the speed of light in vacuum, n is the refractive index and σ is the emission and absorption cross-sections of donor and acceptor ions, respectively. The critical interaction distance, R_0 at which half of the donor excitation energy is transferred to the acceptor can be determined from C_{DA} through the formula,

$$C_{DA} = \frac{R_0^s}{\tau_0} \tag{4}$$

where τ_0 is intrinsic radiative rate (lifetime of donor ions at a low concentration where ion-ion interactions are negligible). The probability of energy transfer in a donor-acceptor system shows a linear relationship with spectral overlap function but, it decreases with the increase in inter-ionic separation. The energy transfer rate, W_{ET} in a uniformly distributed solid medium having inter-ionic separation, R_{Exp} can be defined as (Lakowicz et al. 1990),

$$W_{ET} \propto \left(\frac{R_0}{R_{Exp}} \right)^s \tag{5}$$

For some RE^{3+} ions, the spectral overlap of electronic transitions among donor and acceptor ions does not exist or remains quite poor due to the non-resonance of energy levels participating in the interaction. In such cases, the host matrix helps in bridging the energy gap with the help of the creation or annihilation of one or more host phonons. So, if the host phonons initiate the energy transfer process, its contribution must be taken into consideration while estimating the microparameters related to energy transfer. It is done by making the phonon sidebands (PSB) of stokes to the absorption and emission cross-sectional spectra by adopting the Auzel's exponential law, to use them in the calculation of spectral overlap function (Nostrand et al., 2001),

$$\sigma_{Stokes} = \sigma_{elect} \exp(-\alpha_S \Delta E) \tag{6}$$

where ΔE is known as the energy mismatch between the transitions of electronic and vibronic in nature and α_S is the parameter which is host-dependent for Stokes transitions. This parameter can be represented as,

$$\alpha_S = (h\nu_{max})^{-1} \left(\ln\{ (\overline{N}/S_0)[1 - \exp(-h\nu_{max}/kT)] \} - 1 \right) \tag{7}$$

where $\overline{N}$ stands for the phonon numbers necessary for bridging the energy gap, S_0 is the constant of electron-phonon coupling (≈ 0.04), $h\nu_{max}$ is the highest phonon energy of host, k is known as the Boltzmann constant and T stands for the absolute temperature. The probability of energy transfer for such phonon assisted energy transfer processes becomes (Jaque et al., 2003),

$$P_{ET} \propto I(E_{ph}) = \frac{e^{E_{ph}/k_B T}}{e^{E_{ph}/k_B T} - 1} \int \frac{f_D(E - E_{ph}) f_A(E)}{E^2} dE \tag{8}$$

where $f_D(E)$ and $f_A(E)$ are the line-shape functions of the donor's emission and the acceptor's absorption respectively, and E_{ph} is the host phonon energy. The C_{DA} or $R = 0$ are the host properties and independent of the concentration of activator ions, whereas the energy transfer rate W_{ET} and thus, in turn, the energy transfer efficiency, η_{ET} changes as a function of both donor and acceptor concentrations due to the reduction of inter-ionic separation. The energy transfer rate and efficiency are obtainable from donor fluorescence decay time, as well as from the sensitized luminescence spectra of coactivated samples using the relation (Caldino et al. 2008),

$$W_{ET} = \frac{1}{\tau_D} - \frac{1}{\tau_{DA}} \tag{9}$$

$$\eta_{ET} = 1 - \frac{\tau_D}{\tau_{DA}} \tag{10}$$

where τ_D, τ_{DA} are the lifetimes of donor ions in the presence and absence of acceptor ions respectively.

J-O intensity parameters: The Judd-Ofelt (J-O) intensity parameters, Ω_2, Ω_4, Ω_6 are vital to analyze the quantitative spectroscopic behavior of various RE^{3+} ions situated in various hosts (Judd 1962; Ofelt 1962). These parameters are profound to the host environment of RE ions. The nephelauxetic effect, i.e., the change in the absorption band position of RE ion in the host environment to that of the aqueous solution increases the Ω_2 parameter in the positive side for more asymmetry and covalency at the RE site which in turn influences by the crystal-field effects. Meanwhile, Ω_4 and Ω_6 are related to the rigidity of the network. These parameters show a stronger tendency on the variation of electron density of the oxide ion i.e., higher the density lower the values and vice versa.

The J-O parameters can be calculated from the estimated electric dipole line strength values using the following equations (Kaminskii 1980). The calculations need knowledge of absorption cross-sections (σ_{ij}) of individual transitions and the material's refractive index (n). Equation (11) gives an empirical relation between experimental absorption cross-section and the oscillator strength (f) or line strength (S) respectively (Kaminskii 1980).

$$\frac{mc^2}{\pi e^2 \lambda_{ij}^2} \int \sigma_{ij} d\lambda = f_{ij} = \frac{8\pi^2 mc}{3hg_i \lambda_{ij}} \left[\chi_{ed} S_{ij}^{ed} + \chi_{md} S_{ij}^{md} \right] \tag{11}$$

$$\sigma_{ij} = \frac{2.303 a_{ij}}{N_{RE} l} \tag{12}$$

$\chi_{ed} = \dfrac{(n^2+2)^2}{9n}$ and $\chi_{md} = n^2$ are the dielectric correction factors, λ_{ij} is the wavelength of transition, m and e can be recognized as the mass and charge of an electron, a_{ij} is the optical density, N_{RE} is the RE ion concentration, and l is known as the optical path length. The values of line strength for magnetic dipole transitions can be obtained from the corresponding reports of Carnall et al. ($f_{ij}^{md} = nf_{ij}'$ and $f_{ij} = f_{ij}^{ed} + f_{ij}^{md}$) (Kaminskii 1980). The line strength, S_{ij} for electric dipole transition can be represented as,

$$S_{ij}^{ed} = \sum_{t=2,4,6} \Omega_t \left| \langle i \| U^{(t)} \| j \rangle \right|^2 \tag{13}$$

where $\left| \langle i \| U^{(t)} \| j \rangle \right|^2$ is the squared reduced matrix elements corresponding to

the i→j transitions. The calculated values of squared reduced matrix elements have also been reported by Carnall et al. for all the rare-earth elements in LaF$_3$ matrix (Kaminskii 1980). These values can be used for the estimation of transition line strengths in any host matrix since the unit matrix elements show negligible dependence on the host matrix. The intensity parameters, Ω_t

can now be obtained by running the least square fitting over the experimental line strength to get the best set of Ω_t parameters. The accuracy or root mean square (rms) deviation (δ) can be estimated as follows (Kaminskii 1980),

$$\delta = \left[\sum_{n=1}^{q} (S_{cal} - S_{exp})^2 \frac{1}{q-p} \right]^{1/2} \tag{14}$$

where q is the total number of transitions used for the analysis, and p is the number of parameters evaluated (p = 3). Once the J-O intensity parameters are achieved, the radiative properties such as spontaneous emission probability, $A_{ji'}$ for $j \to i'$ transition, its branching ratio, $\beta_{ji'}$ and the radiative decay time, $\tau_{ji'}$ can be estimated using the following relations (Kaminskii 1980)

$$A_{ji'} = \frac{64\pi^4 e^2}{3h\lambda_{ji'}^3 g_j} \left[\chi'_{ed} S'_{ed} + \chi'_{md} S'_{md} \right] \tag{15}$$

where $S'_{ed} = \sum_{t=2,4,6} \Omega_t \left| \langle (j \| U^{(t)} \| i' \rangle \right|^2$ $\chi'_{ed} = \frac{n(n^2+2)^2}{9}$ and $\chi'_{md} = n^3$

$$\beta_{ji'} = \frac{A_{ji'}}{\sum_{i'} A_{ji'}} \tag{16}$$

$$\tau_{ji'} = \frac{1}{\sum_{i'} A_{ji'}} \tag{17}$$

where $\left| \langle j \| U^{(t)} \| i' \rangle \right|^2$ is the squared reduced matrix elements corresponding to the respective emission transitions and $S_{ji'}^{md}$ can be obtained from the literature (Kaminskii 1980). From the calculated transition probability, using Füchtbauer-Ladenburg relation (Kaminskii 1980; Fowler and Dexter 1962), the stimulated emission cross-section, $\sigma_{ji'}$ can be calculated estimated using the formula:

$$\sigma_{ji'} = \frac{\lambda_{ji'}^4}{8\pi n^2 c \Delta\lambda_{ji'}} A_{ji'} \tag{18}$$

where $\Delta\lambda_{ji'}$ is the transition bandwidth taken as the Full Width at Half Maximum (FWHM). The Füchtbauer-Ladenburg relation can also be applied for full spectrum cross-section by considering the line shape function, $I(\lambda_{ji'})$ of emission transition,

$$\sigma_{ji'} = \frac{\lambda_{ji'}^5}{8\pi n^2 c} A_{ji'} \frac{I(\lambda_{ji'})}{\int \lambda_{ji'} I(\lambda_{ji'}) d\lambda_{ji'}} \tag{19}$$

Another method for calculation of emission cross-section is based on the measured absorption cross-section proposed by McCumber (1964),

$$\sigma_{ji} = \sigma_{ij} \frac{g_i}{g_j} \exp\left[\frac{E_0 - h\nu_{ij}}{k_B T}\right] \qquad (20)$$

where E_0 is the zero-line energy corresponding to the difference between the lowest components of i and j levels, k_B is Boltzmann constant and T is the absolute temperature.

The radiative properties can be easily evaluated from J-O analysis as described above and can correlate with the experimental values for better results. For an overview, the J-O intensity parameters of Eu^{3+} activator in different glasses are given in Table 4.2. As seen from Table 4.2, the Ω_2 parameter in $CaO-La_2O_3-B_2O_3$ glass is smaller than that in other glasses, which indicates Eu^{3+} ions are surrounded by a strong asymmetrical and covalent environment in this glass. The parameter Ω_4 would increase with the decrease in ligand electron density. Meanwhile, in $Na_2O-ZnO-PbO-GeO_2-TeO_2$ glasses, the J-O intensity parameter Ω_6 is absent owing to the nonexistence of the $^5D_0 \rightarrow {}^7F_6$ transition.

Table 4.2. Ω_t (10^{-20} cm^2) parameters of Eu^{3+} in different glasses

Glass	Ω_2	Ω_4	Ω_6	Reference
MgO – PbO - B_2O_3 - SiO_2	6.01	1.96	-	Rao et al. 2013
CaO - La_2O_3 - B_2O_3	5.73	5.42	0.42	Chakrabarti et al. 2007
K_2O – SrO - Al_2O_3 - P_2O_5	6.34	5.67	0.68	Linganna and Jayasankar 2012
Na_2O – ZnO – PbO -GeO_2- TeO_2	6.25	1.77	-	Wang et al. 2013

While spectral positions are insensitive to the host matrix, radiative properties are highly dependent. Most importantly, the emission line strengths and its probabilities are very important to predict/interpret the energy transfer and decay analysis of RE^{3+} ions for varied concentrations as well as multi RE ion interactions. The radiative properties are mainly required for some kind of computer simulations to predict the different energy transfer mechanisms involved among RE ions.

Some Examples to Realize White Light Emission from RE^{3+} Activated Glasses

White emission from a single RE^{3+} activated system: The blue emission, $^6H_{15/2} \rightarrow {}^6F_{11/2}$ of Dy^{3+} is hypersensitive in nature. Therefore, tuning the blue emission and the grouping of blue, yellow as well as red emission bands in the Dy^{3+} activated glass systems can generate white emission. The emission of Dy^{3+} contains dominant emissions at around 485 and 575 nm because of the transitions from $^4F_{9/2}$ to $^6H_{15/2}$ and to $^6H_{13/2}$, respectively. The yellow peak is prime because of the low symmetric ionic sites of Dy^{3+} ions which are non-inversion centers. As elaborated in the energy level scheme of Dy^{3+} ions (Fig. 4.10a), under the UV light excitations, the Dy^{3+} ions in low energy levels are excited to its higher states from which the non-radiative relaxations to the lower energy levels occur via the phonon-assisted processes until they reach to the $^4F_{9/2}$ level. After that, the radiatively deactivates produce blue and yellow emissions including other weak red end emission lines. With the adjustment in the Dy^{3+} content, the above mentioned emissions can be altered. Henceforth, white light may be produced from a glass contained with Dy^{3+} ions via controlling the Dy^{3+} conc. and with subsequent control in the blue vs yellow emission ratio (Reddy et al. 2013; Das et al. 2011).

White emission from a RE^{3+}–RE^{3+} coactivated system via mixing the emission color: The intense red orange photoluminescence from Eu^{3+} and high green emission from Tb^{3+} are useful to generate white light. The incorporation and enhancement of Eu^{3+} ions lead to enhance the red light component. Whereas the green emission from Tb^{3+} ions (Tb^{3+}: $^5D_4 \rightarrow {}^7F_5$) decreased simultaneously. In the tricolor combination of red, green and blue, the appropriate mixing of at least two colors can produce white light. Hence, the right alteration in Eu^{3+} and Tb^{3+} coactivation, the white light can be produced because of the equivalent influence of intensities of green and reddish lights. In the Eu^{3+}-Tb^{3+} coactivated systems, the relaxation of Tb^{3+} ions to the ^7F_J states lead to an energy transfer to ground states of Eu^{3+} with the subsequent excitation of electrons from the ground states to $^5D_1$ (showing by the dashed line in Fig. 4.10b). The succeeding mechanism describes the energy transfer process: $^5D_4/Tb^{3+} + {}^7F_0/Eu^{3+} \rightarrow {}^7F_5/Tb^{3+} + {}^5D_1/Eu^{3+}$ (Reddy et al. 2013).

White emission from a RE^{3+}-RE^{3+} coactivated system via the energy transfer: In Dy^{3+} the yellow and blue (483 and 574 nm, $^4F_{9/2} \rightarrow {}^6H_{13/2}$ and $^6H_{15/2}$) emission peaks and in Eu^{3+}, red and orange (616 and 594 nm of $^5D_0 \rightarrow {}^7F_2$ and $^7F_1$) emission peaks are intense. The hypersensitive emission having electric dipole in nature peaked at 616 nm ($^5D_0 \rightarrow {}^7F_2$ of Eu^{3+}) is enhanced in a systematic way with the rise of Dy^{3+}, which can be possible owing to the energy transfer from Dy^{3+} to Eu^{3+}, as shown in Fig. 4.10c (Fig. 2c of S. Das et al. 2011). Though the energy of the $^4F_{9/2}$ of Dy^{3+} is slightly higher than that of the $^5D_0$ of Eu^{3+} ions, the ET is favorable owing to the phonon-aided non-radiative relaxation from $^4F_{9/2}$ of Dy^{3+} to the $^5D_0$ of Eu^{3+}. The yellow vs blue ($^4F_{9/2} \rightarrow {}^6H_{13/2}$ / $^4F_{9/2} \rightarrow {}^6H_{15/2}$ of Dy^{3+}) and red vs orange ($^5D_0 \rightarrow {}^7F_2$ / $^5D_0 \rightarrow {}^7F_1$ of Eu^{3+}) emission intensity ratios usually have different values for varying conc. of Dy^{3+} and Eu^{3+}. Therefore, suitable alteration of Dy^{3+}-Eu^{3+} ion combinations may obtain white emission in a Eu^{3+}-Dy^{3+} coactivated system (S. Das, pers. comm..).

General Configuration of White Emitting Glass LEDs

In the phosphor-converted LEDs, the layer of phosphor particles is usually coated on the suitable LED via mixing the phosphor powders with organic

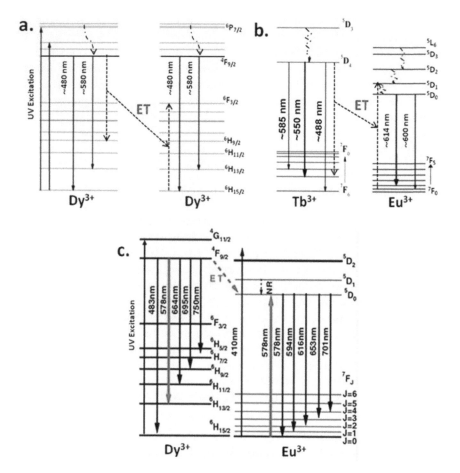

Fig. 4.10. Energy level (ET) diagrams and the energy transfer process in (a) Dy^{3+} ions. (b) Energy level scheme and energy transfer process inTb^{3+} and Eu^{3+} ions. (c) Energy level scheme and energy transfer process in Dy^{3+} and Eu^{3+} ions.

resins to produce a paste. But, such organic resins weaken owing to the lights with short wavelengths and high energies of LEDs, which is known as the photodecomposition. Such photodecomposition results in fading the efficiency of lighting devices with time. Meanwhile, if glasses of appropriate thickness are used, the issue related to the photodecomposition can be simply eliminated, since the glasses constituted of non-decomposable materials are weather-resistant in nature. The schematic illustrations of the design of white emitting LEDs constructed using glasses are shown in Fig. 4.11.

The high correlated color temperature usually around 6000 K is a major disadvantage of the YAG phosphor for making white LEDs for illuminating sources. For reducing the CCT of any white light source, the color coordinate has to move towards the red zone in the chromaticity diagram. This can be achieved via changing the FWHM of the spectral band within the visible

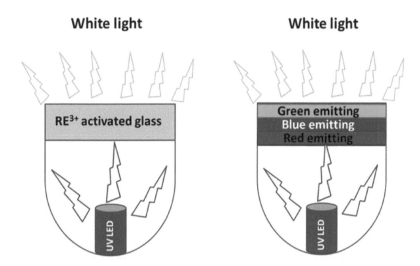

Fig. 4.11. Schematic illustration of a glass converted white LED device.

range, increasing the red component intensity with the subsequent decrease of the blue component in the overall emission. The above mentioned spectral modifications are very difficult to achieve by changing the phosphor powder amount in regular white-LED packaging. But, such adjustments are relatively easy in glass materials, which can be obtained simply by adjusting the glass plate thickness or its compositions.

Conclusions

This chapter elaborates a consolidated impression on the luminescent parameters related to trivalent rare-earth activated or co-activated glasses suitable for solid state type lighting applications, mainly for generating white colored light. Meanwhile, a concise theoretical and experimental outline of the potential rare-earth ions and related photoluminescence parameters suitable for analyzing glasses for lighting applications are also presented. The contribution of Ln ions in the photonic research and development is immeasurable. Considerable research activity is going on RE^{3+} activated glasses for solid-state lighting. Several RE ion has its characteristic emission of different wavelengths in different visible zone, and they are relevant electronic and optical characteristics, for example spectral forms, cross-sections, emission as well as absorption peaks and FWHMs, the excitation state lifetimes, inter-ionic and ion-lattice interactions etc. Meanwhile, the above mentioned properties are strongly influenced by the main host system. Therefore, apart from the selection of RE, ion selection of the host material is also a critical aspect in the development of lighting devices. Glasses are exceptional in their aptitude to incorporate a wide range and very high concentration of activators. This is connected with the random network structure of a glass. The critical issue that influence the rare-earth emission characteristics during the synthesis of glass when doped in high concentration is cluster formation. This chapter could enable the reader on designing potential glasses for solid-state lighting materials.

References

Allu, A.R., Das, S., Som, S., Vardhan, H., Maraka, R., Balaji, S., Santos, L.F., Hönninger, I.M., Jubera, V., Ferreira, J.M.F. 2017. Dependence of Eu^{3+} photoluminescence properties on structural transformations in diopside-based glass-ceramics. J. Alloys Compd. 699, 856-865.

Almeida, A.D., Santos, B. Paolo, B., Quicheron, M. 2014. Solid state lighting review – Potential and challenges in Europe. Renew. Sustain. Ener. Rev. 34, 30-48.

Amjad, R.J., Sahar, M.R., Ghoshal, S.K., Dousti, M.R., Arifin, R. 2013. Synthesis and characterization of Dy^{3+} doped zinc-lead-phosphate glass. Opt. Mater. 35, 1103-1108.

Andriessen, J., Dorenbos, P., van Eijk, C.W.E. 2005. Ab initio calculation of the contribution from anion dipole polarization and dynamic correlation to 4f-5d excitations of Ce^{3+} in ionic compounds. Phys. Rev. B 72, 045129-045142.

Azeem, P.A., Balaji, S., Reddy, R.R. 2008. Spectroscopic properties of Dy^{3+} ions in NaF-B_2O_3-Al_2O_3 glasses. Spectrochimica. Acta. A 69, 183-188.

Babu, B.H., Kumar, V.V.R.K. 2014. White light generation in Ce^{3+}-Tb^{3+}-Sm^{3+} codoped oxyfluoroborate glasses. J. Lumin. 154, 334-338.

Bachmann, V., Ronda, C., Meijerink, A. 2009. Temperature quenching of yellow Ce^{3+} luminescence in YAG:Ce. Chem. Mater. 21, 2077-2084.

Balaji, S., Abdul Azeem, A., Reddy, R.R. 2007. Absorption and emission properties of Eu^{3+} ions in sodium fluoroborate glasses. Physica B: Condensed Matter 394, 62-68.

Blasse, G. 1986. Energy transfer between inequivalent Eu^{2+} ions. J. Solid State Chem. 62, 207-211.

Blasse, G., Bril, A. 1969. Luminescence of phosphors based on host lattices ABO_4 (A is Sc, In; B is P, V, Nb). J. Chem. Phys. 50, 2974.

Braud, A., Girald, S., Doualan, J.L., Moncorge, R. 1998. Spectroscopy and fluorescence dynamics of (Tm^{3+}, Tb^{3+}) and (Tm^{3+}, Eu^{3+}) doped LiYF4 single crystals for 1.5-μm laser operation. IEEE J. Quan. Elect. 34, 2246-2255.

Caird, J.A., Ramponi, A.J., Staver, P.R. 1991. Quantum efficiency and excited-state relaxation dynamics in neodymium-doped phosphate laser glasses. J. Opt. Soc. Am. B. 8, 1391-1403.

Caldiño, U., Jaque, D., Martín-Rodríguez, E., Ramírez, M.O., García Solé, J., Speghini, A., Bettinelli, M. 2008. Nd^{3+} → Yb^{3+} resonant energy transfer in the ferroelectric $Sr_{0.6}Ba_{0.4}Nb_2O_6$ laser crystal. Phys. Rev. B. 77, 075121-075129.

Chakrabarti, R., Das, M., Karmakar, B., Annapurna, K., Buddhudu, S. 2007. Emission analysis of Eu^{3+}:CaO-La_2O_3-B_2O_3 glass. J. Non-Cryst. Solids 353, 1422-1426.

Chen, Q., Dai, N., Liu, Z., Chu, Y., Ye, B., Li, H., Peng, J., Jiang, Z., Li, J., Wang, F., Yang, L. 2013. White light luminous properties and energy transfer mechanism of rare earth ions in Ce^{3+}/Tb^{3+}/Sm^{3+} co-doped glasses. Appl. Phys. A 115, 1159-1166.

Chen, G.H., Yao, L., Zhong, H.J., Cui, S.C. 2016. Luminescence properties and energy transfer behavior between Tm^{3+} and Dy^{3+} ions in co-doped phosphate glasses for white LEDs. J. Lumin. 178, 6-12.

Chen, Y., Liu, X., Chen, G., Yang, T., Yuan, C. 2017. Luminescent characteristics of $Tm^{3+}/Tb^{3+}/Eu^{3+}$ tri-doped borophosphate glasses for LED applications. J. Mater. Sci.: Mater. Electron. 28, 5592-5596.

Das, S., Reddy, A.A., Babu, S.S., Prakash, G.V. 2011. Controllable white light emission from Dy^{3+}-Eu^{3+} co-doped $KCaBO_3$ phosphor. J. Mater. Sci. 46, 7770-7775.

Dexter, D.L., Schulman, J.H. 1954. Theory of concentration quenching in inorganic phosphors. J. Chem. Phys. 22, 1063-1070.

Dorenbos, P. 2000. 5d-level energies of Ce^{3+} and the crystalline environment. I. Fluoride compounds. Phys. Rev. B 62, 15640-15649.

Dorenbos, P. 2002. 5d-level energies of Ce^{3+} and the crystalline environment. IV. Aluminates and "simple" oxides. J. Lumin. 99, 283-299.

Duffy, J.A. 1996. Redox equilibria in glass. J. Non-Cryt. Solids 196, 45-50.

Fowler, W.B., Dexter, D.L. 1962. Relation between absorption and emission probabilities in luminescent centres in ionic solids. Phys. Rev. 128, 2154-2165.

Fuxi, G. 1991. Optical and Spectroscopic Properties of Glass. Springer, Verlag.

Fuxi, G. 1995. Laser Materials. World Scientific Publication Ltd., Singapore.

Gao, G., Wondraczek, L. 2014. Spectral asymmetry and deep red photoluminescence in Eu^{3+} activated $Na_3YSi_3O_9$ glass ceramics. Opt. Mater. Exp. 4, 476-485.

Ghosh, D., Biswas, K., Balaji, S., Annapurna, K. 2017. Tunable white light generation from Ce^{3+}-Tb^{3+}-Mn^{2+} doped metaphosphate glass for LED and solar cell applications. J. Lumin. 183, 143-149.

Hong, T.T., Yen, P.D.H., Quang, V.X., Dung, P.T. 2015. Luminescence properties of Ce/Tb/Sm co-doped tellurite glass for white LEDs application. Mater. Trans. 56, 1419-1421.

Jana, S.K. 2012. Studies on the energy transfer in rare-earth activated glass and glass-ceramics for IR luminescence and visible upconversion applications (PhD Thesis). Jadavpur University, India.

Jaque, D., Ramirez, M.O., Bausà, L.E., García Solé, J., Cavalli, E., Speghini, A., Bettinelli, M. 2003. Nd^{3+}- Yb^{3+} energy transfer in the $YAl_3(BO_3)_4$ nonliner laser crystal. Phys. Rev. B. 68, 035118-9.

Judd, B.R. 1962. Optical absorption intensities of rare-earth ions. Phys. Rev. 127, 750-761.

Kaewnuam, E., Wantana, N., Kim, H.J., Kaewkhao, J. 2017. Development of lithium yttrium borate glass doped with Dy^{3+} for laser medium, W-LEDs and scintillation materials applications. J. Non-Cryst. Solids 464, 96-103.

Kaminskii, A.A. 1980. Laser Crystals – Their physics and properties. 2nd ed. Springer, Verlag Publications, Berlin.

Keller, S.P., Cheroff, G. 1958. Quenching, stimulation, and exhaustion studies on some infrared stimulable phosphors. Phys. Rev. 111, 1533-1539.

Kiss, Z.J. 1965. Energy levels of Dy^{2+} in the cubic hosts of CaF_2, SrF_2, and BaF_2. Phys. Rev. 137, A1749.

Kojima, K., Shimanuki, S., Kojima, T. 1971. Studies on Ag^- centers in alkali halides. II. Ag^-(Na) centers in KCl. J. Phys. Soc. Jpn. 30, 1380-1388.

Kumari, P., Manam, J. 2015. Structural, optical and special spectral changes of Dy^{3+} emissions in orthovanadates. RSC Adv. 5, 107575-107584.

Kumari, P., Manam, J. 2016. Effects of morphology on the structural and photoluminescence properties of co-precipitation derived $GdVO_4:Dy^{3+}$. Chem. Phy. Lett. 662, 56-61.

Lakowicz, J.R., Wiczk, W., Gryczynski, I., Szmacinski, H., Johnson, M.L. 1990. Influence of end-to-end diffusion on intramolecular energy transfer as observed by frequency-domain fluorometry. Biophysical Chem. 38, 99-109.

Langar, A., Bouzidi, C., Elhouichet, H., Gelloz, B., Ferid, M. 2017. Investigation of spectroscopic properties of Sm-Eu codoped phosphate glasses. Displays 48, 61-67.

Li, J.L., Deng, C.Y., Cui, R.R. 2014. Photoluminescence properties of $CaBi_2Ta_2O_9$:RE^{3+} (RE = Sm, Tb, and Tm) phosphors. Opt. Commun. 326, 6-9.

Linganna, K., Jayasankar, C.K. 2012. Optical properties of Eu^{3+} ions in phosphate glasses. Spectrochim. Acta A 97, 788-797.

Liu, S., Zhao, G., Ying, H., Wang, J., Han, G. 2008. Eu/Dy ions co-doped white light luminescence zinc-aluminoborosilicate glasses for white LED. Opt. Mater. 31, 47-50.

Liu, S., Zhao, G., Lin, X., Ying, H., Liu, J., Wang, J., Han, G. 2008. White luminescence of Tm-Dy ions co-doped aluminoborosilicate glasses under UV light excitation. J. Solid State Chem. 181, 2725-2730.

Luewarasirikul, N., Kim, H.J., Meejitpaisan, P., Kaewkhao, J. 2017. White light emission of dysprosium doped lanthanum calcium phosphate oxide and oxyfluoride glasses. Opt. Mater. 66, 559-566.

Luo, Q., Qiao, X., Fan, X., Zhang, X. 2010. Preparation and luminescence properties of Ce^{3+} and Tb^{3+} co-doped glasses and glass ceramics containing SrF_2 nanocrystals. J. Non-Cryst. Solids 356, 2875-2879.

Luo, Q., Qiao, X., Fan, X., Yang, H., Zhang, X., Cui, S., Wang, L. and Wang, G. 2009. Luminescence behavior of Ce^{3+} and Dy^{3+} codoped oxyfluoride glasses and glass ceramics containing LaF_3 nanocrystals. J. Appl. Phys. 105, 043506-043511.

Ma, H., Ye, R., Gao, Y., Hua, Y., Deng, D., Yang, Q., Xu, S. 2014. Luminescence properties and energy transfer mechanism of Ce^{3+}/Sm^{3+} co-doped Na_2O-Y_2O_3-SiO_2 glass for white light emission. J. Non-Cryst. Solids 402, 231-235.

McCumber, D.E. 1964. Einstein relations connecting broadband emission and absorption spectra. Phys. Rev. 136, A954-A957.

Miniscalco, W.J. 1991. Erbium-doped glasses for fiber amplifiers at 1500 nm. J. Lightwave Technol. 9, 234-250.

Murata, T., Sato, M., Yoshida, H., Morinaga, K. 2005. Compositional dependence of ultraviolet fluorescence intensity of Ce^{3+} in silicate, borate, and phosphate glasses. J. Non-Cryst. Solids 351, 312-316.

Nakazawa, E., Shionoya, S. 1967. Energy transfer between trivalent rare earth ions in inorganic solids. J. Chem. Phys. 47, 3211-3219.

Naresh, V., Gupta, K., Reddy, C.P., Ham, B.S. 2017. Energy transfer and colour tunability in UV light induced Tm^{3+}/Tb^{3+}/Eu^{3+}:ZnB glasses generating white light emission. Spectrochim. Acta A 175, 43-50.

Northwest, E. 2011. Introductory guide to LED lighting. Warrington, UK: Envirolink Northwest.

Nostrand, M.C., Page, R.H., Payne, S.A., Isaenko, L.I., Yelisseyev, A.P. 2001. Optical properties of Dy^{3+}- and Nd^{3+}-doped KPb_2Cl_5. J. Opt. Soc. Am. B 18, 264-276.

Ofelt, G.S. 1962. Intensities of crystal spectra of rare earth ions. J. Chem. Phys. 37, 511-520.

Pieterson, L. van, Heeroma, M., de Heer, E., Meijerink, A. 2000. Charge transfer luminescence of Yb^{3+}. J. Lumin. 91, 177-193.

Piper, W.W., Deluca, J.A., Ham, F.S. 1974. Cascade fluorescent decay in Pr^{3+}-doped fluorides: Achievement of a quantum yield greater than unity for emission of visible light. J. Lumin. 8, 344-348.

Rao, T.G.V.M., Kumar, A.R., Neeraja, K., Veeraiah, N., Rami Reddy, M. 2013. Optical and structural investigation of Eu^{3+} ions in Nd^{3+} co-doped magnesium lead borosilicate glasses. J. Alloy. Comp. 557, 209-217.

Reddy, A.A., Pradeesh, K., Sekhar, M.C., Babu, S.S., Prakash, G.V. 2011. Optical properties of Dy^{3+}-doped sodium–aluminum–phosphate glasses. J. Mater. Sci. 46, 2018-2023.

Reddy, A.A., Babu, S.S., Pradeesh, K., Otton, C.J., Prakash, G.V. 2011. Optical properties of highly Er^{3+}-doped sodium-aluminium-phosphate glasses for broadband 1.5 μm emission. J. Alloys Compd. 509, 4047-4052.

Reddy, A.A., Das, S., Goel, A., Sen, R., Siegel, R., Mafra, L., Prakash, G.V., Ferreira, J.M.F. 2013. $KCa_4(BO_3)_3$:Ln^{3+} (Ln = Dy, Eu, Tb) phosphors for near UV excited white-light-emitting diodes. AIP Adv. 3, 022126-022139.

Reddy, A.A., Tulyaganov, D.U., Kharton, V.V., Ferreira, J.M.F. 2015. Development of bilayer glass ceramic SOFC sealants via optimizing the chemical composition of glasses – A review. J. Solid State Electrochem. 19, 2899.

Ronda, C.R. 2008. Luminescence: From Theory to Applications. Wiley, VCH.

Schubert, E.F., Kim, J.K. 2005. Solid-state light sources getting smart. Science 308, 1274-1278.

Shamshad, L., Rooh, G., Kirdsiri, K., Srisittipokakun, N., Damdee, B., Kim, H.J., Kaewkhao, J. 2017. Effect of alkaline earth oxides on the physical and spectroscopic properties of Dy^{3+}-doped Li_2O-B_2O_3 glasses for white emitting material application. Opt. Mater. 64, 268-275.

Shisina, S., Das, S., Som, S., Ahmad, S., Vinduja, V., Merin, P., Gopalan, N.K., Kumari, P., Pandey, M.K. 2019. Structural and optoelectronic properties of palmierite structured $Ba_2Y_{0.67}\delta_{0.33}V_2O_8$: Eu^{3+} red phosphors for n-UV and blue diode based warm white light systems. J. Alloy. Compd. 802, 723-732.

Solé, J.G., Bausá, L.E., Jaque, D. 2005. Optical Spectroscopy of Inorganic Solids. Wiley Publications. Hoboken, New Jersey, United States.

Sontakke, A.D., Ueda, J., Tanabe, S. 2016. Effect of synthesis conditions on Ce^{3+} luminescence in borate glasses. J. Non-Cryst. Solids 431, 150-153.

Sun, X.Y., Wu, S., Liu, X., Gao, P., Huang, S.M. 2013. Intensive white light emission from Dy^{3+}-doped $Li_2B_4O_7$ glasses. J. Non-Cryst. Solids 368, 51-54.

Vedda, A. 2006. Insights into microstructural features governing Ce^{3+} luminescence efficiency in sol-gel silica glasses. Chem. Mater. 18, 6178-6185.

Vijayakumar, M., Marimuthu, K. 2015. Structural and luminescence properties of Dy^{3+} doped oxyfluoro-borophosphate glasses for lasing materials and white LEDs. J. Alloys. Compd. 629, 230-241.

Wang, F., Shen, L.F., Chen, B.J., Pun, E.Y.B., Lin, H. 2013. Broadband fluorescence emission of Eu^{3+} doped germanotellurite glasses for fiber-based irradiation light sources. Opt. Mater. Exp. 3, 1931-1943.

Watts, R.K., Richter, H.J. 1972. Diffusion and transfer of optical excitation in YF_3:Yb, Ho. Physical Review B 6, 1584-1589.

Wybourne, B.G., Smentek, L. 2007. Optical Spectroscopy of Lanthanides. CRC Press.

Yen, W.M., Shionoya, S., Yanamoto, H. 2006. Phosphors Hand Book, 2nd edition. CRC Press.

Zhang, Y., Zhu, Z., Zhang, W., Qiao, Y. 2013. Photoluminescence properties of Sm^{3+} ions doped oxyfluoride calcium borosilicate glasses. J. Alloys Compd. 566, 164-167.

Zhong, H.J., Chen, G.H., Chen, J.S., Yang, Y., Yuan, C.-L., Zhou, C.-R. 2015. Tm^{3+}/Tb^{3+}/Sm^{3+} tri-doped transparent glass ceramic for enhanced white-light-emitting material. J. Mater. Sci.: Mater. Electron. 26, 8143-8146.

Zhong, H., Chen, G., Yao, L., Wang, J., Yang, Y., Zhang, R. 2015. The white light emission properties of Tm^{3+}/Tb^{3+}/Sm^{3+} triply doped SrO-ZnO-P_2O_5 glass. J. Non-Cryst. Solids 427, 10-15.

Zhou, Y., Zhu, C., Yang, X. 2017. Near white light emission of Ce^{3+}-, Ho^{3+}-, and Sm^{3+}-doped oxyfluoride silicate glasses. J. Am. Ceram. Soc. 100, 2059-2068.

Zhu, C.F., Chaussedent, S., Liu, S.J., Zhang, Y.F., Monteil, A., Gaumer, N., Yue, Y.Z. 2013. Composition dependence of luminescence of Eu and Eu/Tb doped silicate glasses for LED applications. J. Alloys Compd. 555, 232-236.

Zhu, C., Wu, D., Liu, J., Zhang, M., Zhang, Y. 2017. Color-tunable luminescence in Ce-, Dy-, and Eu-doped oxyfluoride aluminoborosilicate glasses. J. Lumin. 183, 32-38.

Zhu, C., Wang, J., Zhang, M., Ren, X., Shen, J., Yue, Y. 2014. Eu-, Tb-, and Dy-doped oxyfluoride silicate glasses for LED applications. J. Am. Ceram. Soc. 97, 854-861.

Zu, C., Wang, Y., Chen, J., Han, B., Tao, H. 2011. Luminescent properties and applications of Tb^{3+} doped silicate glasses with industrial scales. J. Non-Cryst. Solids 357, 2435-2439.

Effect of CaZrO₃ Doping by Gd³⁺ on Phototherapy Lamp Phosphor Performance

Neha Dubey[1], Marta Michalska-Domańska[2,3*], Jagjeet Kaur Saluja[1], Janita Saji[4] and Vikas Dubey[5]

[1] Department of Physics, Govt. V.Y.T.PG. Auto. College, Durg - 491001, Chhattisgarh, India
[2] Department of Materials Science and Engineering, Delft University of Technology, Mekelweg 2, 2628 CD Delft, Netherlands
[3] Institute of Optoelectronics, Military University of Technology, 2 Urbanowicza Str., 00-908 Warsaw, Poland
[4] Department of Sciences and Humanities, Faculty of Engineering, CHRIST (Deemed to be University), Bengaluru - 560074
[5] Department of Physics, Bhilai Institute of Technology, Raipur - 493661, India

Introduction

Ultraviolet light B therapy (UVB therapy, phototherapy) is a useful tool for treating dermatological diseases (Hoare et al. 2000; Sonekar et al. 2007). It has been reported in the literature that ultraviolet therapy is useful for treating more than 40 types of skin diseases, such as psoriasis, eczema, atopic skin disorder, vitiligo etc. (Hoare et al. 2000; Sonekar et al. 2007). The Commission International de I'Eclairage classification subdivides the ultraviolet (UV) wavelength band (200–400 nm) into three regions: UVA, UVB and UVC. UVA (320–400 nm) or long wave UV is responsible for skin tanning effects, UVB (280–320 nm) is known as the sun burn (erythemal) region, and UVC (200–280 nm) is the germicidal region. The UVA region is further subdivided into UVA1 (340–400 nm) and UVA2 (320–340 nm). Biological molecules that absorb UV Radiation (UVR) are called chromophores. UVR induces a wide variety of biological responses depending upon the 'action

---
*Corresponding author: m.e.michalska-domanska@tudelft.nl, marta.michalska@wat.edu.pl

spectra' (wavelength dependence responses) and 'absorption spectra' of the chromophore in human skin. Therefore, selection of the proper wavelength☐ emitting phosphor is of prime importance. The action spectra of the phosphor should be in accordance with the biological absorption spectra of the skin (i.e. chromophore) damaged by the particular disease. Two methods are generally used in the treatment of skin diseases, phototherapy with narrow band UVB (NB-UVB; 311 nm) and photochemotherapy with UVA (365 nm) combined psoralen as a photosensitizer (PUVA). Both forms of therapy have a place in dermatology (Gruijl, 2000). The basis for phototherapy is believed to be the direct interaction of certain wavelengths of light with tissue, leading to a change in the immune response (Pinton et al. 1998). Typically, both UVA and UVB have been used in the treatment of various dermatoses; wavelengths in the range 300–320 nm for psoriasis, vitiligo, polymorphic light eruption, lichen sclerosus, pruritus in polycythemia vera, actinicprurigo, seborrheic dermatitis, atopic dermatitis, chronic actinic dermatitis, mycosis fungoides, erythropoietic protoporphyria, eosinophilicpustular folliculitis (Ofuji's disease), lupus erythematous, and wavelengths in the range 360–370 nm for PUVA, balneophototherapy for plaque parapsoriasis, morphea (scleroderma), lichen sclerosus, graft versus-host disease, contact dermatitis, mycosis fungoides, necrobiosislipoidica, acne vulgaris, localized scleroderma and many more dermatoses (Ullrich 1997; Osmancevic et al. 2010; Parrish et al. 1974).

Both phototherapy and photochemotherapy are used. UVB therapy consists of irradiation of patients with UV radiation band closely matching UV wavelength from the sun light, which is provided by using a fluorescent bulb with a strict design of output frequency of ultraviolet. It was found that during phototherapy of psoriasis the light from longer wavelengths of UVB region was more effective while the shorter wavelengths were much less effective or even harmful (Parrish and Jaenicke 1981). Researchers and scientists are still making a lot of efforts to develop new UVB-emitting phosphors with better performances. The interest in ultraviolet phosphor materials is increasing because the global demand for these phosphors in skin diseases therapy is ever increasing. Further for medical applications, it is desirable that the ultraviolet emission should be kept at the minimum to protect the tissues from radiation damage. Thus, there is need to investigate new substances which emit ultraviolet radiations with desired characteristics when excited by very short ultraviolet radiation or X-rays.

It is well known that ultraviolet-emitting phosphors in the wavelength range 200–400 nm has been extensively used in a variety of applications such as examination of counterfeit bank notes, production of vitamin D, fluorescent displays, photochemical reactions, water purifiers, luminescence dosimetry, photocopying and phototherapy (Weelden et al. 1988; Ferguson 1999; Vecht et al. 1998; Holick et al. 1997; Becker 1967; Shie et al. 2008; Scherschun et al. 2001. Moreover, ultraviolet (UV) radiation has been widely used in phototherapy. During phototherapy investigations, it was found that UV in the narrow region from 300 nm to 320 nm showed remarkable good therapeutic effects (MacKie 2000). UV emission in a narrow band centred at about 311 nm is thus required. A narrow UVB source emitting around 311 nm was developed in 1988 (Parrish and Jaenicke 1981), that revolutionized the UVB phototherapy (Weelden et al. 1988). There are a lot of methods to create a narrow UV band emission at around 311 nm, such as the combustion synthesis method, solid state method, sol-gel synthesis method (Hemne et al. 2017).

Though Gd^{3+} doped phosphors are well known for emitting in the ultraviolet region, reports on Gd^{3+} doped ultraviolet-emitting phosphors are limited. Abdullah et al. (Scherschun et al. 2001) studied ultraviolet luminescence in Y$_2$O$_3$ doped with Gd^{3+} ions, and found that strong emission line at 3.96 eV (313 nm) corresponding to the transitions from the $^6P_{7/2}$ to $^8S_{7/2}$ states of Gd^{3+} ions is observed. YAl$_3$(BO$_3$)$_4$:Gd^{3+} phosphor is due to the transitions from O 2 p states to mixed states of mainly B 2p and Y 4d orbitals. Yoshida et al. (2006) proposed an efficient ultraviolet emitting phosphor for the ultraviolet lamps, which was based on YAl$_3$(BO$_3$)$_4$:Gd^{3+}. Ultraviolet-emitting properties of Pr^{3+}-sensitized LaPO$_4$:Gd^{3+} phosphor under vacuum ultraviolet excitation was investigated by Okamoto et al. (2009). They found that luminescent properties of Pr^{3+}-sensitized LaPO$_4$:Gd^{3+} under vacuum-ultraviolet (vuv) light excitation have been investigated. The energy transfer probably occurs from the 5d levels in Pr^{3+} ions to Gd^{3+} ions under 172 nm light excitation. LaPO$_4$:Gd^{3+},Pr^{3+} shows efficient ultraviolet-B (uv-B) emission at 312 nm (Okamoto et al. 2009). We have recently examined Gd^{3+}-doped Y$_2$O$_3$ powder which exhibits photoluminescence at 314 nm (Singh et al. 2011; Singh et al. 2015; Gawande et al. 2016; Dubey et al. 2013; Dubey et al. 2016; Singh et al. 2016; Dubey and Tiwari 2016; Tiwari et al. 2016; Parganiha et al. 2015; Tiwari and Dubey 2016; Dubey et al. 2015; Jeong et al. 2018). In the present chapter the effect of Gd^{3+} ion in addition to CaZrO$_3$ phosphor on its optical properties are shown. Its crystallographic structure and morphology are also studied in detail. The CaZrO$_3$:Gd^{3+} phosphor gives emission in the UVB region and may be useful for phototherapy lamp phosphor as well as in other application in lighting industry.

Experimental

The powder samples of Gd^{3+} activated CaZrO$_3$ with varying concentrations of Gd^{3+} ion (from 0.5 mol% to 2.5 mol% with increase in every 0.5 mol%) were prepared via high temperature modified solid state diffusion method. The stoichiometric amounts starting materials CaCO$_3$, ZrO$_2$, Gd$_2$O$_3$ and H$_3$BO$_3$ (as a flux) in molar ratio (from 0.5% to 2.5% of Gd^{3+}) are taken in alumina crucible and is fired in air at 1000 °C for 2 hours in a muffle furnace. Each heating is followed by intermediate grinding using an agate mortar and pestle. The mixtures of reagents were ground together for 45 minutes to obtain a homogeneous powder. The powder was transferred to alumina crucible, and then heated in a muffle furnace at 1350 °C for 3 hours (Singh et al. 2016; Dubey and Tiwari 2016; Tiwari et al. 2016; Parganiha et al. 2015;

Tiwari and Dubey 2016; Dubey et al. 2015). The phosphor materials were then cooled to room temperature naturally inside the furnace.

The morphology and crystallographic structure of $CaZrO_3$:Gd^{3+} samples were characterized by using SEM, TEM, XRD and Photoluminescence (PL) analysis. Particle morphology was investigated by FEGSEM (field emission gun scanning electron microscope) (JEOL JSM-6360). TEM measurements was executed by PHILIPS CM 200 Operating voltages: 20-200 kV Resolution: 2.4 Å. The XRD measurements were carried out using Bruker D8 Advance X-ray diffractometer. The X-rays were produced using a sealed tube and the wavelength of X-ray was 0.154 nm (Cu K-alpha). The X-rays were detected using a fast counting detector based on Silicon strip technology (Bruker LynxEye detector). The photoluminescence (PL) emission and excitation spectra were recorded at room temperature by use of a Shimadzu RF-5301 PC spectrofluorophotometer. As the excitation source was used a xenon lamp (Singh et al. 2016, Dubey and Tiwari 2016, Tiwari .,et al. 2016, Parganiha, et al. 2015, Tiwari and Dubey 2016, Dubey et al. 2015).

Results and Discussion

Structural and Morphological Characterizations

The XRD analysis was carried out for comprehensive knowledge of phase formation and structural parameters.

In Fig. 5.1 (a) the comparison of the observed, calculated XRD pattern of $CaZrO_3$:Gd^{3+} having 2 mol% i.e. optimized concentration of dopant (Gd^{3+}) and the standard Celref version 3 (Match!2) software. XRD pattern of $CaZrO_3$

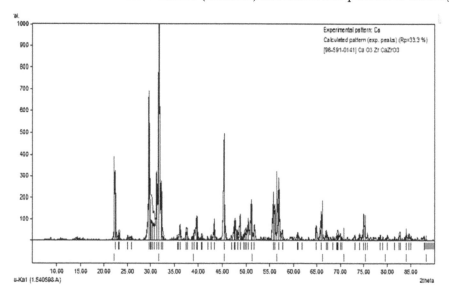

Fig. 5.1. XRD plot and Rietveld refinement of Gd3+ doped CaZrO3 samples (2.0 mol % is optimized concentration of dopant).

was presented. Match! v. 2.3.1 software was used to match the standard (Crystallographic open database) and experimental XRD pattern. It was clearly observed that the experimental XRD pattern with good agreement matched the Crystallographic Open Database (COD) card No. 96-591-0141 with Figure of Merit (FoM) of 0.76.

The datasheet provided by Match! version 2.3.1 expressed that the prepared sample was cubic with an P m -3 m (221) space group. Table 5.1 presents the refined lattice parameters of the phosphor $CaZrO_3:Gd^{3+}$ having a 2 mol% of Gd^{3+} ions. The values of the lattice parameters of the phosphor were obtained from the COD card No. 96-591-0141. Refinement was done using Celref version 3 (Match!2) software (Collaborative Computational Project No. 14 [CCP14] for Single Crystal and Powder Diffraction Birkbeck University of London and Daresbury Laboratory, London, UK). The

Table 5.1. Indexing and refined lattice parameters

Crystal structure							
Crystallographic data							
Space group	P m -3 m (221)						
Crystal system	Cubic						
Cell parameters	a = 3.9900 Å						
Atom coordinates	*Element*	*Oxid.*	*x*	*y*	*z*	*Bi*	*Focc*
	Ca		0.000	0.000	0.000	1.000000	1.000000
	Zr		0.500	0.500	0.500	1.000000	1.000000
	O		0.500	0.500	0.500	1.000000	1.000000
Diffraction lines	*d [Å]*	*Int.*	*h*	*k*	*l*	*Mult.*	
	3.9900	411.1	1	0	0	6	
	2.8214	1000.0	1	0	1	12	
	2.3036	0.1	1	1	1	8	
	1.9950	395.9	2	0	0	6	
	1.7844	148.1	2	0	1	24	
	1.6289	336.5	2	1	1	24	
	1.4107	197.9	2	0	2	12	
	1.3300	60.7	2	1	2	24	
	1.2617	126.1	3	0	1	24	
	1.2030	3.1	3	1	1	24	
	1.1518	57.4	2	2	2	8	
	1.1066	24.9	2	3	0	24	
	1.0664	143.1	3	2	1	48	

calculated spectra confirmed the presence of cubic crystal structure of $CaZrO_3$. The calculated lattice parameters are shown in Table 5.1. They were slightly different than the values obtained in case of the standard COD card. This may be due to the presence of Gd^{3+} (dopant) ions having greater ionic radius than in the case of the Ca^{3+} and Zr ions.

The morphology of the as-synthesized $CaZrO_3$:Gd^{3+} (2 mol% of Gd^{3+}) sample is studied by SEM and the results are shown in Fig. 5.2 (a). The studied samples demonstrated granular morphology and are categorized into two parts: one of them exhibits fine particles having dimensions in the range of ~0.1-2 μm (1st Group) while the other shows the bigger grains comprised of tightly packed fine particles having sizes in the range of ~10-50 μm (2nd Group).

To understand the morphology of as-prepared samples more precisely, TEM studies are performed and the results are shown in Fig. 5.2 (b). The TEM studies revealed the granular agglomerates having dimensions in the range

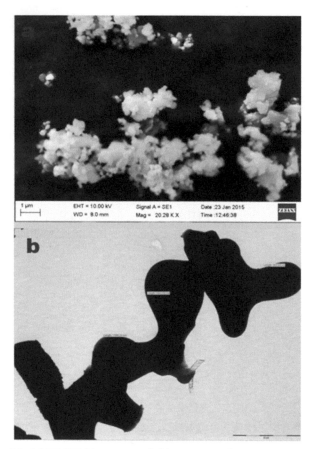

Fig. 5.2. (a) FE SEM images and (b) corresponding TEM images of CaZrO3:Gd3+ (2 mol%) sample.

~0.1-2 μm, which are similar with the results obtained from SEM studies for 1ˢᵗ Group of particles.

Photoluminescence Studies

To examine the optical behavior of as-prepared Gd³⁺ activated CaZrO₃ samples, the room temperature photoluminescence (PL) excitation spectra was recorded. The emission wavelength was kept fixed at 312 nm. The results of PL are depicted in Fig. 5.3 (a). The prominent and sharp excitation bands at around 242 and 274.5 nm are attributed to the $^8S_{7/2} \rightarrow ^6D_{5/2}$ and $^8S_{7/2} \rightarrow ^6I_{11/2}$ transitions of Gd³⁺ ions, respectively.

The photoluminescence emission spectra of solid state diffusion derived CaZrO₃:Gd³⁺ samples were acquired under 274.5 nm UV excitation wavelength as a function of Gd³⁺ concentrations. The results of PL measurements are shown in Fig. 5.3 (b). The sharp emission lines at 312 nm

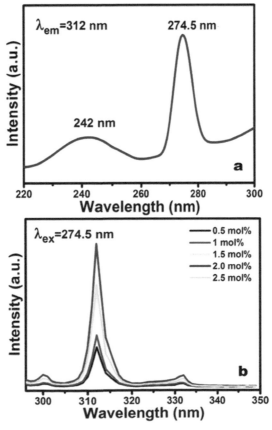

Fig. 5.3. (a) The photoluminescence excitation (2 mol% of Gd3+) and (b) emission spectra of as-prepared CaZrO3:Gd3+ samples with different concentration of Gd3+ (0.1 to 2.5 mol%).

is attributed to $^6P_{7/2} \rightarrow ^8S_{7/2}$ transitions while the less intense peak around 300 nm is attributed to the $^6P_{5/2} \rightarrow ^8S_{7/2}$ transitions of Gd^{3+} ions. According to the previously published scientific papers (Parrish and Jaenicke 1981; Abdullah et al. 2008; Yoshida et al. 2006), the emissions obtained due to Gd^{3+} ions at 312 nm is immoderately beneficial in prospective of medical science. The dose of narrow-band UVB is extremely significant in the cure of early-stage psoriasis, mycosis fungoides and vitiligo. Moreover, this eliminates the possibility of side effects and causes less irritation in comparison to broadband UVB (Hemne et al. 2017). Based on photoluminescence studies it was found that the as-prepared $CaZrO_3:Gd^{3+}$ can be a suitable candidate as a source of narrow-band UVB light for the above mentioned treatments.

    The photoluminescence efficiency of the samples significantly relies on the doping contents of Gd^{3+}. It is clearly visible in Fig. 5.3b that the luminescence efficiency of Gd^{3+} grows rapidly with increasing contents of gadolinium ions up to 2 mol% content of Gd^{3+} doping ions. This is the concentration quenching phenomenon, which relies on decrease of luminescence efficiency if the activator concentration exceeds the specific value known as critical concentration (Jeong et al. 2018; Singh et al. 2018; Wyckoff 1931). In the present chapter the concentration quenching phenomenon was observed above 2 mol% of gadolinium doping contents. The concentration quenching phenomenon takes place due to the reduced distance between the nearest Gd^{3+} ions and the interactions between neighboring Gd^{3+} ions augmenting the probability of non-radiative energy transfer which quenched the luminescence intensity (Fig. 5.4). The energy

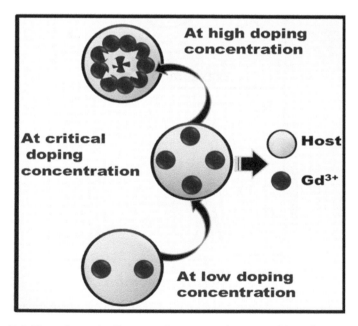

Fig. 5.4. The schematic diagram of concentration quenching phenomenon.

transfer phenomenon responsible for concentration quenching may occur via one of the following mechanism: an exchange interaction or multipole-multipole interaction. To understand and identify the type of interaction mechanism, Blasse introduced a parameter called critical distance (R_c). If the value of R_c is found to be greater than 5Å, then the multipole interaction dominates. Otherwise if the R_c value is less than 5Å, the exchange interaction will be the prime mechanism responsible for energy transfer. The value of R_c can be calculated by using the following expression:

$$R_c = 2\left[\frac{3V}{4\pi X_c N}\right]^{1/3}$$

where V is the unit cell volume, N stands for the effective number of Gd³⁺ sites per unit cell, and X_c represents the critical concentration of the activator. For CaZrO₃:Gd³⁺ system, N = 8, V = 259.91 Å³, and X_c = 0.02. The value of R_c for the present system is determined to be 14.58 Å, which is significantly higher than the 5Å.

The above result suggests the dominance of multipole-multipole interaction over exchange interaction mechanism of concentration quenching phenomenon in CaZrO₃:Gd³⁺ system.

Conclusions

In conclusion, Gd³⁺ activated CaZrO₃ phosphor were successfully prepared via the high temperature modified solid state diffusion method. The structural characterizations revealed the formation of a cubic phase having space group Pm-3m of as-prepared CaZrO₃:Gd³⁺ samples. The morphological studies show granular agglomerates with dimensions in the range of 0.1-50 μm. The photoluminescence emission spectrum of CaZrO₃:Gd³⁺ micro-crystals exhibited the prominent spectral line at around 312 nm attributed to the UVB emission of Gd³⁺ ions. The maximum luminescence intensity was obtained for the CaZrO₃ samples with 2 mol% of Gd³⁺ contents. Above 2 mol% of Gd³⁺ ions addition the concentration quenching phenomenon in tested phosphor takes place. The multipole-multipole interaction was estimated as the main mechanism responsible for luminescence quenching. The narrow-band UVB emission of Gd³⁺ activated CaZrO₃ samples validates its practical applicability as a promising candidate for phototherapy lamps.

References

Abdullah Mikrajuddin, Khairurrijal, Waris Abdul, Sutrisno Widayani, Iis Nurhasanah, Aunuddin S. Vioktalamo. 2008. An ultraviolet phosphor from submicrometer-sized particles of gadolonium-doped yttrium oxide prepared by heating of precursors in a polymer solution. Powder Technol. 183, 297.

Becker K. 1967. Radiophotoluminescent dosimetry. At. Energy Rev. 5, 43.

Dubey, N., Dubey, V. 2016. Synthesis and characterization of europium doped zirconium based phosphor for display applications. Reviews in Fluorescence, 155.

Dubey, V., Kaur, J., Agrawal, S. 2015. Effect of europium concentration on photoluminescence and thermoluminescence behavior of Y_2O_3:Eu^{3+} phosphor. Research on Chemical Intermediates 41(7), 4727.

Dubey, V., Kaur, J., Agrawal, S., Suryanarayana, N.S., Murthy, K.V.R. 2013. Synthesis and characterization of Eu^{3+} doped SrY_2O_4 phosphor. Optik – Int. J. Light Electron Opt. doi 10.1016/j.ijleo.2013.03.153.

Dubey, V., Tiwari, N. 2016. Structural and optical analysis on europium doped $AZrO_3$ (A=Ba, Ca, Sr) phosphor for display devices application. AIP Conference Proceedings 1728, 020002. doi: http://dx.doi.org/10.1063/1.4946052.

Dubey, V., Tiwari R., Tamrakar, R.K., Kaur, J., Dutta, S., Das, S., Visser, H.G., Som, S.J. 2016. Estimation of spectroscopic parameters and colour purity of the red-light-emitting $YBa_3B_9O_{18}$ phosphor: Judd–Ofelt approach. Lumin 180, 169.

Ferguson, J. 1999. The use of narrowband UV-B (tube lamp) in the management of skin disease. Arch. Dermatol. 135, 589.

Gawande, A.B., Sonekar, R.P., Omanwar, S.K. 2007. Combustion synthesis of narrow-band UVB emitting borate phosphors LaB_3O_6:Bi, Gd and $YBaB_9O_{16}$: Bi, Gd for phototherapy applications, 30, 622.

Gawande, A.B., Sonekar, R.P., Omanwar, S.K. 2016. Combustion synthesis of narrow-band UVB emitting borate phosphors LaB_3O_6:Bi, Gd and $YBaB_9O_{16}$: Bi, Gd for phototherapy applications. Optik 127, 3925.

Gruijl, F.R. De. 2000. Biological action spectra. Radiat. Prot. Dosim. 91, 57.

Gruijl, P. Calzavara-x. 1998. Narrow band UVB (311 nm) phototherapy and PUVA photochemotherapy: A combination. J. Am. Acad. Dermatol. 38, 687.

Hartman, R.A. 2002. Apparatus and method for targeted UV phototherapy of skin disorders. US Patent 6,413,268.

Hemne, S., Kunghatkar, R.G., Dhoble, S.J., Moharil, S.V., Singh, V. 2017. Phosphor for phototherapy: Review on psoriasis. Luminescence 32, 260-270.

Hoare, C., Li, A., Wan, Po, Williams, H. 2000. Systematic review of treatments for atopic eczema. Health Technol. Assess. 4, 88.

Holick, M.F. 1997. Hotobiology of vitamin D. pp. 33. In: Feldman, D., Glorieux, F.H., Pike, J.W. (eds.). Vitamin D. Academic Press, San Diego, CA.

Jeong, H., Singh, N., Pathak, M.S., Watanabe, S., Rao, T.K.G., Dubey, V., Singh, V. 2018. Investigations of the ESR and PL characteristics of ultraviolet-emitting gadolinium-doped $ZnMgAl_{10}O_{17}$ phosphors. Optik – International Journal for Light and Electron Optics 157, 1199.

MacKie, R.M. 2000. Effects of ultraviolet radiation on human health. Radiat. Prot. Dosim. 91, 15.

Morison, W.L. 1999. Phototherapy and photochemotherapy: An update. Semin. Cutan. Med. Surg. 18, 297.

Okamoto Shinji, Uchino Rika, Keisuke Kobayashi, Hajime Yamamoto. 2009. Luminescent properties of sensitized ultraviolet-B phosphor under vacuum-ultraviolet light excitation. J. Appl. Phys. 106, 013522.

Osmancevic, A., Wilhelmsen, K. Landin, Larko, O., Krogstad, A.L. 2010. Vitamin D status in psoriasis patients during different treatments with phototherapy. J. Photoch. Photobio. B 101, 117.

Paolo Pinton, Tullio Pozzan, Rosario Rizzuto. 1998. The Golgi apparatus is an inositol 1,4,5-trisphosphate-sensitive Ca^{2+} store, with functional properties distinct from those of the endoplasmic reticulum. EMBO J 17, 5298-5308.

Parganiha, Y., Kaur, J., Dubey, V., Shrivastava, R., Dhoble, S.J. 2015. Synthesis and luminescence study of $BaZrO_3:Eu^{3+}$ phosphor. Superlattices and Microstruc. 88, 262.

Parrish, J.A., Fitzpatrick, T.B., Tanenbaum, L., Pathak, M.A. 1974. Photochemotherapy of psoriasis with oral methoxsalen and longwave ultraviolet light. New Engl. J. Med. 291, 1207.

Parrish, J.A., Jaenicke, K.F. 1981. Action spectrum for phototherapy of psoriasis. J. Investig. Dermatol. 76, 359.

Sacher, C., Konig, C., Scharffetter-Kochanek, K., Krieg, T., Hunzelmann, N. 2001. Bullous pemphigoid in a patient treated with UVA-1 phototherapy for disseminated morphea. Dermatology 202, 54.

Scherschun, L., Kim, J.J., Lim, W.H. 2001. Narrow-band ultraviolet B is a useful and well-tolerated treatment for vitiligo. J. Am. Acad. Dermatol. 44, 999.

Shie, J.L., Lee, C.H., Chiou, C.S., Chang, C.T., Chang, C.C., Chang, C.Y.J. 2008. Photodegradation kinetics of formaldehyde using light sources of UVA, UVC and UVLED in the presence of composed silver titanium oxide photocatalyst. Hazard. Mater. 155, 164.

Singh, V., Singh, N., Pathak, M.S., Watanabe, S., Rao, T.K.G., Singh, P.K., Dubey, V. 2018. UV emission from Gd^{3+} ions in $LaAl_{11}O_{18}$ phosphors. Optik – International Journal for Light and Electron Optics, 157, 1391-1396.

Singh, V., Borkotoky, S., Murali, A., Rao, J.L., Gundu Rao, T.K., Dhoble, S.J. 2015. Electron paramagnetic resonance and photoluminescence investigation on ultraviolet-emitting gadolinium-ion-doped $CaAl_{12}O_{19}$ phosphors. Spectrochimica Acta Part A: Molecular and Biomolecular Spectroscopy 139, 1.

Singh, V., Chakradhar, R.P.S., Rao, J.L., Ledoux-Rak, Isabelle, Kwak, Ho-Young. 2011. Luminescence and EPR studies of $Y_2O_3:Gd^{3+}$ phosphors prepared via solution combustion method. J. Mater. Sci. 46, 1038.

Singh, V.K., Tripathi, S., Mishra, M.K., Tiwari, R., Dubey, V., Tiwari, N. 2016. Optical studies of erbium and ytterbium doped $Gd_2Zr_2O_7$ phosphor for display and optical communication applications. Journal of Display Technology 12(10), 1224.

Tiwari, N., Dubey, V., Kuraria, R.K. 2016. Mechanoluminescence study of europium doped $CaZrO_3$ phosphor. Journal of Fluorescence 26(4), 309.

Tiwari, N., Dubey, V. 2016. Luminescence studies and infrared emission of erbium-doped calcium zirconate phosphor. Luminescence 31, 837.

Tiwari, N., Dubey, V., Dewangan, J., Jain, N. 2016. Near UV-blue emission from cerium doped zirconium dioxide phosphor for display and sensing applications. Journal of Display Technology 12(9), 933.

Vecht, A., Newport, A.C., Bayley, P.A., Crossland, W.A. 1998. Narrow band 390 nm emitting phosphors for photoluminescent liquid crystal displays. J. Appl. Phys. 84, 3827.

Weelden, H. van, Faille, H. Baart de la, Young, E., Leun, J.C. van der. 1988. A new development in UVB phototherapy of psoriasis. Br. J. Dermatol. 119, 11.

Wyckoff, R.W.G. 1931. Structure of Crystals. 2nd edn. The Chemical Catalog Company, Inc. New York. http://jcrystal.com/steffenweber/gallery/StructureTypes/st4.html.

Yoshida, H., Yoshimatsu, R., Watanabe, S., Ogasawara, K. 2006. Optical transitions near the fundamental absorption edge and electronic structures of $YAl_3(BO_3)_4$: Gd^{3+}. Jpn. J. Appl. Phys., Part - 1 45, 146.

Investigations on Tunable Blue Light Emitting P-Acetyl Biphenyl-DPQ Phosphor for OLED Applications

S.Y. Mullemwar[1], G.D. Zade[2], N. Thejo Kalyani[3]*, S.J. Dhoble[4] and Xiaoyong Huang[5]

[1] D. D. Bhoyar College of Arts and Science Mouda, Nagpur - 441104, India
[2] J. N. Arts, Commerce and Science College, Wadi, Nagpur - 440023, India
[3] Department of Applied Physics, Laxminarayan Institute of Technology, Nagpur - 440033, India
[4] Department of Physics, RTM, Nagpur University, Nagpur - 440033, India
[5] College of Physics and Optoelectronics, Taiyuan University of Technology, Taiyuan 030024, P.R. China

Introduction

The era of Organic Light-Emitting Diodes (OLEDs) and Polymeric Light Emitting Diodes (PLEDs) has evolved tremendously from the time when preliminary reports by (Tang and Van Slyke 1987; Burroughes et al. 1990) were accepted. Though researchers are determined to improve the quantum efficiency of photoluminescence (PL) and electroluminescence (EL) OLEDs, challenges still exist. (Zhu et al. 2003). Universally, the blends of three primary colors (red, green and blue) or complementary colours (blue and orange) give rise to white emission. Amongst all, the luminous efficiency of blue OLEDs needs improvement (Kato et al. 2015). Hence, it is imperative to come-up with novel blue light emitting materials, which can compete with their red and green light emissive counterpart materials with respect to luminous efficiency, life time so as to harvest stable white light emission from these three RGB materials. In this regard, organic phosphors based on quinoline comprise an imperative class of a heterocyclic group and hence create substantial awareness amongst researchers worldwide. Poly(quinoline)s were principally reported (Stille 1981) by using Friedlander

---
*Corresponding author: thejokalyani@rediffmail.com

condensation as a polymerization step, so as to stipulate thermally stable and instinctively strong polymers. They show evidence of adaptable properties such as elevated thermal stability with glass-transition temperatures, even greater than 200 °C and onset thermal decomposition temperatures >400 °C, elevated electrical conductivity, and exceptional film forming property (Pelter et al. 1990; Tunney et al. 1987). These distinctive features make them appealing for opto-electronic devices. Prior state of art reveals extensive research on distinguishing properties of poly(quinoline)s such as charge-transfer, photoluminescence (PL) and Electro Luminescence (EL) and electron transporting properties (Kim et al. 1992; Lee and Rhee 1995; Chen et al. 1996; Jen et al. 1998; Mullemwar et al. 2016; Ghate et al. 2017; Kim et al. 2000; Zhang et al. 1998; Kwon et al. 2004) for their potential applications in OLEDs devices (Zhang and Jenekhe 2000; Zhang et al. 1999; Jenekhe et al. 2001; Tonzola et al. 2004). These complexes can be integrated into numerous matrices including zeolites, semiconductors (Alam and Jenekhe 2001; Maas et al. 2002), porous materials (Xu et al. 2002; Minoofar et al. 2002), polymer thin films (Wang et al. 2002) and a range of aqueous solutions (Kalyani et al. 2010) to avoid luminescent concentration quenching, elevate the processing ability and enhance thermal as well as mechanical stabilities. This chapter describes the synthesis and characterization of a novel phosphor (Acetyl biphenyl substituted diphenyl quinolines) 2-([1, 1'-biphenyl]-4-yl)-4-phenylquinoline (P-Acetyl biphenyl-DPQ) in various organic solvents and polystyrene at different wt%.

Experimental

Reagent and Solvents

Starting materials used for the synthesis of 2-([1,1'-biphenyl]-4-yl)-4-phenylquinoline ($C_{27}H_{19}N$) (P-Acetyl biphenyl-DPQ) complex are 4 Acetylbiphenyl ($C_{14}H_{12}O$) [Himedia] 2 Amino Benzophenone ($C_{13}H_{11}NO$) [Himedia] Diphenylphosphate ($C_6H_5O)_2P(O)OH$) [Sigma Aldrich], mcresol [$CH_3C_6H_4OH$], Dichloromethane (CH_2Cl_2) [Fisher scientific], Sodium hydroxide (NaOH) [Fisher scientific], Hexane ($CH_3(CH_2)_4CH_3$) [Loba chemiecals] Chloroform ($CHCl_3$), [Fisher scientific], Polystyrene [CH_2CH (C_6H_5)]$_n$, [Sigma Aldrich] and double distilled water.

Synthesis of P-Acetyl Biphenyl-DPQ

Synthesis scheme of P-Acetyl biphenyl-DPQ is adopted from the literature (Zade et al. 2011) and successive reactions are shown in Fig. 6.1.

Preparation of Blended Films

The most popular polymer, namely polystyrene (PS) was chosen for blending the synthesized complex at known wt% such as 10 and 5 wt% and the blended films were made according to literature methods (Mullemwar et al. 2016) . Preparation of blended thin films is schematically represented in Fig. 6.2.

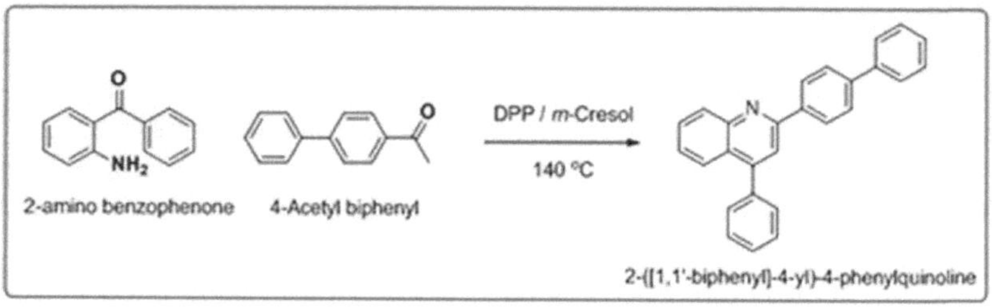

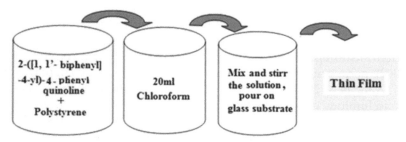

Fig. 6.1. Synthesis scheme of 2-([1,1′-biphenyl]-4-yl)-4-phenylquinoline (P-Acetyl biphenyl-DPQ).

Fig. 6.2. Schematic representation: Preparation of blended thin films.

Results and Discussion

Physical, chemical and optical properties of the synthesized P-Acetyl biphenyl-DPQ phosphor were investigated as follows: ^{1}H-NMR, ^{13}C-NMR on Bruker Benchtop instrument, Fourier Transform Infrared (FTIR) spectra on Bruker-Alpha at room temperature, and Thermo gravimetric/Differential Thermal Analysis (TGA/DTA) on Perkin Elmer diamond. Absorption and photo luminescence measurements were carried out on SPECORD 50 spectrophotometer and Humamatsu F-4500 spectrofluorometer, respectively. CIE coordinates were calculated by CIE1931 (Commission International d'Eclairage) system.

NMR Spectroscopy

^{1}H–NMR spectrum of P-Acetyl biphenyl-DPQ: ^{1}H NMR (400 MHz, chloroform-d($CDCl_3$)) δ (ppm) 8.35 (d, 3H), 7.97 (d, 1H), 7.92 (s, 1H), 7.81 (m, 2H), 7.74 (d, 2H), 7.62 (m, 4H), 7.53 (t, 4H), 7.43 (d, 2H). This spectrum confirms the presence of 19 H-atom in the synthesized compound. These peaks can be assigned to the aromatic protons (Fig. 6.3), which correlates with the structure as described in Fig. 6.1. Chemical shifts from ^{13}C–NMR spectrum: ^{13}C-NMR (100 MHz, $CDCl_3$) δ (ppm) 156.43, 149.26, 148.90, 142.16, 140.63, 138.55, 130.21, 129.65, 128.90, 128.70, 128.47, 128.05, 127.61, 127.22, 126.40, 125.88, 125.69, 119.27 as shown in Fig. 6.4. Each carbon nucleus experiences different magnetic fields according to their electronic

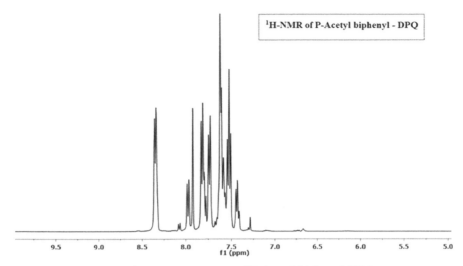

Fig. 6.3. ¹H-NMR spectrum of P-Acetyl biphenyl-DPQ.

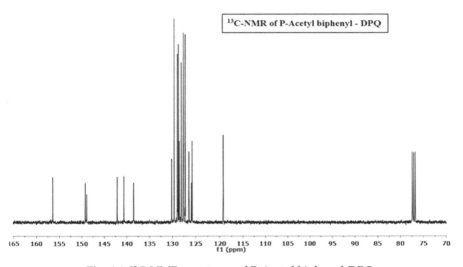

Fig. 6.4. ¹³C-NMR spectrum of P-Acetyl biphenyl-DPQ.

environment. In the range of 115–160 ppm, P-Acetyl biphenyl-DPQ confirms 27 aromatic C-nuclei. Due to the symmetrically related nature of maximum C-nuclei, the ¹³C spectrum displays 18 signals for a total number of 27 nuclei. The signals between 70 and 80 ppm are due to the $CDCl_3$ solvent.

Fourier Transform Infrared (FTIR) Spectra

The molecular structure of P-Acetyl biphenyl-DPQ chromophore are confirmed by FT-IR spectra over the range 4000–400 cm⁻¹ by averaging 500 scans at a maximum resolution of 20 cm⁻¹ as shown in Fig. 6.5. Peaks recorded

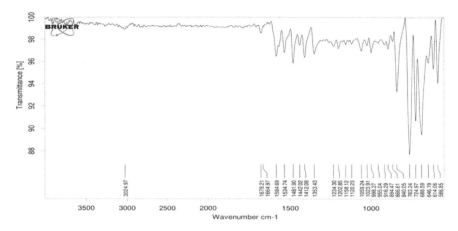

Fig. 6.5. FTIR spectrum of P-Acetyl biphenyl DPQ.

at 916.29, and 763.24 cm⁻¹ may be attributed to phenyl ring substituted bands, while the peaks at 1400 cm⁻¹ confirms (C=N) group, revealing the presence of the quinoline ring. Peaks at 866.61, 763.24, 724.97 and 688.59 cm⁻¹ can be assigned to the benzene ring. Conjugated C=O stretching vibrations are clearly observed at 1664.97 cm⁻¹, conjugated C=C stretching vibrations are portrayed at 1584.69 and 1534.74 cm⁻¹, thereby declaring the complete structural formation of P-Acetyl biphenyl-DPQ phosphor.

Thermogravimetric and Differential Thermal Analysis (TGA/DTA)

Thermogram of P-Acetyl biphenyl-DPQ phosphor displays a horizontal plateau, till 300 °C as indicated in Fig. 6.6. With further increase in temperature, the thermogram takes a curved portion, indicating decomposition or weight

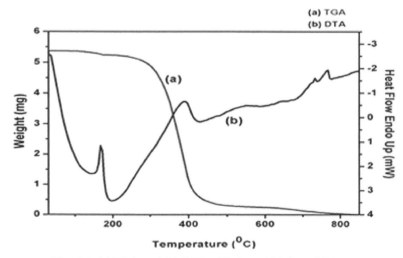

Fig. 6.6. (a) TGA and (b) DTA of P-Acetyl biphenyl-DPQ.

loss of the sample due to heating. Hence, the synthesized complex can be operated up till 300 °C without any degradation in its properties.

DTA curve of P-Acetyl biphenyl-DPQ displays two endothermic peaks, one centred at 25 °C and the other at 181 °C corresponding to the distortion of water and evaporation of residual moisture from the complex. Exothermic peaks roughly at 189 °C and 429 °C in the DTA curve can be accredited to the breakdown process of the residual organic molecules.

UV-visible Absorption Spectra

UV-visible absorption spectroscopy was carried out to investigate the optical absorption peaks and the role of polarity index of solvent on the absorption spectra of P-Acetyl biphenyl-DPQ phosphor in chloroform, dichloromethane, acetic acid and formic acid at 10^{-3} mol/L. Two absorption peaks: one attributed to the π - π^* transition (at the lower wavelength with high energy) of the conjugated polymer main chains; another is attributed to the n - π^* transition (at a higher wavelength with low energy) of the conjugated side chains were observed. However, the peak position and optical absorption density were found to be different for different solvents as shown in Fig. 6.7. Solvated DPQ in formic acid displays a strong absorption peak at 366 nm with a shoulder at 262 nm, while in acetic acid, it displays strong absorption peak at 364 nm with a weak shoulder at 258 nm.

In both the cases, the peak at the lower wavelength is due to their large molar extinction coefficients of the conjugated polymer main chains,

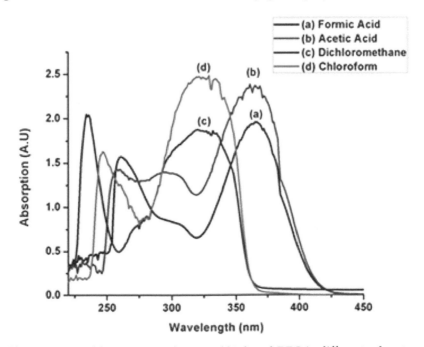

Fig. 6.7. UV-visible spectrum of P-Acetyl biphenyl-DPQ in different solvents.

while the peak at the higher wavelength is attributed to conjugated side chains. Similarly in dichloromethane and chloroform absorption peaks were observed at 327 and 332 nm with a weak shoulder at 234 and 245 nm, respectively. Hypsochromic shift was monitored in the absorption spectra of P-Acetyl biphenyl-DPQ, when the solvent is changed from formic/acetic acid to chloroform and dichloromethane. These changes may be due to (i) protonation of the imine nitrogen of the quinoline ring to form the quinolium ion (Kalyani et al. 2012). Poor optical densities were observed in formic and acetic acid, indicating sparse decomposition of the ligand in acidic conditions.

Determination of Optical Band Gap

The optical band gap of P-Acetyl biphenyl-DPQ in various solvents was calculated by adopting the method given by Morita et al. (1995). These values were found to be 3.02, 3.07, 3.46 and 3.44 eV in formic acid, acetic acid, dichloromethane and chloroform, respectively as shown in Fig. 6.8.

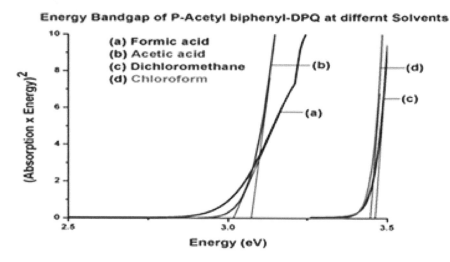

Fig. 6.8. Energy band gap of P-Acetyl biphenyl-DPQ in various solvents.

Photoluminescence (PL) Spectra in Organic Solvents

Consecutively, to explore the spectral features of P-Acetyl biphenyl-DPQ in a solid state, various organic solvents and blended polystyrene films, PL spectra was carried out individually. The PL spectra of P-Acetyl biphenyl-DPQ in a solid state demonstrate intense blue light emission at 388 nm, when excited at 362 nm as shown in Fig. 6.9.

The PL spectra of solvated P-Acetyl biphenyl-DPQ in formic acid, acetic acid, chloroform and dichloromethane are shown in Figs. 6.10 (a), (b), (c) and (d), respectively. Solvated P-Acetyl biphenyl-DPQ in formic acid peaks at 485 nm when excited at 412 nm, while in formic acid the emission peak shifted to 478 nm when excited at 409 nm. Thus a hypsochromic shift of 7 nm was

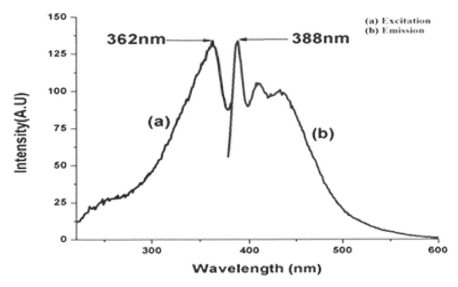

Fig. 6.9. PL spectra of P-Acetyl biphenyl-DPQ in powder form.

observed when the change in solvent from acetic to formic acid, while in dichloromethane emission peak was observed at 466 nm when excited at 369 nm, while in chloroform, the emission peaked at 381 nm when excited at 357 nm. Thus a considerable hypsochromic shift of 85 nm was observed when the solvent is changed from DCM to chloroform. However, all the solvated fluorophores emits intense blue light, which can be tuned with the change in solvent.

Photoluminescence (PL) Spectra in Polystyrene

The PL spectra of molecularly doped P -Acetyl biphenyl-DPQ with polystyrene (PS) was investigated for different wt %. In case of P Acetyl biphenyl-DPQ+PS at 10 wt %, the PL spectrum displays emission at 384 nm, when excited at 360 nm and for 5 wt%, the polymeric chromophore emits intense blue emission at wavelength 381 nm, when excited at 357 nm as shown in Fig. 6.11. When these results are compared with the PL results of P-Acetyl biphenyl-DPQ in a solid state and in various solvents, the emission intensity was observed in the order of I_{emi} in blended films > I_{emi} in solvents > I_{emi} in a solid state. This infers that the synthesized phosphor form aggregates in PS matrix as well as in solvents and hence enhancement in intensity is observed in solvated P-Acetyl biphenyl-DPQ and its blended films (Kin et al. 2009; Raut et al. 2016).

Determination of Stokes Shift

The difference between the reciprocal of absorption to the emission wavelength (Stokes shift) was calculated by using the relation

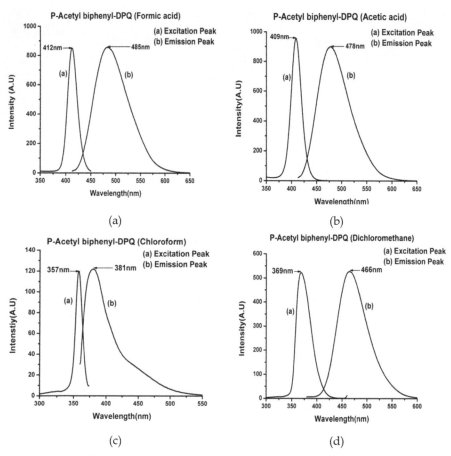

Fig. 6.10. PL spectra of phosphor in (a) Formic acid, (b) Acetic acid, (c) Chloroform and (d) Dichloromethane.

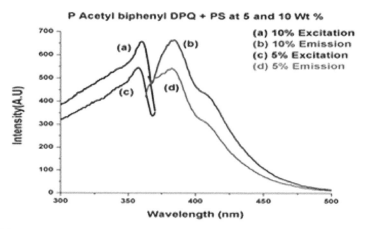

Fig. 6.11. PL spectra of P-Acetyl biphenyl-DPQ+PS (5%) and (10%) Film.

$$\text{Stock's shift } (\Delta v) = \left[\frac{1}{\lambda_{abs}} - \frac{1}{\lambda_{emi}} \right] \times 10^7 \, cm^{-1}$$

The synthesized phosphors display a Stokes shift of about 67, 65, 91, 38 nm in formic acid, acetic acid, dichloromethane and chloroform as shown in Figs. 6.12 (a), (b), (c) and (d), respectively.

Commission International d'Eclairage (CIE) Coordinates

CIE coordinates of P-Acetyl biphenyl-DPQ in various solvents were calculated on the CIE 1931 (Commission International d'Eclairage) system software (Dahule et al. 2015). They were found to be (0.154, 0.323) in formic

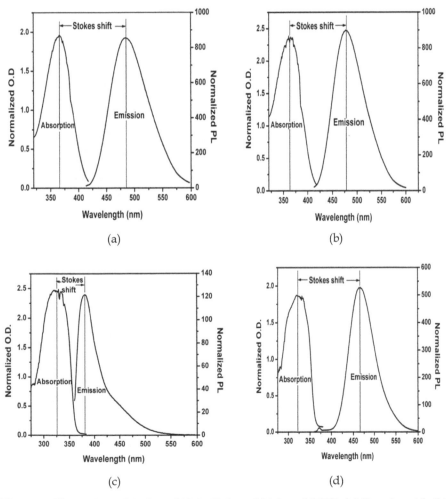

Fig. 6.12. Illustration of Stokes shift in P-Acetyl biphenyl-DPQ: (a) Formic acid, (b) Acetic acid, (c) Chloroform and (d) Dichloromethane.

acid, (0.143, 0.255) in acetic acid, (0.139, 0.160) in dichloromethane and (0.162, 0.063) in chloroform, respectively as shown in Fig. 6.13. A summary of optical and photometric properties of P-Acetyl biphenyl-DPQ in various environments is tabulated in Table 6.1.

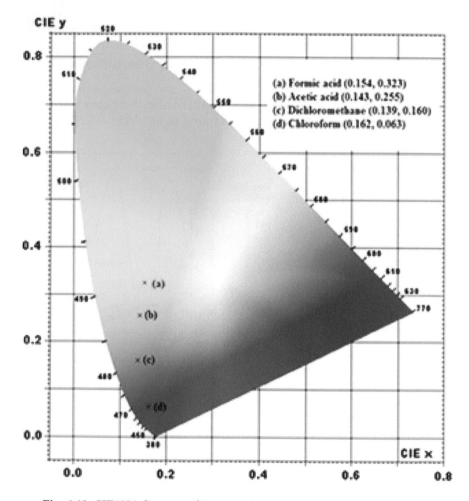

Fig. 6.13. CIE1931 diagram of P-Acetyl biphenyl-DPQ in various solvents.

Conclusions

P-Acetyl biphenyl-DPQ was synthesized by Friedlander condensation reaction at 140 °C for 4 hours. ^{1}H–NMR spectrum of P-Acetyl biphenyl-DPQ confirms the presence of the desired 19 H-atoms that can be assigned to the aromatic protons, which were found to be correlated with the structure. FTIR spectrum shows molecular confirmation of P-Acetyl biphenyl-DPQ chromophore. TGA/DTA result infers that the complex has thermal stability

Table 6.1. Summary of photo physical properties of P-Acetyl biphenyl-DPQ in various environments

Solvent	Polarity index	Absorption		Optical density		E_g(eV)	Molar absorbitivity ($\varepsilon = A/CL$)		Emission parameter		FWHM (nm)	Stoke's shift (cm^{-1})	CIE Coordinate (x,y)
		Band I (λ_{max})	Band II (λ_{max})	Band I	Band II		Band I	Band II	λ_{ext} (nm)	λ_{ext} (nm)			
-		-	-	-	-	-	-	-	-	-	362	-	-
Spectral properties of P-Acetyl biphenyl-DPQ in Different solvents													
Formic acid	58	262	366	1.56		1.95		3.02	4.19	5.31	412	485	(0.154, 0.323)
Acetic acid	6.2	258	364	1.43		2.34		3.07	3.91	6.54	409	478	(0.143, 0.255)
Dichloromethane	9.1	234	327	2.04		1.85		3.46	5.59	5.03	369	466	(0.139, 0.160)
Chloroform	4.81	245	332	1.58		2.42		3.44	4.19	6.77	357	381	(0.162, 0.063)
Spectral properties of P-Acetyl biphenyl-DPQ in Blended film with PS													
10%	-	-	-	-	-	-	-	-	-	-	360		
5%	-	-	-	-	-	-	-	-	-	-	357		

till 300 °C. Two endothermic peaks, centred at 25 °C and 181 °C were observed in the DTA curve. UV-Vis absorption spectra reveal that the optical density of the complex in acetic acid and formic acid are much smaller than in chloroform and dichloromethane. The energy band gap of the synthesized complex was found to be 3.02, 3.07, 3.46 and 3.44 eV in formic acid, acetic acid, dichloromethane and chloroform, respectively. PL spectra of P-Acetyl biphenyl DPQ in solid state and blended films reveals tunable blue light emission within the range of 381 nm to 485 nm with narrow Full Width at Half-Maximum (FWHM). The synthesized phosphors display a Stokes shift of about 67, 65, 91, 38 nm in formic acid, acetic acid, dichloromethane and chloroform, demonstrating its prospective as a tunable emissive material for OLED fabrication by solution processed techniques.

References

Alam, M.M., Jenekhe, S.A. 2001. Nanolayered heterojunctions of donor and acceptor conjugated polymers of interest in light emitting and photovoltaic devices: Photoinduced electron transfer at polythiophene/polyquinoline interfaces. J. Phys. Chem. B 105(13), 2479-2482.

Burroughes, J.H., Bradley, D.D.C., Brown, A.R., Marks, R.N., Mackay, K., Friend, R.H., Burns, P.L., Holmes, A.B. 1990. Light-emitting diodes based on conjugated polymers. Nature 347, 539.

Chen, T.A., Jen, A.K.Y., Cai, Y. M. 1996. A novel class of nonlinear optical side-chain polymer: Polyquinolines with large second-order nonlinearity and thermal stability. Chem. Mater. 8(3), 607-609.

Color Calculator, Version 2. 2007. A software from Radiant Imaging, Inc.

Dahule, H.K., Kalyani, N. Thejo, Dhoble, S.J. 2015. Novel Br-DPQ blue light-emitting phosphors for OLED. Luminescence 30, 405-410.

Ghate, M., Dahule, H.K., Kalyani, N. Thejo, Dhoble, S.J. 2017. Deep blue light emitting Cyno-DPQ phosphor with large stokes shift and high thermal stability for OLEDs and display applications. Optik 149, 198-205.

Ghate, M., Dahule, H.K., Kalyani, N. Thejo, Dhoble, S.J. 2017. Synthesis and characterization of high quantum yield and oscillator strength 6-chloro-2-(4-cynophenyl)–4–phenyl quinoline (cl-CN-DPQ) organic phosphor for solid-state lighting. Luminescence 1, 1-8.

Jen, A.K.Y., Wu, X.M., Ma, H. 1998. High-performance polyquinolines with pendent high-temperature chromophores for second-order nonlinear optics. Chem. Mater. 10(2), 471-473.

Jenekhe, S.A., Lu, L.D., Alam, M.M. 2001. New conjugated polymers with donor-acceptor architectures: Synthesis and photophysics of carbazole-quinoline and phenothiazine-quinoline copolymers and oligomers exhibiting large intramolecular charge transfer. Macromolecules 34(21), 7315-7324.

Kalyani, N. Thejo, Dhoble, S.J., Pode, R.B. 2010. Optical properties of EuxRe(1- x) (TTA)3Phen organic complexes in different solvents. J. Korean Physical Soc. 57(4), 746-751.

Kalyani, N. Thejo, Dhoble, S.J., Pode, R.B. 2012. Enhancement of photoluminescence in various Eu xRe(1−x)TTA3Phen (Re = Y, Tb) complexes molecularly doped in PMMA. Indian J. Phys. 86(7), 613-618.

Kato, Kazuki & Iwasaki, Toshihiko & Tsujimura, Takatoshi. 2015. Over 130 lm/W all-phosphorescent white OLEDs for next-generation lighting. J. Photopolym. Sci. Tec. 28, 335-340.

Kim, J.L., Kim, J.K., Cho, H.N. 2000. New polyquinoline copolymers: Synthesis, optical, luminescent, and hole-blocking/electron-transporting properties. Macromolecules 33, 5880-5885.

Kim, K.A., Park, S.Y., Kim, Y.J. 1992. Synthesis and photoelectrical properties of poly[2,6-(p-phenoxy)-4-phenylquinoline]. J. Appl. Polym. Sci. 46, 1-7.

Kin, E., Fukuda, T., Kato, S., Honda, Z., Kamata, N. 2009. pH and concentration dependence of luminescent characteristics in glass-encapsulated Eu-complex. J. Sol-Gel Sci. Technol. 50(3), 409-414.

Kwon, T.W., Alam, M.M., Jenekhe, S.A. 2004. n-Type conjugated dendrimers: Convergent synthesis, photophysics, electroluminescence, and use as electron-transport materials for light-emitting diodes. Chem. Mater. 16(23), 4657-4666.

Lee, S.H., Rhee, B.K. 1995. Nonlinear optical transmission of polyquinoline thin film at 532 nm. Opt. Quant. Electron 27, 371-377.

Maas, H., Currao, A., Calzaferri, G. 2002. Encapsulated lanthanides as luminescent materials. Angew. Chem. Int. Ed. 41, 2495-2497.

Minoofar, P.N., Hernandez, R., Chia, S., Dunn, B., Zink, J.I., Franville, A.-C. 2002. Placement and characterization of pairs of luminescent molecules in spatially separated regions of nanostructured thin films. J. Am. Chem. Soc. 124(48), 14388-14396.

Morita, S., Akashi, T., Fujii, A., Yoshida, M., Ohmori, Y., Yoshimoto, K., Kawai, T., Zakhidrov, A.A., Lee, S.B., Yoshino, K. 1995. Unique electrical and optical characteristics in poly (p-phenylen)-C60 system. Synth. Met. 69, 433-434.

Mullemwar, S.Y., Zade, G.D., Kalyani, N. Thejo, Dhoble, S.J. 2016. Synthesis and characterization of novel 2-(4-bromophenyl)-6-chloro-4-phenylquinoline, blue light emitting organic phosphor. I.J.L.A. 6(1), 68-72.

Mullemwar, S.Y., Zade, G.D., Kalyani, N. Thejo, Dhoble, S.J. 2016. Blue light emitting P-hydroxy DPQ phosphor for OLEDs. Optik 127, 10546-10553.

Pelter, M.W., Stille, J.K. 1990. Thermal rigidification of polyquinolines by thermolytic elimination of ethylene from a 9,10-dihydro-9,10-ethanoanthracene unit. Macromolecules 23(9), 2418-2422.

Raut, Soniya, Kalyani, N. Thejo, Dhoble, S.J. 2016. Organic Light Emitting Diodes (OLED) Materials, Technology and Advantages. NOVA Science Publishers, New York. 41-92. Edited by Douglas Rivera New York. ISBN: 978-1-63484001-9.

Stille, J.K. 1981. Macromolecules. Polyquinolines 14, 870-880.

Tang, C.W., Van Slyke, S.A. 1987. Organic electroluminescent diodes. Appl. Phys. Lett. 51, 913-915.

Tonzola, C.J., Alam, M.M., Bean, B.A., Jenekhe, S.A. 2004. New soluble n-type conjugated polymers for use as electron transport materials in light-emitting diodes. Macromolecules 37(10), 3554-3563.

Tunney, S.E., Suenaga, J., Stille. 1987. Conducting polyquinolines. Macromolecules 20, 258-264.

Wang, L.-H., Wang, W., Zhang, W.-G., Kang, E.-T., Huang, W. 2002. Synthesis and luminescence properties of novel Eu-containing copolymers consisting of Eu(III)−Acrylate−β-Diketonate complex monomers and methyl methacrylate. Chem. Mater. 12(8), 2212-2218.

Xu, Q., Li, L., Liu, X., Xu, R. 2002. Incorporation of rare-earth complex Eu(TTA)$_4$C$_5$H$_5$NC$_{16}$H$_{33}$ into surface-modified Si-MCM-41 and its photophysical properties. Chem. Mater. 14(2), 549-555.

Zade, G.D., Dhoble, S.J., Raut, S.B., Pode, R.B. 2011. Synthesis and characterization of chlorine-methoxy-diphenylquinoline (Cl-MO-DPQ) and chlorine-methyl-diphenylquinoline (Cl-M-DPQ) blue emitting organic phosphors. J. Mod. Phy. 2(12), 1523-1529.

Zhang, X.J., Jenekhe, S.A. 2000. Electroluminescence of multicomponent conjugated polymers. 1. Roles of polymer/polymer interfaces in emission enhancement and voltage-tunable multicolor emission in semiconducting polymer/polymer heterojunctions. Macromolecules 33(6), 2069-2082.

Zhang, X., Shetty, A.S., Jenekhe, S.A. 1998. Efficient electroluminescence from a new n-type conjugated polyquinoline. Acta. Polym. 49, 52-55.

Zhang, X.J., Shetty, A.S., Jenekhe, S.A. 1999. Electroluminescence and photophysical properties of polyquinolines. Macromolecules 32(22), 7422-7429.

Zhu, Yan, Alam, Maksudul M., Jenekhe, Samson A. 2003. Regioregular head-to-tail poly(4-alkylquinoline)s: Synthesis, characterization, self-organization, photophysics, and electroluminescence of new n-type conjugated polymers. Macromolecules 36(24), 8958-8968.

Phosphors in Role of Magnetic Resonance, Medical Imaging and Drug Delivery Applications: A Review

Neha Dubey[1], Vikas Dubey[2], Jagjeet Kaur[1], Dhananjay Kumar Deshmukh[3]* and K.V.R. Murthy[4]

[1] Department of Physics, Govt. V.Y.T.PG. Auto. College, Durg - 491001, India
[2] Department of Physics, Bhilai Institute of Technology Raipur, Kendri - 493661, India
[3] Chubu Institute for Advance Studies, Chubu University, Kasugai - 487-8501, Japan
[4] Fellow Luminescence Society of India, President, Luminescence Society of India, Applied Physics Department, Faculty of Technology and Engineering, M. S. University of Baroda, Baroda - 390001, India

Introduction

Metastasis—which accounts for 90% of cancer-related deaths (Chaffer and Weinberg 2011), and occurs when cancer cells detach from their primary site and home in distant organs—can be detected through non-invasive clinical-imaging modalities such as Magnetic Resonance Imaging (MRI), X-ray Computed Tomography (CT) and Positron Emission Tomography (PET) (O'Connor et al. 2011; Heinzmann et al. 2017). Contrast enhanced MRI is typically the preferred choice because of its higher sensitivity and specificity; yet CT, which is highly accessible and has lower operating costs, is used more widely. Although these imaging modalities are capable of detecting large metastases, they do not offer the resolution necessary to detect the early spread of metastatic tumor cells. Rare-earth-doped nanoprobes emitting short-wavelength infrared light enable the detection of metastatic lesions in multiple organs (Zhang 2017).

In vivo fluorescence imaging in the near-infrared region between 1500–1700 nm (NIR-IIb window) affords high spatial resolution, deep-tissue penetration and diminished auto-fluorescence due to the suppressed

*Corresponding author: deshmukhdhananjay@gmail.com

scattering of long-wavelength photons and large fluorophore Stokes shifts (Zhong et al. 2017).

In vivo fluorescence-based optical imaging provides high spatial and temporal resolution, giving researchers the unique ability to visualize biological processes in real-time (30 frames per second) down to the cellular level (Choi et al. 2013; Hong et al. 2014; Zhu et al. 2017). For decades, one-photon fluorescence imaging in the visible (400–700 nm) and traditional near-infrared (NIR-I; 750–900 nm) regions of the electromagnetic spectrum have been plagued by the inability to clearly resolve deep-tissue structures and physiological dynamics (Hong et al. 2012). As a recent development, NIR-emissive fluorescent probes in the second near-infrared window (NIR-II, 1000–1700 nm) have led to improved *in vivo* fluorescence imaging quality owing to suppressed scattering of photons and diminished autofluorescence (Welsher et al. 2011; Hong et al. 2014; Zhang et al. 2012). Several classes of fluorescent NIR-II probes have been reported including carbon nanotubes (Hong et al. 2014), conjugated polymers (Tao et al. 2013), small molecular dyes (Antaris et al. 2016) and inorganic-based nanoparticles of quantum dots (Zhang et al. 2012; Tao et al. 2013; Franke et al. 2016), and rare-earth nanocrystals (Naczynski et al. 2013). Indeed, progress has been made in NIR-II in vivo biological imaging owing to the development of various NIR-II fluorescent probes (Sun et al. 2016; Li et al. 2014; Dong et al. 2013). Still, bright probes with emission in the long end of the NIR-II region remain scarce and are desired in order to further reduce scattering of emitted photons and maximize *in vivo* fluorescence imaging depth and clarity.

Cancer is one of the most fatal diseases in today's world that kills millions of people every year. It is one of the major health concerns of the 21st century which does not have any boundary and can affect any organ of people from any place (Bharali and Mousa 2010). Cancer, the uncontrolled proliferation of cells where apoptosis disappears, requires a very complex process of treatment. Because of complexity in genetic and phenotypic levels, it shows clinical diversity and therapeutic resistance. A variety of approaches are being practised for the treatment of cancer each of which has some significant limitations and side effects (Zhao and Rodriguez 2013). Cancer treatment includes surgical removal, chemotherapy, radiation and hormone therapy. Chemotherapy, a very common treatment, delivers anticancer drugs systemically to patients for quenching the uncontrolled proliferation of cancerous cells (Jabir et al. 2012). Unfortunately, due to nonspecific targeting by anticancer agents, many side effects occur and poor drug delivery of those agents cannot bring out the desired outcome in most of the cases. Cancer drug development involves a very complex procedure which is associated with advanced polymer chemistry and electronic engineering. The main challenge of cancer therapeutics is to differentiate the cancerous cells and the normal body cells. That is why the main objective becomes engineering the drug in such a way that it can identify the cancer cells to diminish their growth and proliferation.

Conventional chemotherapy fails to target the cancerous cells selectively without interacting with the normal body cells. Thus they cause serious

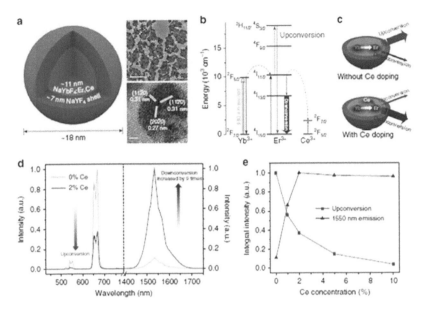

Fig. 7.1. Ce^{3+} doped rare-earth nanoparticles with enhanced NIR-IIb luminescence. (a) Schematic design of a $NaYbF_4$:Er,Ce@$NaYF_4$ core-shell nanoparticle (left) with corresponding large scale TEM image (upper right, scale bar = 200 nm) and HRTEM image (lower right, scale bar = 2 nm). (b) Simplified energy-level diagrams depicting the energy transfer between Yb^{3+}, Er^{3+}, and Ce^{3+} ions. (c) Schematic illustration of the proposed energy-transfer mechanisms in Er-RENPs with and without Ce^{3+} doping. (d) Upconversion and downconversion luminescence spectra of the Er-RENPs with 0 and 2% Ce^{3+} doping. (e) Schematic representation of Ce^{3+} doping concentration and corresponding upconversion and downconversion emission intensity of the Er-RENPs upon 980 nm excitation (Zhong et al. 2017).

side effects including organ damage resulting in impaired treatment with lower dose and ultimately low survival rates (Mousa and Bharali 2011). Nanotechnology is the science that usually deals with the size range from a few nanometres (nm) to hundred nm, depending on their intended use (Peer et al. 2007). It has been an area of interest over the last decade for developing precise drug delivery systems as it offers numerous benefits to overcome the limitations of conventional formulations (Malam et al. 2009).

Nanoparticles are rapidly being developed and trailed to overcome several limitations of traditional drug delivery systems and are coming up as a distinct therapeutics for cancer treatment. Conventional chemotherapeutics possess some serious side effects including damage of the immune system and other organs with rapidly proliferating cells due to nonspecific targeting, lack of solubility and inability to enter the core of the tumors resulting in impaired treatment with a reduced dose and with low survival rate. Nanotechnology has provided the opportunity to get direct access of the cancerous cells selectively with increased drug localization and cellular uptake. Nanoparticles can be programmed for recognizing the cancerous cells and giving selective and accurate drug delivery avoiding interaction with the healthy cells. This chapter mainly discusses the cell's recognizing ability of nanoparticles by various strategies having unique identifying properties that distinguish them from previous anticancer therapies. It also discusses specific drug delivery by nanoparticles inside the cells illustrating many successful researches and how nanoparticles remove the side effects of conventional therapies with tailored cancer treatment (Sutradhar and Amin 2014).

Phosphors for Drug Delivery, Medical Imaging and Magnetic Response Comparative Study

Figure 7.2 illustrates the process of active targeting. Nanoparticles can also target cancer through passive targeting. As apoptosis is stopped in cancerous cells, they continue sucking nutritious agents abnormally through the blood vessels forming wide and leaky blood vessels around the cells induced by angiogenesis. Leaky blood vessels are formed due to basement membrane abnormalities and decreased numbers of pericytes lining rapidly proliferating endothelial cells (Hobbs and Seymour 1998). Hence, the permeability of molecules to pass through the vessel wall into the interstitium surrounding tumor cells is increased. The size of the pores in leaky endothelial cells ranges from 100 to 780 nm (Hobes et al. 1998; Rubin and Casarett 1966; Shubik 1982). Thus nanoparticles below that size can easily pass through the pores (Jain and Stylianopoulos 2010; Jang et al. 2003). As a result, it facilitates to efflux the nanoparticles to cluster around the neoplastic cells. Nanoparticles can be targeted to a specific area of capillary endothelium, to concentrate the drug within a particular organ and perforate the tumor cells by passive diffusion or convection. Lack of lymphatic drainage eases the diffusion process. The tumor interstitium is composed of a collagen network and a gel-like fluid. The fluid has high interstitial pressures which resist the inward flux of molecules. Tumors also lack well-defined lymphatic networks having leaky vasculature. Therefore, drugs that enter the interstitial area may have extended retention times in the tumor interstitium. This feature is called

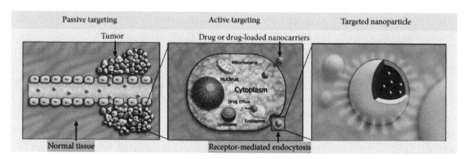

Fig. 7.2. Active and passive targeting by nanoparticles. (Sutradhar and Amin 2014) Copyright Hindawi Publication House 2014.

Table 7.1. Comparative study of literature related to drug delivery, MR and medical imaging

S. No.	Author	Synthesis method	Study	Application	Remarks
1.	Grebenik et al. 2014		Biological tissues is an important focus area of present-day medical diagnostics	Pathogenic tissues, including malignant tumors	New approach to the development of targeted constructs on the basis of UCNP and 4D5 scFv specific to the HER tumor marker
2.	Kantammeni et al. 2017		Real-time surveillance of lesions in multiple organs should facilitate pre- and post-therapy monitoring in preclinical settings	Major challenge in cancer diagnostics and therapy	
3.	Gmeiner and Ghosh 2015		Materials on the nanoscale are increasingly being targeted to cancer cells with great specificity through both active and passive targeting	Use of nanotechnology for cancer treatment with an emphasis on targeted drug delivery	
4.	Wang et al. 2013		NIR triggered drug and gene delivery, as well as several other UCNP-based cancer therapeutic	Upconversion nanoparticles for photodynamic therapy and other cancer therapeutics	NIR-excited PDT or other NIR-triggered theranostics using UCNPs

(Contd.)

Table 7.1. (*Contd.*)

S. No.	Author	Synthesis method	Study	Application	Remarks
5.	Gupta et al. 2012	Sol-gel technique	Europium-doped yttrium oxide nanophosphors	Nanophosphors in biomedical studies	High-contrast cellular and tissue imaging with high sensitivity, magnetic tracking capability and low toxicity
6.	Feng et al. 2013		Multi-modality bioimaging, and NIR light-induced therapy	Developing field, and provide guidance to design and to fabricate new nanocomposites based on upconversion nanophosphors	
7.	Yuan et al. 2012		Great potential to achieving better therapeutic effects in cancer treatment.	Zwitterionic polymer for enhanced drug delivery to tumor	
8.	Chen et al. 2013		Mono dispersed biocompatible Yb/Er or Yb/Tm doped β-NaGdF4 upconversion phosphors	Bright luminescence under 1 cm chicken breast tissue	Nanophosphors for deep tissue and dual MRI imaging
9.	Budijono et al. 2010	One-step cothermolysis utilizing oleic acid (OA) and trioctyl phosphine (TOP) ligands	Rare earth ion-doped nanophosphors (NaYF4: Yb^{3+}, Er^{3+}) opens new possibilities for improved biolabelling	Applications in bioimaging and photodynamic therapy	

No.	Author	Synthesis	Material	Application	Notes
10.	Hou et al. 2011		IBU–loaded α-NaYF$_4$:Yb^{3+}, Er^{3+}@silica fibre nanocomposites show UC emission of Er^{3+} under 980 nm NIR laser excitation	Drug delivery and disease therapy based on its bioactive, luminescent, and porous properties	Upconversion (UC) luminescent porous silica fibres decorated with NaYF$_4$:Yb^{3+}, Er^{3+} nanocrystals (NCs)
11.	Chen et al. 2013		Drug release as a function of nanoparticle size, shape, surface chemistry, and tissue type	One of the greatest challenges in cancer therapy is to develop methods to deliver chemotherapy agents to tumor cells while reducing systemic toxicity to noncancerous cells	
12.	Xu et al. 2011	Solution-phase synthesis	Drug carrier system varies with the released amount of ibuprofen	Eu^{3+}-doped GdPO$_4$ hollow spheres exhibit strong orange-red emission	Biocompatibility test on L929 fibroblast cells using MTT assay reveals low cytotoxicity of the system
13.	Tian et al. 2011	Large scale via a template-directed method using hydrothermal carbon spheres as sacrificed templates	RE ions (Yb/Er) into the Gd$_2$O$_3$ host matrix, these NPs emitted bright multicolored upconversion	MR/fluorescent imaging and therapeutic applications	Ibuprofen (IBU) was selected as a model drug to study the drug storage and release properties of this system.

(Contd.)

Table 7.1. (*Contd.*)

S. No.	Author	Synthesis method	Study	Application	Remarks
14.	Chen et al. 2016		For safe and effective therapy, drugs must be delivered efficiently and with minimal systemic side effects	Inorganic drug carriers for cancer therapy	Nanostructures are novel entrants to the world of drug delivery systems. The past decade has witnessed the rapid development of novel nanostructures and hybrid nanostructures in the field of nanomedicine.
15.	Kang et al. 2011	$NaYF_4:Yb^{3+}/Er^{3+}$ nanoparticles via a simple two-step sol-gel process	Nanospheres emit green ($^2H_{11/2}$ and $^4S_{3/2}$-$^4I_{15/2}$) and red ($^4F_{9/2}$-$^4I_{15/2}$) fluorescence of Er^{3+} even after the loading of IBU	Core shell structured $NaYF_4:Yb^{3+}/Er^{3+}$@$nSiO_2$@m-SiO_2 nanospheres are a promising material for controlled drug release	
16.	Yi et al. 2014	Hydrothermal conditions using the $Gd(OH)CO_3$:Ce/Tb precursor	Three-dimensional (3D) in vivo X-ray bioimaging of the mouse can provide the accurate location from multiple directions	Promising application in targeted therapy of tumors	Ce/Tb co-doped $GdPO_4$ hollow spheres

the Enhanced Permeability and Retention (EPR) effect and facilitates tumor interstitial drug accumulation (Fig. 7.2) (Maeda 2001; Maeda et al. 2000). Nanoparticles can easily accumulate selectively by enhanced permeability and retention effect and then diffuse into the cells (Yuan 1998).

Very recently, Hyeon and co-workers (Park et al. 2012) reported in vivo PDT effect through the systemic administration of UCNP–Ce6 followed by the 980-nm irradiation (Fig. 7.3). NaGdY4-based UCNPs after PEGylation were loaded with Ce6 molecules by both physical adsorption and chemical conjugation, yielding a UCNP-Ce6 complex with >103 Ce6 molecules per particle. The blood circulation half-life of UCNP–Ce6 was determined to be 21.6 min in BABL/C mice after intravenous injection. Nude mice bearing U87MG tumors were injected with UCNP–Ce6 through the tail vein (0.1 mg of rare earth per mouse). Obvious tumour accumulation of UCNP–Ce6 nanoparticles was revealed by dual-modal upconversion luminescence imaging and T1-weighted MR imaging (Fig. 7.3 a and b), and could likely be attributed to the Enhanced Permeability and Retention (EPR) effect of cancerous tumors. Under the 980 nm irradiation, tumor growth of UCNP-Ce6 injected mice was significantly inhibited compared with other control groups (Fig. 7.3c). These results clearly indicate the great potential of using

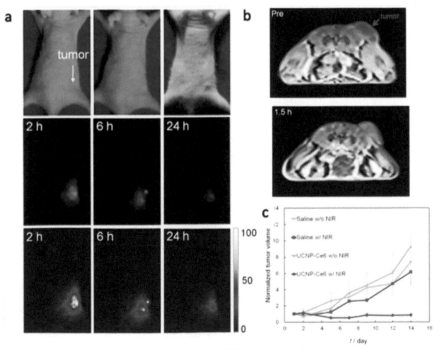

Fig. 7.3. *In vivo* imaging-guided PDT. (a) UCL images of nude mice bearing tumors after intravenous injection of UCNP–Ce6. (b) T1-weighted MR images of a tumour-bearing nude mouse before and after 1.5 hours intravenous injection of UCNP–Ce6. (c) Growth of tumors after various treatments indicated for efficient imaging guided PDT therapy. Copyright 2012 Wiley-VCH (Park et al. 2012; Wang et al. 2013).

UCNPs for multi-modal imaging guided PDT. Remarkably, this study is the first report to demonstrate UCNP-based *in vivo* PDT through the systemic administration of UCNP-PS complexes (Wang et al. 2013).

Recently, Zhang and co-workers (Jayakumar et al. 2012) reported the use of NIR-to-UV UCNPs for photo-controllable gene expression (Fig. 7.4). Plasmid DNA encoding Green Fluorescence Protein (GFP) and small interfering RNA (siRNA) target GFP mRNA were both caged with 4, 5-dimethoxy-2-nitroacetophenone DMNPE to block their respective functions. After NIR light treatment, they were uncaged by the energy transferred from UCNPs, inducing controlled gene expression and specific gene silencing, respectively. The NIR-to-UV UCNPs overcome the drawback of current photo-responsive systems in which UV light is needed to activate

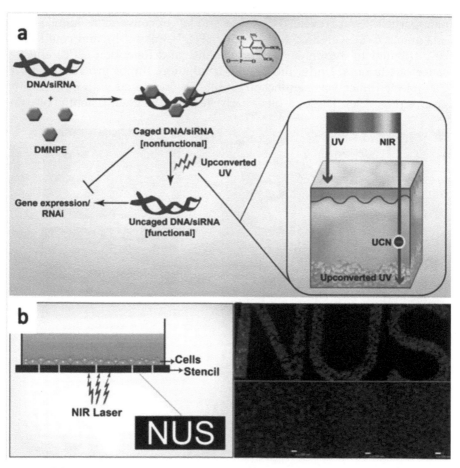

Fig. 7.4. Schematic illustration of NIR triggered gene release using NIR-UV UCNPs. (a) Plasmid DNA or siRNA are caged with DMNPE and then uncaged by upconverted UV light emitted from NIR-to-UV UCNPs. Inset shows the penetration depth of UV and NIR light in the skin. (b) Photoactivation and patterned activation of caged GFP nucleic acids in cells. Copyright 2012 Highwire Press PNAS. (Jayakumar et al. 2012).

DMNPE caged nucleic acids. Besides, the upconverted UV produced by the irradiated UCNPs was also found to be relatively safe for the cells, under the applied nanoparticle dosage and duration of NIR laser irradiation. This system was then further used to activate photocaged nucleic acids in animal models. Cells transfected with UCNPs containing photocaged GFP plasmid were loaded into a poly-dimethyl-siloxane (PDMS) device, which was transplanted into mice. Efficient GFP activation was observed even at the deep tissues. These results prove that this technique has enormous potential in a number of fields including gene therapy for controlled and specific gene delivery/knockdown, developmental biology for site-specific gene knockdown, and patterning of biomolecules using safe NIR light (Wang et al. 2013).

Bioluminescence imaging is a technology that allows for the non-invasive study of ongoing biological processes in small laboratory animals. Xing and co-workers (Yang et al. 2012) reported NIR light controlled uncaging of d-luciferin and bioluminescence imaging *in vivo* using NIR-to-UV UCNP probes. The core-shell NIR-to-UV UCNPs were coated with thiolated silane molecules and subsequently coupled to d-luciferin that was caged with a 1-(2-nitrophenyl) ethyl group. UV light emitted from UCNPs under NIR irradiation could activate caged d-luciferin to release d-luciferin molecules which was an active substrate of luciferase used in bioluminescence imaging. Cell viability assays showed no obvious cytotoxicity for C6 glioma cells treated with the d-luciferin/UCNP conjugate after two hours of irradiation with NIR light. In marked contrast, UV irradiation resulted in significant cellular damage after a short exposure time. Importantly, strong bioluminescence signals were detected in the mouse injected d-luciferin/UCNP conjugate after NIR-light induced photo uncaging. While under UV irradiation, no notable bioluminescence was detected in the mouse owing to the poor tissue penetration of UV light (Fig. 7.5).

Drug Carrier with Upconversion Indicator

Drug Carrier for Photo-thermal Therapy: The design of upconversion nanocomposites for photo-thermal therapy is based on an agent that absorbs excitation radiation and transfers it to heat energy. Because of the low absorption ability of UCNPs, nanocomposites can possess the photo-thermal therapy effect only by introducing an additional component with high absorption efficiency at the excitation wavelength, otherwise the UCNP can only act as an upconversion indicator. Au and Ag nanoparticles, with strong extinction bands originating from the surface plasmon, can be introduced as a photo-thermal therapy agent. Song et al. have designed and synthesized a core-shell $NaYF_4$:Yb,Er@Ag nanocomposite for UCL imaging and therapeutic applications (Dong et al. 2011). Thioglycolic acid (TGA) was chosen as the link to coordinate on the surface of UCNPs, and its thiol group performs as the active site during the formation of Ag nanocrystals. The Ag shell has the ability to transfer the incident 808 nm radiation to heat energy,

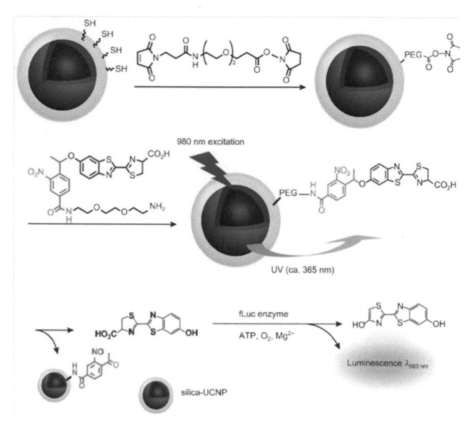

Fig. 7.5. Experimental design for NIR activated bioluminescence imaging. Caged luciferin conjugated on NIR-to-UV UCNPs was activated after 980-nm laser irradiation, releasing un-caged luciferin as an active substrate in bioluminescence imaging. Copyright 2012 Wiley-VCH (Yang et al. 2012).

and such light-to-heat conversion could be used to kill cancer cells effectively. Liu et al. incorporated Au nanoparticles onto the surface of UCNPs with the assistance of an intermediate layer composed of iron oxide (Cheng et al. 2012).

Upconversion Nanocomposites for Detection Application

Detection applications are an important research field for upconversion nanocomposites. Because the luminescent properties of rare earth ions are mainly decided by the host lattice, and the quenching centers exist inside or on the surface of nanocrystals, UCNPs usually respond little to changes in the surroundings. This makes UCNPs difficult to use as direct sensors for detection applications.

Fortunately, energy transfer processes provide a chance to introduce an analyte-responsive chromophore as the optical detection component, and the wavelength conversion ability of UCNPs can extend the working wavelength of this chromophore from the original short wavelength region to the NIR range.

In nanocomposites designed for detection based on the FRET process, the distance between the energy donor and acceptor and the spectral overlap are factors that can be controlled to influence the energy-transfer process (Fig. 7.6). The energy acceptor chosen for upconversion detection nanocomposites on the basis of the FRET process should have absorption corresponding to the upconversion emission of UCNPs. The distance between the donor and the acceptor in the FRET process should be small enough to ensure highly efficient energy transfer. The upconversion emission is usually chosen as the detection signal, which will be turned on or off when the FRET process is turned off or on, respectively. Some detection examples based on upconversion FRET are introduced below (Feng et ,al. 2013).

Chen et al. (2013) developed and fabricated mono-dispersed biocompatible Yb/Er or Yb/Tm doped β-NaGdF$_4$ upconversion phosphors using polyelectrolytes to prevent irreversible particle aggregation during conversion of the precursor,

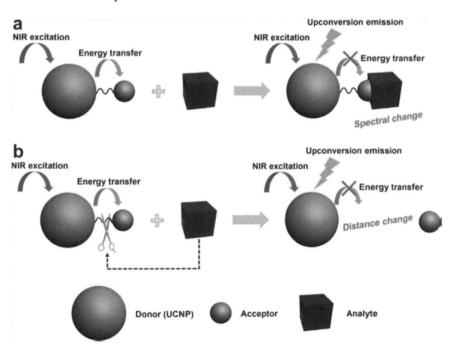

Fig. 7.6. Design principle of the upconversion nanocomposite for UC-FRET detection based on changing the spectra overlap (a) and alternation of the distance between donor and acceptor (b) (Feng et al. 2013). Copyright Wiley-VCH Verlag GmbH & Co. KGaA.

$Gd_2O(CO_3)_2 \cdot H_2O:Yb/Er$ or Yb/Tm, to β-NaGdF$_4$:Yb/Er or Yb/Tm. The polyelectrolyte on the outer surface of nanophosphors also provided an amine tag for PEGylation. This method is also used to fabricate PEGylated magnetic upconversion phosphors with Fe_3O_4 as the core and β-NaGdF$_4$ as a shell. These magnetic upconversion nanophosphors have relatively high saturation magnetization (7.0 emu g^{-1}) and magnetic susceptibility (1.7 $\times$ 10^{-2} emu g^{-1} Oe^{-1}), providing them with large magneto-phoretic mobilities. The magnetic properties for separation and controlled release in flow, their optical properties for cell labelling, deep tissue imaging, and their T$_1$- and T$_2$-weighted Magnetic Resonance Imaging (MRI) relaxivities are studied. The magnetic upconversion phosphors display both strong magnetophoresis,

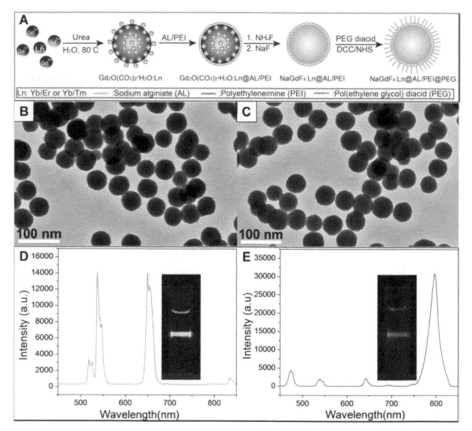

Fig. 7.7. (A) Schematic figure showing the synthesis route of PEGylated β-NaGdF$_4$: Yb/Er and β-NaGdF$_4$:Yb/Tm nanophosphors. (B) TEM image of PEGylated β-NaGdF$_4$:Yb/Er. (C) TEM image of PEGylated β-NaGdF$_4$:Yb/Tm nanophosphors. Upconversion luminescence spectra for (D) PEGylated β-NaGdF$_4$:Yb/Er and (E) PEGylated β-NaGdF$_4$:Yb/Tm nanophosphors. The insets show photographs of the luminescence from the respective particle solutions in a 1 cm plastic cuvette during irradiation with a 980 nm laser (Chen et al. 2013). Copyright Wiley-VCH Verlag GmbH & Co. KGaA

dual MRI imaging (r_1 = 2.9 mM^{-1} s^{-1}, r^2 = 204 mM^{-1} s^{-1}), and bright luminescence under 1 cm chicken breast tissue.

Rare earth doped upconversion nanophosphors are promising bio-imaging agents because of their narrow and distinct spectral emission peaks, the long optical penetration depth of their near-infrared excitation and red/near-infrared emission, the negligible auto-fluorescence backgrounds generated at these wavelengths, and the particles' excellent photo and chemical stability, and low toxicity (Mader et al. 2010; Zhou et al. 2012; Wang et al. 2010). The luminescence of upconversion nano-phosphors is generated by a sequential absorption of multiple low energy photons (e.g. 980 nm near infrared light) followed by fluorescent emission of higher energy photons from the excited states. They have been applied in immunoassays (Soukka et al. 2008), chemical sensing (Achatz et al. 2011), cell labelling (Wang et al. 2009) and small animal imaging (Zhou et al. 2012). Upconversion nanophosphors are primarily synthesized by hydrothermal and thermal decomposition methods (Zhou et al. 2012). Traditional hydrothermal synthesis methods usually generate relatively mono-dispersed particles with controlled size and shape by incubating rare-earth precursors with fluoride salts in aqueous solutions at high temperature (150 ~ 300 °C) (Wang et al. 2007). The thermal

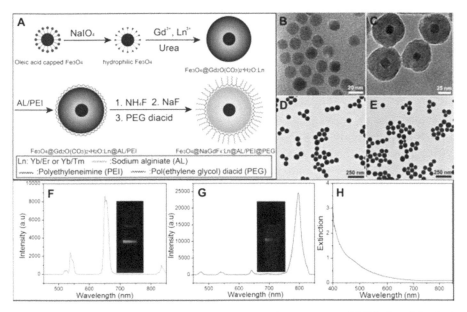

Fig. 7.8. (A) Schematic figure showing the synthesis route of PEGylated Fe_3O_4@ β-NaGdF$_4$:Yb/Er or Yb/Tm nanophosphors. TEM image of (B) Fe_3O_4 seeds after surface modification, (C) Fe_3O_4 @Gd$_2$O(CO$_3$)$_2$•H$_2$O:Yb/Er@AL/PEI, (D) NaGdF$_4$:Yb/Tm. Upconversion luminescent spectra and images for (F) PEGlayted Fe_3O_4@ β-NaGdF$_4$:Yb/Er, (G) PEGlayted Fe_3O_4@ β-NaGdF$_4$:Yb/Tm nanophosphor. (H) Extinction spectrum of a 0.18 mg/mL colloidal solution of oleic acid-coated magnetite cores in a 1 cm path length cuvette (Chen et al. 2013). Copyright Wiley-VCH Verlag GmbH & Co. KGaA

decomposition method involves the decomposition of organometallic reagents (e.g. rare-earth tri-fluoro-acetates) with hydrophobic organic ligands (e.g., oleic acid, oleylamine) in a high boiling point organic solvent (Mai et al. 2006). In order to apply these hydrophobic upconversion nanophosphors in bioimaging, their surface usually needs to be functionalized with hydrophilic ligands. This method also has been demonstrated to yield upconversion nanophosphors with well dispersed size and shape (Yi et al. 2006). Magnetic nanoparticles are another attractive material for bioimaging and actuation because they can be manipulated using external magnetic fields. They play an important role in biomedical applications such as bio-magnetic separation, drug delivery, MRI, and hyperthermia (Hu et al. 2006). Much effort has been applied to designing and fabricating multifunctional magnetic upconversion nanophosphors for multimodal imaging, separation and sensing. Previous work on magnetic upconversion nanophosphor fabrication usually involved chemically conjugating or physically attaching, magnetic particles to upconversion nanophosphors. In 2010, Shen and coworkers fabricated magnetic upconversion phosphor by conjugating iron oxide nanoparticles on the surface of upconversion phosphors by organic ligands (Shen et al. 2010). However, the magnetic nanophosphors synthesized in this work were relatively large (~ 200 nm). In 2012, Zhang and co-workers reported using magnetic upconversion nanophosphors (Fe_3O_4@α-$NaYF_4$:Yb/Er, 115 ± 20 nm) as drug carrier for doxorubicin (Zhang et al. 2011). However, the upconversion nano-shell is cubic phase $NaYF_4$:Yb/Er (α-$NaYF_4$:Yb/Er) which exhibits lower upconversion luminescent efficiency than hexagonal phase nanophosphors (β-$NaYF_4$:Yb/Er) (He et al. 2011).

Cai et al. reported a type of magnetic upconversion nanophosphor (~80 nm) with a Fe_3O_4 core and β-$NaGdF_4$:Yb/Er shell to serve as a drug carrier by attaching upconversion phosphors onto iron oxide nanoparticles (Gai et al. 2010). However, the iron oxide core needed to be coated with a thick porous silica shell in order to load small α-$NaGdF_4$:Yb/Er nanoparticles into the porous silica shell. Meanwhile, calcination at 400 °C in H_2 atmosphere was required to convert α-$NaGdF_4$:Yb/Er to brighter luminescent β-$NaGdF_4$:Yb/Er. Liu and co-workers recently report a method to obtain Fe_3O_4@ β-$NaGdF_4$:Yb/Er by thermal decomposition of organometallic reagents on iron oxide nanoparticles (Zhong et al. 2012). The obtained magnetic upconversion nanoparticles have a narrow size distribution and good size range (~30 nm) to achieve long blood circulation time. However, the non-water soluble surface limits its bio-applications. To obtain biocompatible magnetic upconversion nanoparticles with a long blood circulation time, the size of particles should be large enough (>5 nm) to avoid renal filtration but small enough to minimize opsonization and reticuloendothelial system (RES) clearance (Yoo et al. 2010). It remains a challenge to prepare to appropriately sized (5–100 nm) mono-dispersed multifunctional nanophosphors with both magnetic (e.g. Fe_3O_4) and brightly luminescent components (β-$NaYF_4$:Yb/Er or β-NaYF$_4$:Yb/Tm) because a thick upconversion nanoshell or a separation layer (e.g. SiO$_2$) is needed to compensate luminescence absorption and quenching by a magnetic core. Herein, we report a novel method to fabricate monodispersed PEGylated hexagonal phase nanophosphors (β-NaGdF$_4$:Yb/Er and β-NaGdF$_4$:Yb/Tm). This novel method is further applied to synthesize magnetic upconversion nanophosphors with iron oxide core and hexagonal phase nanophosphors (β-NaGdF$_4$:Yb/Er and β-NaGdF$_4$:Yb/Tm). The resultant magnetic upconversion nano-phosphors display bright upconversion luminescence with low quenching by the iron oxide core. In addition, a new technique was used to magnetically capture these upconversion nanophosphors from a flowing solution onto a magnetized surface and release them with a rotating magnetic field for drug eluting stent applications. Furthermore, the magnetic upconversion nanophosphors were successfully used as cell labelling reagents and imaged through 1 cm chicken breast with low dose (0.2 mL, 1 mg/mL) under 980 nm laser excitation. Last but not least, these magnetic nanophosphors with gadolinium and iron contents are showed good dual-modal (T$_1$ and T$_2$-weighted) MRI contrast agents.

Conclusion

Our aim is to compare some of the luminescent materials for therapeutic use in detail. It is concluded from the above article Nano Particles are useful for Magnetic Resonance (MR), imaging, drug delivery host carriers and drug storage/release properties. Some models are described in detail for upconversion behavior of NPs which were employed for MR/fluorescent imaging and therapeutic applications. Here NaGdF$_4$:Yb/Er phosphor are mostly used for drug delivery application apart from that Yb/Er doped Gd$_2$O$_3$ NPs are studied for MR and medical imaging purpose.

Acknowledgement

One of the authors Dr. Neha Dubey is very thankful to department of science and technology funding through project WOS-A by DST file No. SR/WOS-A/PM-14/2018 (G) entitled "Synthesis and characterization of rare earth active zirconium based phosphor for biomedical applications". The authors are also grateful to Wiley Publication House and high wire press PNAS for permission to reuse the figures. Also very thankful to Dr. R.N. Singh, Principal, Govt. V.Y.T.PG. Auto. College Durg, for providing laboratory facilities and department of physics for kind support. Author Dr. Vikas Dubey is very thankful to TEQIP-III CSVTU Bhilai for funding through Collaborative Research Project Ref. No. CSVTU/CRP/TEQIP-III/16 date 17/08/2019.

References

Achatz, D., Ali, R., Wolfbeis, O. 2011. Luminescence applied in sensor science. pp. 29. *In*: L. Prodi, M. Montalti, N. Zaccheroni (Eds.). Luminescent Chemical Sensing, Biosensing, and Screening Using Upconverting Nanoparticles. Vol. 300. Springer, Berlin/Heidelberg.

Antaris, A.L., Hao Chen, Kai Cheng, Yao Sun, Guosong Hong, Chunrong Qu, Shuo Diao, Zixin Deng, Xianming Hu, Bo Zhang, Xiaodong Zhang, Omar K. Yaghi, Zita R. Alamparambil, Xuechuan Hong, Zhen Cheng & Hongjie Dai. 2016. A small-molecule dye for NIR-II imaging. Nat. Mater. 15, 235-242.

Baban, D.F., Seymour, L.W. 1998. Control of tumour vascular permeability. Advan. Drug Del. Rev. 34(1), 109-119.

Bharali, D.J., Mousa, S.A. 2010. Emerging nanomedicines for early cancer detection and improved treatment: Current perspective and future promise. Pharmacology & Therapeutics 128(2), 324-335.

Budijono, Stephanie J., Jingning Shan, Nan Yao, Yutaka Miura, Thomas Hoye, Robert H. Austin, Yiguang Ju, Robert K. Prud'homme. 2010. Synthesis of stable block-copolymer-protected $NaYF_4$:Yb^{3+}, Er^{3+} up-converting phosphor nanoparticles. Chem. Mater. 22, 311-318.

Chaffer, C.L., Weinberg, R.A. 2011. Synthesis of a novel magnetic drug delivery system composed of doxorubicin-conjugated Fe_3O_4 nanoparticle cores and a PEG-functionalized porous silica shell. Science 331, 1559-1564.

Chen, F.-H., Zhang, L.-M., Chen, Q.-T., Zhang, Y., Zhang, Z.-J. 2010. A perspective on cancer cell metastasis. Chem. Commun. 46, 8633.

Chen, G., Ohulchanskyy, T.Y., Kumar, R., Ågren, H., Prasad, P.N. 2010. Ultrasmall monodisperse $NaYF_4$:Yb^{3+}/Tm^{3+} nanocrystals with enhanced near-infrared to near-infrared upconversion photoluminescence. ACS Nano 4, 3163.

Chen, Hongyu, Bin, Qi. Thomas Moore, Daniel C. Colvin, Thomas, Crawford, John C. Gore, Frank Alexis, O. Thompson Mefford, Jeffrey N. Anker. 2013. Synthesis of brightly PEGylated luminescent magnetic upconversion nanophosphors for deep tissue and dual MRI imaging. Small DOI: 10.1002/smll.201300828.

Chen, Hongyu, Thomas Moore, Bin Qi, Daniel C. Colvin, Erika K. Jelen, Dale A. Hitchcock, Jian He, O. Thompson Mefford, John C. Gore, Frank Alexis, Jeffrey N. Anker. 2013. Monitoring pH-triggered drug release from radioluminescent nanocapsules with X-ray excited, optical luminescence. ACS Nano 7(2), 1178-1187.

Chen, Shizhu, Xiaohong Hao, Xingjie Liang, Qun Zhang, Cuimiao Zhang, Guoqiang Zhou, Shigang Shen, Guang Jia, Jinchao Zhang. 2016. Inorganic nanomaterials as carriers for drug delivery. J. Biomed. Nanotechnol. 12(1), 1550-7033.

Cheng, L., Yang, K., Li, Y.G., Zeng, X., Shao, M.W., Lee, S.T., Liu, Z. 2012. Multifunctional nanoparticles for upconversion luminescence/MR multimodal imaging and magnetically targeted photothermal therapy. Biomaterials 33, 2215.

Choi, H.S., Summer, L. Gibbs, Jeong Heon Lee, Soon Hee Kim, Yoshitomo Ashitate, Fangbing Liu, Hoon Hyun, GwangLi Park, Yang Xie, Soochan Bae, Maged Henary, and John V Frangioni. 2013. Targeted zwitterionic near-infrared fluorophores for improved optical imaging. Nat. Biotechnol. 31, 148-153.

Dong, B., Xu, S., Sun, J., Bi, S., Li, D., Bai, X., Wang, Y., Wang, L.P., Song, H.W. 2011. J. Mater. Chem. 21, 6193.

Dou, Q.Q., Idris, N.M., Zhang, Y. 2013. Biomaterials 34, 1722.

Feng, Wei, Han, Chunmiao, Li, Fuyou. 2013. Upconversion-nanophosphor-based functional nanocomposites. Adv. Mater. 25, 5287-5303.

Franke, D., Daniel, K. Harris, Ou Chen, Oliver T. Bruns, Jessica, A. Carr, Mark, W.B. Wilson, Moungi G. Bawendi. 2016. Continuous injection synthesis of indium arsenide quantum dots emissive in the short-wavelength infrared. Nat. Commun. 7, 12749.

Gai, S., Yang, P., Li, C., Wang, W., Dai, Y., Niu, N., Lin, J. 2010. Synthesis of magnetic, up-conversion luminescent, and mesoporous core–shell-structured nanocomposites as drug carriers. Adv. Funct. Mater. 20, 1166.

Guardia, P., Di Corato, R., Lartigue, L., Wilhelm, C., Espinosa, A., Garcia-Hernandez, M., Gazeau, F., Manna, L., Pellegrino, T. 2012. Water-soluble iron oxide nanocubes with high values of specific absorption rate for cancer cell hyperthermia treatment. ACS Nano 6, 3080.

He, M., Huang, P., Zhang, C., Hu, H., Bao, C., Gao, G., He, R., Cui, D. 2011. Dual phase-controlled synthesis of uniform lanthanide-doped $NaGdF_4$ upconversion nanocrystals via an OA/ionic liquid two-phase system for in vivo dual-modality imaging. Adv. Funct. Mater. 21, 4470.

Heinzmann, K., Carter, L.M., Lewis, J.S., Aboagye, E.O. 2017. Multiplexed imaging for diagnosis and therapy. Nat. Biomed. Eng. 1, 697-713.

Hobbs, S.K., Hobbs, W.L., Monsky, F. Yuan et al. 1998. Regulation of transport pathways in tumor vessels: Role of tumor type and microenvironment. Proceedings of the National Academy of Sciences of the United States of America 95(8), 4607-4612.

Hong, G., Jerry C. Lee, Joshua T. Robinson, Uwe Raaz, Liming Xie, Ngan F. Huang, John P. Cooke, Hongjie Dai. 2012. Multifunctional in vivo vascular imaging using near-infrared II fluorescence. Nat. Med. 18, 1841-1846.

Hong, G., Yingping Zou, Alexander L. Antaris, Shuo Diao, Di Wu, Kai Cheng, Xiaodong Zhang, Changxin Chen, Bo Liu, Yuehui He, Justin Z. Wu, Jun Yuan, Bo Zhang, Zhimin Tao, Chihiro Fukunaga, Hongjie Dai. 2014. Ultrafast fluorescence imaging in vivo with conjugated polymer fluorophores in the second near-infrared window. Nat. Commun. 5, 4206.

Hong, G., Shuo Diao, Junlei Chang, Alexander L. Antaris, Changxin Chen, Bo Zhang, Su Zhao, Dmitriy N. Atochin, Paul L. Huang, Katrin I. Andreasson, Calvin J. Kuo, Hongjie Dai. 2014. Through-skull fluorescence imaging of the brain in a new near-infrared window. Nat. Photonics 8, 723-730.

Hou Zhiyao, Chunxia Li, Pingan Ma, Guogang Li, Ziyong Cheng, Chong Peng, Dongmei Yang, Piaoping Yang, Jun Lin. 2011. Electrospinning preparation and drug-delivery properties of an up-conversion luminescent porous $NaYF_4$:Yb^{3+}, Er^{3+} @Silica Fiber Nanocomposite. Adv. Funct. Mater. 21, 2356-2365.

Hu, F.Q., Wei, L., Zhou, Z., Ran, Y.L., Li, Z., Gao, M.Y. 2006. Preparation of biocompatible magnetite nanocrystals for in vivo magnetic resonance detection of cancer. Adv. Mater. 18, 2553.

Jabir, Nasimudeen R., Tabrez, Shams, Ashraf, Ghulam Md., Shakil, Shazi, Damanhouri, Ghazi A., Kamal, Mohammad A. (2012). Nanotechnology-based approaches in anticancer research. Int. J. Nanomedicine 7, 4391-4408.

Jain, R.K. and Stylianopoulos, T. 2010. Delivering nanomedicine to solid tumors. Nature Rev. Clin. Oncol. 7(11), 653-664.

Jang, S.H., Wientjes, M.G., Lu, D., Au, J.L.-S. 2003. Drug delivery and transport to solid tumors. Pharma. Res. 20(9), 1337-1350.

Jayakumar, M.K.G., Idris, N.M., Zhang, Y. 2012. Remote activation of biomolecules in deep tissues using near-infrared-to-UV upconversion nanotransducers. P. Natl. Acad. Sci. USA. 109, 8483-8488.

Jun, Y.-W., Huh, Y.-M., Choi, J.-S., Lee, J.-H., Song, H.-T., Kim, S., Yoon, S., Kim, K.-S., Shin, J.-S., Suh, J.-S., Cheon, J. 2005. J. Am. Chem. Soc. 127, 5732.

Kang, Xiaojiao, Ziyong Cheng, Chunxia Li, Dongmei Yang, Mengmeng Shang, Ping'an Ma, Guogang Li, Nian Liu, Jun Lin. 2011. Core-shell structured up-conversion luminescent and mesoporous $NaYF_4:Yb^{3+}/Er^{3+}@nSiO_2@mSiO_2$ nanospheres as carriers for drug delivery. J. Phys. Chem. C 115, 15801-15811.

Kantamneni, H., Margot Zevon, Michael J. Donzanti, Xinyu Zhao, Yang Sheng, Shravani R. Barkund, Lucas H. McCabe, Whitney Banach-Petrosky, Laura M. Higgins, Shridar Ganesan, Richard E. Riman, Charles M. Roth, Mei-Chee Tan, Mark C. Pierce, Vidya Ganapathy, Prabhas V. Moghe. 2017. Surveillance nanotechnology for multi-organ cancer metastases. Nat. Biomed. Eng. 1, 993-1003 https://doi.org/10.1038/s41551-017-0167-9

Katagiri, K., Imai, Y., Koumoto, K., Kaiden, T., Kono, K., Aoshima, S. 2011. Magnetoresponsive on-demand release of hybrid liposomes formed from Fe_3O_4 nanoparticles and thermosensitive block copolymers. Small 7, 1683.

Li, C., Yejun Zhang, Mao Wang, Yan Zhang, Guangcun Chen, Lun Li, Dongmin Wu, Qiangbin Wang. 2014. In vivo real-time visualization of tissue blood flow and angiogenesis using Ag_2S quantum dots in the NIR-II window. Biomaterials 35, 393-400.

Li, Z., Zhang, Y. 2008. An efficient and user-friendly method for the synthesis of hexagonal-phase $NaYF_4:Yb$, Er/Tm nanocrystals with controllable shape and upconversion fluorescence. Nanotechnology 19, 345606.

Mader, H.S., Kele, P., Saleh, S.M., Wolfbeis, O.S. 2010. Curr. Op. Chem. Biol. 14, 582.

Maeda, H. 2001. The enhanced permeability and retention (EPR) effect in tumor vasculature: The key role of tumor-selective macromolecular drug targeting. Adv. Enzyme Regul. 41, 189-207.

Maeda, H., Wu, J., Sawa, T., Matsumura, Y., Hori, K. 2000. Tumor vascular permeability and the EPR effect in macromolecular therapeutics: A review. J. Cont. Release 65(1-2), 271-284.

Mai, H.-X., Zhang, Y.-W., Si, R., Yan, Z.-G., Sun, L.-d., You, L.-P., Yan, C.-H. 2006. J. Am. Chem. Soc. 128, 6426.

Malam, Y., Loizidou, M., Seifalian, A.M. 2009. Liposomes and nanoparticles: nanosized vehicles for drug delivery in cancer. Trends in Pharmacol. Sci. 30(11), 592-599.

Mousa, S.A., Bharali, D.J., 2011. Nanotechnology-based detection and targeted therapy in cancer: Nano-bio paradigms and applications. Cancers 3, 2888-2903.

Naczynski, D.J., Tan, M.C., Zevon, M., Wall, B., Kohl, J., Kulesa, A., Chen, S., Roth, C.M., Riman, R.E., Moghe, P.V., 2013. Rare-earth-doped biological composites as in vivo shortwave infrared reporters. Nat. Commun. 4, 2199.

O'Connor, J.P.B., Eric O. Aboagye, Judith E. Adams, Hugo J.W.L. Aerts, Sally F. Barrington, Ambros J. Beer, Ronald Boellaard, Sarah E. Bohndiek, Michael Brady, Gina Brown, David L. Buckley, Thomas L. Chenevert, Laurence P. Clarke, Sandra Collette, Gary J. Cook, Nandita M. deSouza, John C. Dickson, Caroline Dive, Jeffrey L. Evelhoch, Corinne Faivre-Finn, Ferdia A. Gallagher, Fiona J. Gilbert, Robert J. Gillies, Vicky Goh, John R. Griffiths, Ashley M. Groves, Steve Halligan, Adrian L. Harris, David J. Hawkes, Otto S. Hoekstra, Erich P. Huang, Brian F. Hutton, Edward F. Jackson, Gordon C. Jayson, Andrew Jones, Dow-Mu Koh, Denis Lacombe, Philippe Lambin, Nathalie Lassau, Martin O. Leach, Ting-Yim Lee, Edward L. Leen, Jason S. Lewis, Yan Liu, Mark F. Lythgoe, Prakash Manoharan, Ross J. Maxwell, Kenneth A. Miles, Bruno Morgan, Steve Morris, Tony Ng, Anwar R. Padhani, Geoff J.M. Parker, Mike Partridge, Arvind P. Pathak, Andrew C. Peet, Shonit Punwani, Andrew R. Reynolds, Simon P. Robinson, Lalitha K. Shankar, Ricky A. Sharma, Dmitry Soloviev, Sigrid Stroobants, Daniel C. Sullivan, Stuart A. Taylor, Paul S. Tofts, Gillian M. Tozer, Marcel van Herk, Simon Walker-Samuel, James Wason, Kaye J. Williams, Paul Workman,

Thomas E. Yankeelov, Kevin M. Brindle, Lisa M. McShane, Alan Jackson, John C. Waterton. 2017. Charged-particle therapy in cancer: Clinical uses and future perspectives. Nat. Rev. Clin. Oncol. 14, 169-186.

Park, Y.I., Kim, H.M., Kim, J.H., Moon, K.C., Yoo, B., Lee, K.T., Nohyun Lee, Yoonseok Choi, Wooram Park, Daishun Ling, Kun Na, 2012. Theranostic probe based on lanthanide-doped nanoparticles for simultaneous in vivo dual-modal imaging and photodynamic therapy. Adv. Mater. doi:10.1002/adma.201202433.

Peer, D., Karp, J., Hong, S. 2007. Nanocarriers as an emerging platform for cancer therapy. Nature Nanotech 2, 751–760. https://doi.org/10.1038/nnano.2007.387

Qian, H.S., Zhang, Y. 2008. Synthesis of hexagonal-phase core-shell $NaYF_4$ nanocrystals with tunable upconversion fluorescence. Langmuir 24, 12123.

Rubin, P., Casarett, G. 1966. Microcirculation of tumors Part I: Anatomy, function, and necrosis. Clin. Radiol. 17(3), 220-229.

Shen, J., Sun, L.-D., Zhang, Y.-W., Yan, C.-H. 2010. Superparamagnetic and upconversion emitting $Fe_3O_4/NaYF_4$:Yb, Er hetero-nanoparticles via a crosslinker anchoring strategy. Chem. Commun. 46, 5731.

Shubik, P. 1982. Vascularization of tumors: A review. J. Cancer Res. Clin. Oncol. 103(3), 211- 226.

Soukka, T., Rantanen, T., Kuningas, K. 2008. Photon upconversion in homogeneous fluorescence-based bioanalytical assays. Ann. NY Acad. Sci. 1130, 188.

Yao Sun, Chunrong Qu, Hao Chen, Maomao He, Chu Tang, Kangquan Shou, Suhyun Hong, Meng Yang, Yuxin Jiang, Bingbing Ding, Yuling Xiao, Lei Xing, Xuechuan Hong, Zhen Cheng. 2016. Novel benzo-bis(1,2,5-thiadiazole) fluorophores for *in vivo* NIR-II imaging of cancer. Chem. Sci. 7, 6203-6207.

Sutradhar, Kumar Bishwajit, Amin Md. Lutful. 2014. Nanotechnology in Cancer Drug Delivery and Selective Targeting. Hindawi Publishing Corporation. ISRN Nanotechnology. Volume 2014, Article ID 939378, 12 pages. https://doi.org/10.1155/2014/939378.

Tao, Zhimin, Guosong Hong, Chihiro Shinji, Changxin Chen, Shuo Diao, Alexander L. Antaris, Bo Zhang, Yingping Zou, Hongjie Dai. 2013. Biological imaging using nanoparticles of small organic molecules with fluorescence emission at wavelengths longer than 1000 nm. Angew. Chem. Int. Ed. 52, 13002-13006.

Teng, X., Zhu, Y., Wei, W., Wang, S., Huang, J., Naccache, R., Hu, W., Tok, A.I.Y. Han, Y., Zhang, Q., Fan, Q., Huang, W., Capobianco, J.A., Huang, L. 2012. Lanthanide-Doped $NaxScF^{3+}x$ Nanocrystals: Crystal Structure Evolution and Multicolor Tuning J. Am. Chem. Soc. 134, 8340.

Tian Gan, Zhanjun Gu, Xiaoxiao Liu, Liangjun Zhou, Wenyan Yin, Liang Yan, Shan Jin, Wenlu Ren, Gengmei Xing, Shoujian Li, Yuliang Zhao. 2011. Facile fabrication of rare-earth-doped Gd_2O_3 hollow spheres with upconversion luminescence, magnetic resonance, and drug delivery properties. J. Phys. Chem. C 115, 23790-23796.

Wang, Chao, Cheng Liang, Liu Zhuang. 2013. Upconversion nanoparticles for photodynamic therapy and other cancer therapeutics. Theranostics 3(5): 317-330.

Wang, F., Liu, X. 2009. Recent advances in the chemistry of lanthanide-doped upconversion nanocrystals. Chem. Soc. Rev. 38, 976.

Wang, F., Banerjee, D., Liu, Y., Chen, X., Liu, X. 2010. Upconversion nanoparticles in biological labeling, imaging, and therapy. Analyst 135, 1839.

Wang, L., Li, Y. 2007. Controlled synthesis and luminescence of lanthanide doped $NaYF_4$ nanocrystals. Chem. Mater. 19, 727.

Welsher, K., Sherlock, S.P., Dai, H. 2011. Deep-tissue anatomical imaging of mice using carbon nanotube fluorophores in the second near-infrared window. Proc. Natl. Acad. Sci. USA 108, 8943-8948.

Xu, Zhenhe, Yu Cao, Chunxia Li, Ping'an Ma, Xuefeng Zhai, Shanshan Huang, Xiaojiao Kang, Mengmeng Shang, Dongmei Yang, Yunlu Daiab, Jun Lin. 2011. Urchin-like $GdPO_4$ and $GdPO_4$:Eu^{3+} hollow spheres – hydrothermal synthesis, luminescence and drug-delivery properties. J. Mater. Chem. 21, 3686.

Yang, Y.M., Shao, Q., Deng, R.R., Wang, C., Teng, X., Cheng, K., Zhen Cheng, Ling Huang, Zhuang Liu, Xiaogang Liu, Bengang Xing. 2012. *In vitro* and *in vivo* uncaging and bioluminescence imaging by using photocaged upconversion nanoparticles. Angew. Chem. Int. Edit. 51: 3125-3129.

Yi, G.S., Chow, G.M. 2006. Synthesis of hexagonal-phase $NaYF_4$:Yb,Er and $NaYF_4$:Yb,Tm nanocrystals with efficient up-conversion fluorescence. Adv. Funct. Mater. 16, 2324.

Yi, Zhigao, Wei Lu, Chao Qian, Tianmei Zeng, Lingzhen Yin, Haibo Wang, Ling Rao, Hongrong Liua, Songjun Zeng. 2014. Urchin-like Ce/Tb co-doped $GdPO_4$ hollow spheres for in vivo luminescence/X-ray bioimaging and drug delivery. Biomater. Sci. 2, 1404-1411.

Yoo, J.W., Chambers, E., Mitragotri, S. 2010. Factors that control the circulation time of nanoparticles in blood: Challenges, solutions and future prospects. Curr. Pharmaceut. Design 16, 2298.

Yu, M.K., Jeong, Y.Y., Park, J., Park, S., Kim, J.W., Min, J.J., Kim, K., Jon, S. 2008. Angew. Chem. Int. Ed. 47, 5362.

Yuan, F. 1998. Transvascular drug delivery in solid tumors. Seminars Radia. Oncol. 8(3), 164-175.

Yuan, You-Yong, Cheng-Qiong, Mao, Xiao-Jiao, Du, Jin-Zhi, Du, Feng, Wang, Jun, Wang. 2012. Surface charge switchable nanoparticles based on zwitterionic polymer for enhanced drug delivery to tumor. Adv. Mater. DOI: 10.1002/adma.201202296.

Zhang, F., Braun, G.B., Pallaoro, A., Zhang, Y., Shi, Y., Cui, D., Moskovits, M., Zhao, D., Stucky, G.D. 2011. Mesoporous multifunctional upconversion luminescent and magnetic "nanorattle" materials for targeted chemotherapy. Nano Lett. 12, 61.

Zhang, Miqin. 2017. Early detection of multi-organ metastases. Nature Biomedical Engineering 1, 934-936. doi.org/10.1038/s41551-017-0173-y.

Zhang, Y., Guosong Hong, Yejun Zhang, Guangcun Chen, Feng Li, Hongjie Dai, Qiangbin Wang. 2012. Ag_2S quantum dot: A bright and biocompatible fluorescent nanoprobe in the second near-infrared window. ACS Nano 6, 3695-3702.

Zhao, Gang, Rodriguez, B. Leticia. 2013. Molecular targeting of liposomal nanoparticles to tumor microenvironment. Int. J. Nanomedicine 8, 61-71.

Zhong, C., Yang, P., Li, X., Li, C., Wang, D., Gai, S., Lin, J. 2012. Monodisperse bifunctional Fe_3O_4@ $NaGdF_4$:Yb/Er@ $NaGdF_4$:Yb/Er core-shell nanoparticles. RSC Adv. 2, 3194.

Zhong, Yeteng, Zhuoran Ma, Shoujun Zhu, Jingying Yue, Mingxi Zhang, Alexander L. Antaris, Jie Yuan, Ran Cui, Hao Wan, Ying Zhou, Weizhi Wang, Ngan F. Huang, Jian Luo, Zhiyuan Hu, Hongjie Dai. 2017. Boosting the down-shifting luminescence of rare-earth nanocrystals for biological imaging beyond 1500 nm. Nature Communications 8, 737. DOI: 10.1038/s41467-017-00917-6

Zhou, J., Liu, Z., Li, F. 2012. Upconversion nanophosphors for small-animal imaging. Chem. Soc. Rev. 41, 1323.

Zhu, S. Qinglai Yang, Alexander, L.A., Jingying Yue, Zhuoran Ma, Huasen Wang, Wei Huang, Hao Wan, Joy Wang, Shuo Diao, Bo Zhang, Xiaoyang Li, Yeteng Zhong, Kuai Yu, Guosong Hong, Jian Luo, Yongye Liang, Hongjie Dai. 2017. Molecular imaging of biological systems with a clickable dye in the broad 800-1700 nm near-infrared window. Proc. Natl. Acad. Sci. USA 114, 962-967.

Visible Light Emitting Ln³⁺ Ions (Ln = Eu, Sm, Tb, Dy) in Mg$_2$SiO$_4$ Host Lattice

Ramachandra Naik[1]*, S.C. Prashantha[2]*, H. Nagabhushana[3], Yashwanth V. Naik[3], K.M. Girish[4], H.B. Premkumar[5] and D.M. Jnaneshwara[6]

[1] Department of Physics, New Horizon College of Engineering, Bengaluru - 560103, India
[2] Research Center, Department of Science, East West Institute of Technology, VTU, Bengaluru - 560091, India
[3] Prof. C.N.R. Rao Center for Advanced Materials, Tumkur University, Tumkur - 572103, India
[4] Department of Physics, DSATM, Bangalore - 560082, India
[5] Department of Physics, Ramaiah University of Applied Sciences, Bangalore - 560054, India
[6] Department of Physics, SJB Institute of Technology, Bangalore - 560060, India

Introduction

Among nanomaterials, silicates are widely studied materials for a wide range of applications due to their visible light transparency. In particular, inorganic nanophosphors with the incorporation of trivalent rare earth cations reveal major luminescence effects. Further, various vacancies and defects present in host matrix results in different luminescence features (Prashantha et al. 2011). Enhanced electrical, luminescent and optical properties of nano phosphors were caused by the quantum size effect, which was generated by an increase in the band gap due to a decrease in the quantum allowed state and the high surface-to-volume ratio (Cho et al. 2010). Among the silicate family, the Mg$_2$SiO$_4$ (forsterite) host doped with rare earth ions exhibit some interesting applications such as long lasting phosphor, X-ray imaging, Light Emitting Display (LED), environmental monitoring, etc. Mg$_2$SiO$_4$ is a member of the olivine family of crystals and has an orthorhombic crystalline structure in

*Corresponding authors: rcnaikphysics@gmail.com; scphysics@gmail.com

which Mg^{2+} occupies two non equivalent octahedral sites: one (M1) with inversion symmetry (CI) and the other (M2) with mirror symmetry (CS). The material has a high surface area, low thermal expansion, low electrical conductivity, good chemical stability and excellent insulation properties even at high temperatures (Sunitha et al. 2012; Saberi et al. 2009; Sun et al. 2009).

Solid-state lighting using Light-Emitting Diode (LED) and luminescent phosphor material to generate white light is the current research focus in the lighting industry which depends critically on the design of bright and stable phosphors. The application of WLED (White Light Emitting Diode) as the next generation light source will change the life of humans dramatically. The common WLED device is composed of chips that emit blue light and phosphors that can be excited by blue light to emit other colors. Since WLED devices are characterized by their high emission intensity, the quality of phosphors is vital to the performance of WLED device. WLEDs can save about 70% of the energy and do not need any harmful ingredient in comparison with the conventional light sources, such as incandescence light bulbs and the luminescent tubes. Therefore WLEDs have a great potential to replace them and are considered as the next generation solid state light devices (Naik et al. 2014a).

Rare Earth (RE) ions were characterized by an incompletely filled 4f shell, which can absorb the excitation energy to be at the excited state and then return to the ground state, resulting in emitting in the visible region. Rare earth ions have a number of efficient and narrow emission lines in the visible region whose position was insensitive to their matrices due to the shielding effect of outer 5s and 5p electrons. Forsterite was prepared preferably through solution based methods in order to get high chemical homogeneity and small crystallite size compared to conventional solid state reaction, which needs higher calcination temperatures to obtain phase pure crystals. However, the synthesis of pure and doped nano crystalline forsterite with controlled particle size still remained challenging (Naik et al. 2014b).

Prashantha et al. reported that, nanoparticles of Eu^{3+} doped Mg_2SiO_4 are prepared using low temperature solution combustion technique with metal nitrate as precursor and urea as fuel. The synthesized samples are calcined at 800 °C for 3 hours. The Powder X-ray diffraction (PXRD) patterns of the sample revealed orthorhombic structure with α-phase. These phosphors exhibit a bright red color upon excitation by 256 nm light and showed the characteristic emission of the Eu^{3+} ions. The electronic transition corresponding to $^5D_0 \rightarrow {}^7F_2$ of Eu^{3+} ions (612 nm) is stronger than the magnetic dipole transition corresponding to $^5D_0 \rightarrow {}^7F_1$ of Eu^{3+} ions (590 nm). Enhancement in PL intensity of Eu^{3+} was observed due to the formation of different lattice sites in the host phosphor (Prashantha et al. 2011).

Tabrizi et al. reported that a Mg_2SiO_4:Eu^{3+} nanopowder was synthesized by a polyacrylamide gel method. In this route, the gelation of the solution is achieved by the formation of a polymer network which provides a structural framework for the growth of particles. The densification of the powders was also studied. An amorphous nanopowder was synthesized and crystallized to Mg_2SiO_4 after heat-treatment via a solid-state reaction at a relatively low temperature of about 700 °C. The powders prepared by the polyacrylamide gel method showed better sinterability than the powders synthesized by the conventional sol gel method. The relative density of the sample was 97% at 1500 °C (Hassanzadeh-Tabrizi et al. 2013).

Synthesis and Characterization of Mg_2SiO_4:Ln^{3+} (Ln = Eu, Sm, Tb, Dy)

Synthesis of Mg_2SiO_4: Ln^{3+} (Ln = Eu, Sm, Tb, Dy)

To prepare Mg_2SiO_4 various methods were adopted by different researchers but the solution combustion method is one of the efficient techniques due to its special features such as, energy saving, cost effective and time saving. The detailed steps and procedures to prepare undoped and Ln^{3+} doped Mg_2SiO_4 is given in Fig. 8.1.

Stoichiometric amount of $Mg(NO_3)_2$, SiO_2 (fumed silica), and oxalyl dihydrazide (ODH) as fuel were used as raw materials to prepare an undoped sample. Further, required amount and type of Ln^{3+} (Ln = Eu, Sm, Tb, Dy)

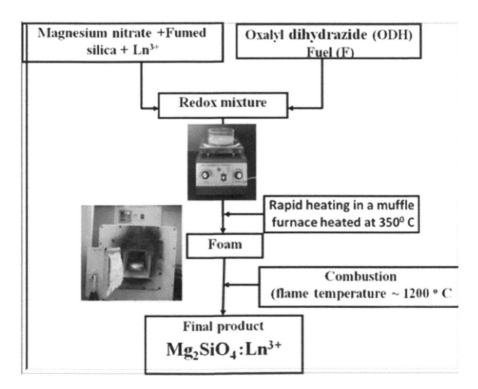

Fig. 8.1. Flow chart of combustion synthesis.

were added to prepare doped samples. Redox mixture was prepared in a petri dish using distilled water and was stirred continuously on a magnetic stirrer to get a uniform solution. The petri dish containing the solution was kept in a preheated muffle furnace with temperature maintained at 350 °C. The solution boiled and underwent dehydration with liberation of gases and turned into a final product in powder form. Final products were ground into a fine powder using agate and a mortar. These powders were used for further characterizations without any post calcinations. All samples were prepared at low temperature and without any calcination was the key feature of this presentation (Naik et al. 2014a).

Characterization of $Mg_2SiO_4:Ln^{3+}$ (Ln = Eu, Sm, Tb, Dy)

Phase Purity and Structure: XRD

Undoped and doped samples were characterized using Powder X-Ray Diffraction (PXRD) technique in order to check its purity and the structure. All samples show orthorhombic structure with JCPDS card No. 78-1371 with space group p_{bnm} (No. 62). No other impurity peaks were detected which indicates prepared samples are pure. Crystallite size (D) is calculated using Scherrer's method (Eq. (1)) (Naik et al. 2014a).

$$D = \frac{k\lambda}{\beta \cos \theta} \tag{1}$$

where 'k' is constant, 'λ' is wavelength of X-rays, and 'β' is Full Width at Half Maximum (FWHM). It was found that, all samples show crystallite size in nano range between 1-100 nm. The D value of undoped sample was found to be ~25 nm but for the doped samples D value changed with respect to concentration. In Eu^{3+} doped samples D values decreased from ~35 nm to ~27 nm. D value decreased in case of Tb^{3+} doped samples from 22 to 18 nm, from 27 to 25 nm in Sm^{3+} doped sample and from 26 to 19 nm in Dy^{3+} doped samples. These variations in crystallite size may be due to temperature variations during combustion process.

Bandgap Analysis: DRS

The measurement of diffused radiation reflected from a surface constitutes the area of spectroscopy known as diffuse reflectance spectroscopy. Diffuse reflectance spectrometry concerns one of the two components of reflected radiation from an irradiated sample, namely specular reflected radiation, R_s and diffusely reflected radiation, R_d. The former component is due to the reflection at the surface of single crystallites while the latter arises from the radiation penetrating into the interior of the solid and re-emerging to the surface after being scattered numerous times. These spectra can exhibit both absorbance and reflectance features due to contributions from transmission, internal and specular reflectance components as well as the scattering phenomenon in the collected radiation. Diffuse Reflectance Spectra (DRS) of doped samples were recorded and analyzed its absorption peaks. It was observed in different doped samples that different peaks were present which are attributed to characteristic transitions of dopant ions. In Eu^{3+} doped samples, the bands observed in the DR spectra at ~393 and ~464 nm were attributed to the intra configurationally 4f–4f transitions from the ground $^7F_0$ level. The Tb^{3+} doped samples show peaks at 323 nm and 390 nm which confirms that absorption takes place at these particular wavelengths. The DR spectra of Sm^{3+} doped Mg_2SiO_4 nanophosphors shows a strong absorption band in the shorter wavelength range 200–300 nm (i.e. higher energy) for all samples may be ascribed to the absorption of the host lattice. When Sm^{3+} ions is introduced into the host lattices, several weak absorption bands in the larger wavelength range 300–480 nm (i.e. low energy) are observed. These electronic bands are the characteristics of Sm^{3+} ions, starting from the ground state $^6H_{5/2}$ to the various excited states of Sm^{3+} ions. In Dy^{3+} doped Mg_2SiO_4 nanophosphors DR spectra show a strong absorption bands at 320, 350 and 390 nm which are attributed to Dy^{3+} ions (Kumar et al. 2014).

Powder sample diffuses the light in a large quantity and thickness is also large, which make the absorption spectrum difficult to interpret. To avoid this difficulty, DRS were used and Schuster–Kubelka–Munk (SKM) relation was used to correlate the diffuse reflectance to absorption coefficient (Kumar et al. 2014). In this case; the Kubelka–Munk equation at any wavelength becomes

$$F(R) = \frac{(1-R)^2}{2R} \qquad (2)$$

where R is the absolute reflectance of the sampled and $F(R)$ is the so-called Kubelka–Munk function. The optical band gap of phosphors exhibits the fundamental absorption, which corresponds to electron excitation from the valance band to conduction band and determined by the relation $(F(R)$ $h\upsilon)^n = A(h\upsilon - Eg)$, where $n = 2$ for a direct allowed transition, and $n = 1/2$ for an indirect allowed transition, 'A' is the absorption coefficient ($A = 4\,k/\lambda$; k is the absorption index or absorbance) and '$h\upsilon$' is the photon energy (Naik et al. 2014a). For Mg_2SiO_4 $n = 2$. The linear part of the curve was extrapolated to $(F(R)h\upsilon)^2 = 0$ to get the direct bandgap energy.

The bandgap of doped samples are as follows: Eu^{3+} doped samples exhibits in the range 5.63-5.87 eV, 4.9-5 eV for Tb^{3+} doped samples, 4.7-5 eV for Sm^{3+} doped samples, and 4.9-5 eV for Dy^{3+} doped samples. It was observed that, upon changing Ln^{3+} ions the bandgap changes because the charge carrier concentration increases. As the doping concentration increased, the density of states also improved and created a continuum of states resulting in the decrease of the bandgap (Naik et al. 2014a). This suggests decrease in the bandgap due to the inclusion of Ln^{3+} ions into Mg_2SiO_4 matrix which changes the electronic structure leading to the form of intermediate energy levels.

Thus, $Mg_2SiO_4:Ln^{3+}$ could be considered as ultraviolet absorption material, in which by varying the Fe content absorption energy can be adjusted.

Luminescence Properties of $Mg_2SiO_4:Ln^{3+}$ (Ln = Eu, Sm, Tb, Dy)

Photoluminescence Mechanism

When a sample is illuminated by light, it absorbs radiation of a certain wavelength to jump to an excited state. A part of the absorbed energy is lost on vibration relaxation, i.e., radiationless transition to the lowest vibrational level takes place in the excited state and finally it returns to the ground state by emitting energy. This process is called fluorescence. Fluorescence continues for a period of 10^{-8} to 10^{-9} s in most cases. Since a part of the radiation absorbed is lost, the fluorescence emitted from the substance has a longer wavelength than the excitation radiation. The emission transition occurring in a solid state is seen as a glow and is recorded in the form of a band in the luminescence spectrum. The position of the band in the luminescence spectrum does not depend on the method of excitation. The luminescence spectra are normally observed with the intensity of luminescence as a function of the emission wavelength. The same instrument can be used to measure the spectral distribution of luminescence (emission spectrum) and the vibration in the emission intensity with excitation wavelength (excitation spectrum) or with activator concentration. Figure 8.2 shows the block diagram of the experimental set-up used to record the luminescence spectrum of samples. The light emitted from the Xe-lamp enters the excitation monochromator. The light emerging from the excitation monochromator is split by the beam splitter and a fraction is directed to the monitor detector. All the driving

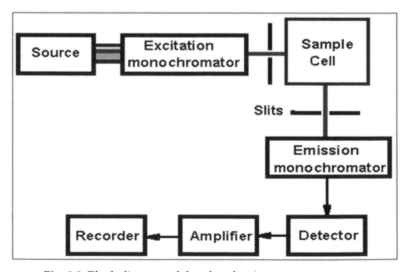

Fig. 8.2. Block diagram of the photoluminescence spectrometer.

components i.e., the wavelength drive motors and slit control motors are operated by the signal sent from the computer. On the other hand, output signal from the monitor detector and fluorescence detector (photomultiplier) are processed by the computer via the A/D converter transmitted to the CRT or graphic plotter.

Mg_2SiO_4:Ln^{3+} (Ln = Eu, Sm, Tb, Dy) samples were analyzed using the photoluminescence technique. Excitation spectrum was recorded for all different dopants to determine excitation wavelength. Eu^{3+} doped samples exhibit a broad excitation band in the wavelength range 300–550 nm, which can be attributed to the intra-configurational (f–f) transitions of Eu^{3+}. The excitation peak observed at 393 nm suggests that the interaction between O^{2-} and Eu^{3+} was stronger, therefore it was selected as excitation wavelength (Naik et al. 2014a).

From the excitation spectrum of Sm^{3+} doped sample, it was observed that the intensity of f–f transition at 315 nm was high compared with the other transitions and has been chosen as the excitation wavelength for the measurement of emission spectra of Mg_2SiO_4:Sm^{3+} phosphors. The most intense peak at 315 nm clearly signifies that these phosphors were effectively excited by near-ultraviolet (NUV) light (Naik et al. 2015).

The excitation spectrum of Tb^{3+} show a series of sharp lines in the region 300–450 nm that belong to f-f transitions of Tb^{3+} ions. The transitions at longer wavelengths pertain to transitions from the $^7F_6$ ground state to the excited levels of the 4f8 configuration. The most intense lines with maxima at 368 and 377 nm can be assigned to the $^7F_6 \rightarrow$ ($^5G_6$, $^5G_5$, $^5L_9$) and $^7F_6 \rightarrow$ ($^5D_3$, $^5L_{10}$) on the basis of the energy levels scheme of Tb^{3+} (Naik et al. 2014b). The excitation peak at 350 nm in Dy^{3+} doped samples is strong and gives intense emission compared to other wavelengths peaks in the spectrum; therefore samples are excited at 350 nm (Naik et al. 2016).

The emission spectra of Eu^{3+} doped sample consist of emission peaks at 577, 590, 612, 650 and 703 nm which were attributed to Eu^{3+} transitions $^5D_0 \rightarrow {}^7F_0$, $^5D_0 \rightarrow {}^7F_1$, $^5D_0 \rightarrow {}^7F_2$, $^5D_0 \rightarrow {}^7F_3$ and $^5D_0 \rightarrow {}^7F_4$ respectively. As dopant concentration increases emission intensity also increases upto 11 mol%. The mechanism of energy transfer between the activators in the Mg_2SiO_4 host, the critical energy transfer distance (R_c) was calculated by the following equation:

$$R_C \approx 2 \left[\frac{3V}{4\pi CN} \right]^{\frac{1}{3}} \tag{3}$$

where C is the critical concentration, N is the number of cation sites in the unit cell and V is the volume of the unit cell. For the present Mg_2SiO_4:Eu^{3+} phosphor the values of N, V and C were 4, 289.92⁻³ and 0.11 respectively (Naik et al. 2014a).

The emission spectra of Sm^{3+} doped sample consisting of four main groups of lines from 550-750 nm was the characteristic of the samarium ions. These emissions can be assigned to the intra 4-f orbital transitions $^4G_{5/2} \rightarrow$

$^6H_{5/2}$ (576 nm), $^4G_{5/2} \rightarrow {^6H_{7/2}}$ (611 nm), $^4G_{5/2} \rightarrow {^6H_{9/2}}$ (656 nm) and $^4G_{5/2} \rightarrow$ $^6H_{11/2}$ (713 nm) because of their consistent luminescence behavior with the Sm^{3+} emission characteristics. As dopant concentration increases emission intensity also increases upto 5 mol% and thereafter it decreases due to concentration quenching. The critical transfer distance of the center Sm^{3+} in Mg$_2$SiO$_4$:Sm^{3+} phosphor by taking the appropriate values of V, N, C and R_c are 290 Å^3, 4, 0.05 and 13.71 Å respectively (Naik et al. 2015).

The emission of Tb^{3+} ions occurs due to the transitions of $^5D_3$ and $^5D_4$ excited states to the ^7F_J ground states. The emission spectrum lines are separated in two groups. The blue emission group is from $^5D_3 \rightarrow {^7F_J}$ (J = 5, 4 and 3) below 480 nm and the green emission group is from $^5D_4 \rightarrow {^7F_J}$ (J = 6, 5, 4 and 3) above 480 nm. The spectra consist of emission peaks at 417, 436, 458 nm (blue emissions) corresponding to $^5D_3 \rightarrow {^7F_5}$, $^5D_3 \rightarrow {^7F_4}$, $^5D_3 \rightarrow {^7F_3}$ transitions and 486, 541, 584, 621 nm (green emissions) which are attributed to the electronic transitions $^5D_4 \rightarrow {^7F_6}$, $^5D_4 \rightarrow {^7F_5}$, $^5D_4 \rightarrow {^7F_4}$ and $^5D_4 \rightarrow {^7F_3}$, respectively. As dopant concentration of Tb^{3+} increases, $^5D_4 \rightarrow {^7F_5}$ transition dominates and the emission intensity increases. This may be attributed to the increase in distortion of the local field around the Tb^{3+} ions. When the Tb^{3+} ions concentration increases more than 3 mol%, PL emission intensity begins to decrease indicating the concentration quenching. The critical transfer distance (R_c) for energy transfer can be calculated using the relation given by Blasse. (3) Where C is the critical concentration, N is the number of Z ions in the unit cell and V is the volume of the unit cell. By taking the experimental and analytical values, V is 289.966^{-3}, N is 4 and X_c is 0.03. The critical energy transfer distance R_c in Mg$_2$SiO$_4$:Tb^{3+} phosphor is calculated to be 16.3 Å (Naik et al. 2014b). The PL spectra of Dy^{3+} doped sample consist of three main emission peaks at 476 nm, 577 nm and 666 nm which is assigned to the electronic transitions of $^4F_{9/2} \rightarrow {^6H_{15/2}}$ (blue), $^4F_{9/2} \rightarrow {^6H_{13/2}}$ (yellow) and $^4F_{9/2} \rightarrow {^6H_{11/2}}$ (red) respectively. The positions of the emission peaks are not influenced by the Dy^{3+} concentration.

The defect reaction equation can be represented in the following way,

$$(1-x)Mg_2SiO_4 + 0.5xDy_2O_3 =$$
$$xDy'_{Mg} + 0.5x\,V''_0 + (1-x)\,Mg^x_{Mg} + (2-0.5x)O^x_o \qquad (4)$$

where Dy$'_{Mg}$ means Dy occupying the site normally occupied by a Mg^{2+} due to replacement by Dy, V$''_0$ is the O^{2-} vacancy, Mg$^x_{Mg}$ represents the rest magnesium in the lattice of Mg$_2$SiO$_4$, and O^x_o is the oxygen in the lattice of Mg$_2$SiO$_4$. By taking the experimental and analytic values of N, V and X_c as 4, 289.96 Å^3 and 0.03 respectively, the estimated R_c value is ~16.3 Å (Naik et al. 2016).

Materials Performance: Chromaticity Coordinates and Color Temperature

The Commission International de l'Eclairage (CIE) 1931 chromaticity coordinates for Mg$_2$SiO$_4$:Eu^{3+} (1-11 mol%) phosphors as a function of Eu^{3+}

concentration for the luminous color was depicted by the PL spectra. According to our knowledge, the CIE coordinates of red emission of Eu^{3+} ions not only depend upon the asymmetric ratio but also depend on the higher energy $^5D_J (= 3, 2, 1)$ emission levels. It was observed that the CIE co-ordinates of 1 to 11 mol% Eu^{3+} activated $Mg_2SiO_4:Eu^{3+}$ phosphor was tuning from (0.4345, 0.5339) to (0.5910, 0.4002) respectively (Naik et al. 2014a).

CIE and Correlated Color Temperature (CCT) were calculated in order to know the change in the photometric characterization of $Mg_2SiO_4:Sm^{3+}$. The X, Y coordinates of the sample (5 mol%) excited under 315 nm were calculated from the photoluminescence spectra and was found to be close to NTSC values of red emission. Also, CCT can be estimated by Planckian locus, which is only a small portion of the (x, y) chromaticity diagram. CCT was calculated by transforming the (x, y) coordinates of the light source to (U', V') and was found to be 1756 K (Naik et al. 2015).

It was observed that the CIE co-ordinates of 3 mol% Tb^{3+} activated Mg_2SiO_4 phosphor is found to lie in the green region ($x = 0.25903$, $y = 0.44203$). Therefore, the present phosphor is highly useful for the production of artificial white light owing to its better spectral overlap and also in the green component in white LEDs (Naik et al. 2014b).

It is observed that the CIE co-ordinates of different concentrations of Dy^{3+} activated Mg_2SiO_4 phosphor shifting towards white region. CCT 9095 K was obtained by transforming the (x, y) coordinates of the light source to (U', V') by using equations. The CCT value of studied phosphors found to vary from 8180 to 10730 K (Naik et al. 2016).

Conclusions

Ln^{3+} doped forsterite phosphors were synthesized by a solution combustion method. The diffraction patterns as-prepared Mg_2SiO_4 phosphors can be indexed to a single orthorhombic phase. The calculated average crystallite was found to be in the range 25-50 nm. The Eu^{3+} doped phosphor showed excellent CIE chromaticity co-ordinates (x, y) and CCT values; as a result it was quite useful for display applications. The orange red emission properties and the estimated CIE chromaticity co-ordinates (0.588, 0.386) of Sm^{3+} doped phosphors was very close to NTSC standard value of red emission of this phosphor and CCT was found to be 1756 K. Therefore, the $Mg_2SiO_4:Sm^{3+}$ phosphors were promising materials in the red region for WLEDs and optical display system applications. The emission spectra of $Mg_2SiO_4:Tb^{3+}$ is related to $^5D_4 \rightarrow {}^7F_J$ (J = 6, 5, 4, 3) transitions of Tb^{3+} ion, the most intense emission of Tb^{3+} is registered for the transition $(^5D_4 \rightarrow {}^7F_5)$ at 541 nm. Further, the excellent green emission properties and the estimated CIE chromaticity co-ordinates suggest that the $Mg_2SiO_4:Tb^{3+}$ phosphors are promising materials in the green region for optical display system applications. $Mg_2SiO_4:Dy^{3+}$ (1-11 mol%) nanophosphors showed three main peaks in blue (magnetic dipole), yellow (forced electric dipole) and red regions. Further, the phosphor showed excellent CIE chromaticity co-ordinates (x, y); as a result it is quite useful for display applications.

References

Cho, S., Lee, R., Lee, H., Kim, J., Moon, C., Nam, S., Park, J. 2010. Synthesis and characterization of Eu^{3+} doped Lu_2O_3 nanophosphor using a solution–combustion method. J. Sol–Gel Sci. Technol. 53, 171-175.

Hassanzadeh-Tabrizi, S.A., Taheri-Nassaj, E. 2013. Polyacrylamide gel synthesis and sintering of Mg_2SiO_4:Eu^{3+} nanopowder. Ceram. Inter. 39, 6313-6317.

Kumar, Vinod, Som, S., Kumar, Vijay, Kumar, Vinay, Ntwaeaborwa, O.M., Coetsee, E., Swart, H.C. 2014. Tunable and white emission from ZnO:Tb^{3+} nanophosphors for solid state lighting applications. Chem. Eng. 255, 541-552.

Naik, Ramachandra, Prashantha, S.C., Nagabhushana, H., Sharma, S.C., Nagabhushana, B.M., Nagaswarupa, H.P., Premakumar, H.B. 2014a. Low temperature synthesis and photoluminescence properties of redemitting Mg_2SiO_4:Eu^{3+} nanophosphor for near UV light emitting diodes. Sens. Actu. B: Chem. 195, 140-149.

Naik, Ramachandra, Prashantha, S.C., Nagabhushana, H., Nagaswarupa, H.P., Anantha Raju, K.S., Sharma, S.C., Nagabhushana, B.M., Premakumar, H.B., Girish, K.M. 2014b. Mg_2SiO_4:Tb^{3+} nanophosphor: Auto ignition route and near UV excited photoluminescence properties for WLEDs J. Alloys Comp. 617, 69-75.

Naik, Ramachandra, Prashantha, S.C., Nagabhushana, H., Sharma, S.C., Nagaswarupa, H.P., Anantharaju, K.S., Nagabhushana, B.M., Premakumar, H.B., Girish, K.M. 2015. A single phase, red emissive Mg_2SiO_4:Sm^{3+} nanophosphor prepared via rapid propellant combustion route. Spectrochim. Acta A 140, 516-523.

Naik, Ramachandra, Prashantha, S.C., Nagabhushana, H., Sharma, S.C., Nagaswarupa, H.P., Anantharaju, K.S., Jnaneshwara, D.M., Girish, K.M. 2016. Tunable white light emissive Mg_2SiO_4:Dy^{3+} nanophosphor: Its photoluminescence, Judd-Ofelt and photocatalytic studies. Dye. Pigm. 127, 25-36.

Prashantha, S.C., Lakshminarasappa, B.N., Nagabhushana, B.M. 2011. Photoluminescence and thermoluminescence studies of Mg_2SiO_4:Eu^{3+} nanophosphor. J. Alloys Compd. 509, 10185-10189.

Saberi, A., Negahdari, Z., Alinejad, B., Golestani-Fard, F. 2009. Synthesis, characterization of nanocrystalline forsterite through citrate–nitrate route. Ceram. Int. 35, 1705-1708.

Sun, H.T., Fujii, M., Nitta, N., Mizuhata, M., Yasuda, H., Deki, S., Hayashi, S. 2009. Molten-salt synthesis and characterization of nickel-doped forsterite nanocrystals. J. Am. Ceram. Soc. 92, 962-966.

Sunitha, D.V., Nagabhushana, H., Singh, F., Nagabhushana, B.M., Sharma, S.C., Chakradhar, R.P.S. 2012. Thermo, iono and photoluminescence properties of 100 MeV Si^{7+} ions bombarded $CaSiO_3$:Eu^{3+} nanophosphor. J. Lumin. 132, 2065-2071.

Synthesis and Luminescence Characteristics of Europium Doped Gadolinium Based Oxide Phosphors for Display and Lighting Applications

Jyoti Singh[1]*, Dirk Poelman[2], Vikas Dubey[3] and Vinay Gautam[4]

[1] School of Basic and Applied Sciences, Galgotias University, Greater Noida, Uttar Pradesh, India
[2] Lumilab, Department of Solid State Sciences, Ghent University, Ghent, Belgium
[3] Department of Physics, Bhilai Institute of Technology, Raipur, India
[4] Department of Applied Sciences, Galgotias College of Engineering and Technology, Greater Noida, Uttar Pradesh, India

Introduction

Phosphors are materials which are capable of emitting radiation when subjected to ultraviolet light, electron bombardment, X-rays or some other form of excitation (Antony et al. 2001). This emission is known as luminescence. Phosphors are mainly categorized into two types: organic and inorganic. A wide range of organic molecules and polymers are found to show luminescence. These so-called fluorophores find applications as luminescent labels for bio-imaging, as dyes, optical brighteners and in organic light emitting diodes (Sasabe and Kido 2013). This chapter is focused on the study of novel inorganic phosphors which are ideal for applications requiring long lifetimes, such as LEDs for lighting and displays. The inorganic phosphor is comprised of two components: (i) host lattice and (ii) the impurity or activator ions (mainly rare-earth ions or transition metals) (Kikuchi 2010; Smet et al. 2010, 2011; Avci et al. 2009; Poelman et al. 1993). Generally, inorganic phosphors are semiconductors or insulators with a wide band gap. Semiconductors with a wide and direct band gap are of great interest in the optical industry and have a wide range

*Corresponding author: jyoti.singh.phy@gmail.com

of applications in the field of optoelectronics, lasers, solid-state lighting, sensors, solar cell etc. (Bridot et al. 2007; Zhou et al. 2011; Vries et al. 2005).

Recently, Rare Earth (RE) activated oxide phosphors have been vastly used in White Light Emitting Diodes (WLEDs) and display devices (Kikuchi 2010; Smet et al. 2010, 2011). Oxide based matrices offer a range of structural and compositional possibilities that makes the oxide host suitable for recent developments in the display technology, such as Field-Emission Displays (FED) and Plasma Display Panels (PDP) (Antony et al. 2001). Moreover, phosphor-converted (pc) W-LEDs have been known for their enormous commercial application potential due to their high luminous efficiency, reliability, eco-friendly nature, energy saving and durability with a long working lifetime.

Normally, the light absorption cross-section of semiconductors is larger as compared to the lanthanide ions which have poor absorption cross-section due to the forbidden nature of f–f transition (parity selection rule). Nevertheless, this forbidden nature of f–f transition can be modified by the crystal field when the lanthanide (Ln^{3+}) ion activators occupy a crystallographic site with a certain symmetry (Kim et al. 2014; Blasse 1966; Sieber et al. 1995; Kolar et al. 1999; Barry and Roy 1967). So the doping of lanthanide ions in the suitable host is one of the prevalent techniques to overcome this problem. The approach of using rare earth ions as activators and semiconductors as the host offers the advantage of combining the optical properties of rare earth ions and the unique qualities of the semiconductors. Rare-earth-doped semiconductor materials, namely phosphors, are widely applied for emissive displays, W-LEDs and fluorescent lamps.

The commercial and most commonly used W-LEDs have been largely obtained via mixing the emission from a blue LED with the broad yellow spectrum of Ce-doped $Y_3Al_5O_{12}$ (YAG:Ce) (Smet et al. 2011). Alternatively, and offering more flexibility in tuning the W-LED emission spectrum, of Red, Green and Blue (RGB) color emitting phosphors can be coated onto blue/ UV chips (Avci et al. 2009; Poelman et al. 1993). In this regard, trivalent europium (Eu^{3+}) ion which is very useful, since it is well-known for strong red and orange emissions of f–f transitions ($^5D_0 \rightarrow {}^7F_2$ and $^7F_1$) and can be doped into various host materials such as tungstates, molybdates and fluorides etc. in order to yield strong red emission for lighting and display usage (Smet et al. 2010, 2011; Avci et al. 2009). Although, Eu^{3+} doped sulphide phosphors such as (Cd, Sr) S, Gd_2O_2S and Y_2O_2S have been commercially used, these materials are chemically instable and suffer from an environment unfriendly fabrication process (Smet et al. 2010, 2011, Avci et al. 2009). In addition, the volatility of sulphur in these sulphide based phosphors are accountable for the cathodes to deteriorate, which results in reduced luminous efficiency and restricts their application in FEDs and LEDs (Bridot et al. 2007; Zhou et al. 2011; Vries et al. 2005).

Most of the properties of wide band gap oxide based matrices such as strontium carbonate $SrCO_3$, gadolinium oxide Gd_2O_3, yttrium oxide Y_2O_3 etc. have been studied or are being studied separately but the studies on mixed oxides such as lanthanide based alkaline earth binary oxides ($SrGd_2O_4$, $BaGd_2O_4$, SrY_2O_4 etc.) have not been done properly yet (Srinivas and Rao 2012; Zhou et al. 2007; Chaker et al. 2003; Li et al. 2009; Karunadasa et al. 2005). The reason behind the study on mixed oxides is because the dual combination of elements may affect the material properties, so one can get different/modified material properties by making these mixed oxides (Park et al. 1999; Fu et al. 2008; Lojpur et al. 2014; Zhang et al. 2013; Pavitra et al. 2012; Dubey et al. 2016; Singh et al. 2016; Raju et al. 2014; Zhou et al. 2007; Mari et al. 2011).

Mixed oxides with the composition of alkaline earth and rare earth elements, manifest potential applications in the technological areas such as solid state lighting and display devices (Som et al. 2014, Dutta et al. 2015, Raju et al. 2014a; Sun et al. 2013a, b; Besara et al. 2014; Wang et al. 2007a, b; Raju et al. 2014b, c; Zhang et al. 2012; Maekawa et al. 2007; Fu et al. 2006). Recently, Maekawa et al. (2007) reported SrY_2O_4 and BaY_2O_4 as host materials and studied their thermo-physical and mechanical properties, which stimulate these host compounds as effective thermal barrier coating materials. ARE_2O_4 compounds such as $SrGd_2O_4$, $BaGd_2O_4$, SrY_2O_4 and BaY_2O_4 etc., consist of one alkaline earth (A) site and two RE sites (Lakshinarasimhan and Varadaraj 2008; Wang and Tian 2011; Sun et al. 2014; Singh et al. 2016; Wang et al. 2007a). Both Rare-Earth (RE) sites are occupied without inversion symmetry with C_s point symmetry, which is appropriate for the introduction of other RE^{3+} (Eu^{3+}, Tb^{3+}, Dy^{3+}, Sm^{3+} and Er^{3+}) ions for efficient characteristic emission (Carnall et al. 1968 a, b; Som et al. 2015; Zhang et al. 2013; Tian et al. 2012; Das et al. 2014; Reddy et al. 2012; Som et al. 2014a; Bandi et al. 2012). In spite of this, studies related to $SrGd_2O_4$ and $BaGd_2O_4$ phosphors have not yet been done in detail for powder samples. Next to their importance for luminescent applications, these compounds could open up other possible applications in bio-imaging thanks to the presence of magnetic Gd-ions and the high average atomic number of the hosts. Indeed, due to their high X-ray mass absorption coefficient and magnetic Gd^{3+} ions, these phosphors could potentially be used as multimodal imaging contrast ageing, for fluorescent imaging, CT as well as MRI (Bridot et al. 2007; Zhou et al. 2011; Vries et al. 2005).

The aim of this chapter is to study the effect of Eu^{3+} ions on the structural and optical properties of the AGd_2O_4 system (A = Sr or Ba). The changes in luminescence characteristics of AGd_2O_4 system occurring due to doping are explored with the help of Judd-Ofelt calculations.

Reported Works on $BaRE_2O_4$ and $SrRE_2O_4$

- Luminescence properties of Eu^{3+} doped $BaGd_2O_4$ was first reported by Blasse 1966. It is reported that the quantum efficiency of the $BaGd_2O_4$: Eu^{3+} phosphor for mercury lamps is about 40%.
- Tb^{3+} and Sm^{3+} co-activated $BaGd_2O_4$ phosphors were developed by Sieber et al. 1995 for highly efficient X-ray storage phosphors.
- Maekawa et al. 2007 reported the thermophysical properties of BaY_2O_4 phosphors which have a potential to be a new candidate material for

thermal barrier coating. The solid state reaction method followed by a spark plasma sintering was adopted to synthesize BaY_2O_4 phosphors.

- Kolar et al. 1999 reported the scintillating properties of $BaGd_2O_4$ phosphors. Both BaO and Gd_2O_3 in $BaGd_2O_4$ contributed to the improvement of host density which is reported to be 7.68 g/cm^3 slightly higher than that of famous lutetium oxyorthosilicate (LSO) crystal (7.4 g/cm^3). Moreover, the $BaGd_2O_4$ host is very stable up to 1860 °C which is of significance for practical applications.

- Luminescence properties of RE doped $BaGd_2O_4$ phosphors were reported by Zhou et al. 2007. It was stated that $BaGd_2O_4$ phosphors can be utilized in the emerging Field Emission Displays (FED) and Plasma Display Panel (PDP) technology fields.

- Barry and Roy 1967 reported a study on the chemical relationships between sesquioxides of certain rare earth ions [La, Gd, Y, Ho, Yb] and the alkaline earth oxides [Ca, Sr and Ba]. CaY_2O_4, $CaHo_2O_4$, SrY_2O_4, $SrHo_2O_4$ $SrYb_2O_4$, $BaGd_2O_4$, BaY_2O_4, $BaHo_2O_4$ and $BaYb_2O_4$ compounds were prepared and the chemical stability was checked. It was reported that for given RE ions, the tendency toward compound formation would be greater for the larger ionic radius of the alkaline earth ion employed.

- Srinivas and Rao 2012 reported photoluminescence and thermo-luminescence properties of Eu^{3+} doped $BaGd_2O_4$ phosphors synthesized by a conventional solid-state reaction method. The samples showed strong red emission at 611 nm under 254 nm UV light. The different kinetic parameters were determined by Chen's method by using the TL glow curve.

- Sun et al. 2013b reported the photoluminescence, X-ray Excited Luminescence (XEL) and thermoluminescence properties of Dy^{3+} doped $BaGd_2O_4$ phosphors in the scintillating field. The strongest XEL intensity of Dy^{3+} ions in $BaGd_2O_4$ host in the blue and yellow region was observed by the naked eyes at room temperature. The TL properties of $BaGd_2O_4$: Dy^{3+} phosphor were studied after irradiation with 254 nm UV light.

- Sun et al. 2014 reported the PL and TL properties of Y^{3+} and Na^+ co-doped $BaGd_2O_4$: Eu^{3+} phosphors. It is reported that incorporation of Y^{3+} ions enhanced the sensitizing effect of Gd^{3+} in $BaGd_2O_4$ which gives intense red emission in $BaGd_2O_4$: Eu^{3+} host.

- Raju et al. 2014a reported the PL and cathodoluminescence properties of Sm^{3+} doped $BaGd_2O_4$ phosphors synthesized via a solvothermal reaction method. The PL studies and cathodoluminescent spectrum revealed the reddish-orange emission for LED and FED applications.

- Besara et al. 2014 described the magnetic properties of the $BaLn_2O_4$ (Ln = lanthanide) family. It is reported that the single crystals of a series of compounds in the $BaLn_2O_4$ family were prepared by using a molten metal flux. A sign of magnetic frustration was observed in all compounds due to the trigonal arrangements of the trivalent lanthanide cations in the structure.

- Chaker et al. 2003 reported on the new phase of $SrGd_2O_4$ samples by using the Rietveld refinement method. They described the structure of $SrGd_2O_4$ as an assembly of bi-octahedra $[Gd_2O_{10}]$ which are linked together by O^{2-} anions and of do-decahedra of SrO_8.

- Li et al. 2009 performed the theoretical researches on the $CaFe_2O_4$-type binary rare earth oxides ARE_2O_4 by using chemical bond theory of dielectric description. The chemical bond properties, thermal expansion properties and compressibility of these crystals were studied in order to understand their contribution to the mechanical properties.

- Karunadasa et al. 2005 described the magnetic properties of the SrL_2O_4 (L = Gd, Dy, Ho, Er, Tm, and Yb) host. The complex crystal structure produces strong geometric frustration for the magnetic system as evidenced in both magnetic susceptibility and neutron-scattering data at low temperatures. Susceptibility measurements for the series, including $SrGd_2O_4$ were also reported.

- Park et al. 1999 reported the PL properties of Eu^{3+} doped SrY_2O_4 phosphors synthesized via a combustion method. The emission spectrum of SrY_2O_4: Eu^{3+} showed two kinds of Eu^{3+} emissions, which could be assigned to Eu^{3+} ions on the Sr site and the Y site and reducing SrY_2O_4: Eu^{3+} yielded a blue emission when excited with 295 nm UV light. The latter could be attributed to crystal defects introduced by the reduction.

- Fu et al. 2008 compared the band structure calculation of bulk and nanocrytalline SrY_2O_4 samples synthesized via combustion and solid-state reaction method, respectively.

- Zhang et al. 2012 reported the photoluminescence properties of RE doped SrY_2O_4 and $SrGd_2O_4$ phosphors prepared via the solid-state reaction method. It is reported that under different excitation such as VUV-UV excitation, X-ray excitation and near-infrared excitation RE doped phosphors emitted various color emissions.

- Lojpur et al. 2014 investigated the temperature sensing behavior of Eu^{3+} doped SrY_2O_4 prepared via the citrate sol-gel method. The emission rise time of Eu^{3+} doped SrY_2O_4 phosphors was investigated in the temperature range of 20-200°C for the application in luminescence thermometry.

- Zhang et al. 2013 reported the photoluminescence and cathodoluminescence (CL) properties of Tb^{3+}, Tm^{3+} and Dy^{3+} doped SrY_2O_4 phosphors prepared via soft chemical synthesis method. A color-tunable emission and white CL in SrY_2O_4 phosphors could be realized by co-doping with Tm^{3+} and Dy^{3+} ions which suggests its applicability in W-LED's and FED devices.

- Pavitra et al. 2012 reported the PL properties of Eu^{3+} or Sm^{3+} ions co-doped SrY_2O_4: Tb^{3+} phosphors prepared via the sol-gel process.

Multiple emissions (green, orange and white) were reported for Eu^{3+} or Sm^{3+} ions co-doped SrY_2O_4: Tb^{3+} phosphors under UV excitation.

- Dubey et al. 2016 investigated the TL properties of Eu^{3+} doped SrY_2O_4 phosphors under UV, beta and gamma irradiation. The study of formation of deep trapping mechanism was studied in detail for Eu^{3+}doped SrY_2O_4 phosphors.

- Singh et al. 2016 performed the comparative study of PL properties of Eu^{3+} doped Gd_2O_3 and AGd_2O_4 (A = Ba or Sr) phosphors prepared by facile gel combustion process using hexamethylenetetramine as an organic fuel.

In this chapter, we have worked on a modified synthesis technique, homogeneous precipitation method followed by a subsequent combustion process, to stabilize the phase purity. These methods are very appropriate to find the pure $SrGd_2O_4$ and $BaGd_2O_4$ phase comparison to the above reported methods.

Significance of the $SrGd_2O_4$ and $BaGd_2O_4$ Host

The AGd_2O_4 (A = Ba or Sr) system exhibits excellent thermal stability, high density of about 7.13 g/cm^3, high melting point (around 2400 °C) and good charge stability, which is significant for practical applications (Raju et al. 2014a, b, c; Zhou et al. 2007; Mari et al. 2011). Moreover, the AGd_2O_4 host can effectively transfer its absorbed energy to the incorporated RE activators via the sensitizing effect of Gd^{3+} ions resulting in the enhancement of emission intensity (Carnall et al. 1968 a, b; Som et al. 2015; Zhang et al. 2012; Tian et al. 2012; Das et al. 2014; Reddy et al. 2012). Additionally, in the AGd_2O_4 host, the Gd^{3+} ions have the same valence state and similar ionic radius as other lanthanide ions. These are beneficial for the introduction of luminescent centers of various rare-earth ions such as Eu^{3+}, Tb^{3+} and Dy^{3+} since they can easily substitute the identical ionic sites of the Gd^{3+} ions.

The AGd_2O_4 system crystallizes in the form of a calcium ferrite ($CaFe_2O_4$)-related structure having space group *Pnam*, which is composed of a double octahedral $Gd_2O_4^{2-}$ framework with alkaline earth ions residing within the framework. The full structure of AGd_2O_4 is shown in Fig. 9.1 (a) and the GdO_6 octahedra are shown in polyhedral representation. It is worth mentioning that the Gd^{3+} ions occupy two crystallographically inequivalent Gd sites; both sites are coordinated by six oxygen atoms with C_s point symmetry (Lakskminarasimhan and Varadaraju 2008; Sun et al. 2014). Figure 9.1 (a) indicates that the Gd (1) site is nearly octahedral while, the Gd (2) site is in a more distorted coordination environment (Fig. 9.1(b)). In the a-b plane, Gd^{3+} ions are linked in a network of hexagons and also linked by triangles along c-axis. Two crystallographically inequivalent Gd^{3+} ions (yellow and green in color) form two different triangular ladders along the c-direction (Singh et al. 2016). The Sr site in this host is composed of eight nearest neighbors exhibiting C_s symmetry. It is well manifested, that an acceptable percentage

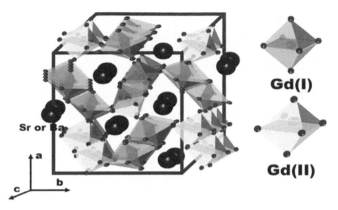

Fig. 9.1. The crystal structure of AGd_2O_4. Two crystallographically inequivalent Gd^{3+} ions, Gd (1) is shown as green color, whereas Gd (2) is shown as yellow color.

difference in ionic radii between doped and substituted host ions should not exceed ±30% (Singh et al. 2016), which suggests that Eu^{3+} ions (0.947 Å, CN = 6) prefer to substitute Gd^{3+} (0.938 Å, CN = 6) ions rather than Sr^{2+} (1.18 Å, CN = 8) ions or Ba^{2+} (1.35 Å, CN = 8) ions (Zhang et al. 2012; Maekawa et al. 2007).

Synthesis Routes and Characterization Techniques

Generally, there are many types of synthesis routes to prepare the different types of the oxide samples for various applications. Among them, one of the key routes to prepare the microparticles is the homogeneous precipitation method followed by subsequent combustion process.

Homogeneous Precipitation Method Followed by Combustion Process

The powder samples of AGd_2O_4: Eu (A = Sr or Ba) were prepared via the homogeneous precipitation method followed by a subsequent combustion process with controlled heating at 1200 °C for 3 hours in air ambiance. The method of Singh et al. 2016 is followed to prepare the phosphors. The procedure is as follows:

(a) First, the stoichiometric amounts of starting materials $Sr(NO_3)_2$ or $Ba(NO_3)_2$ (Otto Chemie Pvt. Limited 99%), $Gd(NO_3)_3.6H_2O$ (Otto Chemie Pvt. Limited, 99.9%) and $Eu(NO_3)_3.xH_2O$ (Sigma-Aldrich, 99.9%) were dissolved in the excess of ethanol.

(b) The above mixture taken in a glass beaker was stirred well for 30 minutes under heating at 80 °C to get a homogeneous clear solution; after that, the quantity of urea (H_2NCONH_2) was added to the nitrate solution. The molar ratio of urea to metal nitrate is fixed to 2.5.

(c) Later the obtained solution was continuously stirred at 100 °C for at least 3 hours until the liquid totally evaporated and white powders were left.

(d) In the next step, the powders were ground properly and kept at 1200 °C for 3 hours in order to investigate the systematic studies.

(e) Interestingly, the combustion and the annealing effect simultaneously occurred during this process. During the combustion process, the material goes through a rapid dehydration and foaming followed by decomposition and generating combustible gases. Thereby, a voluminous solid is yielded when these volatile combustible gases ignite and burn with a flame. The combustion process utilizes the enthalpy of combustion and the whole process is completed in a short duration of time. Along with the combustion synthesis the concurrent annealing process leads to the increase in the crystallinity of the solid obtained.

Characterization Techniques

The X-Ray Diffraction (XRD) patterns of the as-prepared samples were recorded on Bruker-D8 Focus X-ray powder diffractometer with Cu Kα = 1.5406 Å in the range $20° \leq 2\theta \leq 70°$ at a scan rate 1° min^{-1} with 0.02° step size, since all the prominent peaks are available in the concerned range. The morphology of the prepared samples was examined by using a Field-Emission Scanning Electron Microscope (FESEM Supra-55, Germany) images and a Philips CM12 transmission electron microscope. The room-temperature photoluminescence (PL) spectra and life time measurement were recorded on Horiba FL3-21 fluorescence spectrophotometer.

Results and Discussion

Structural and Luminescence Characteristics of SrGd$_2$O$_4$ and BaGd$_2$O$_4$ Phosphors

Phase Identification and Morphological Characterizations

The XRD studies were carried out to investigate the structural information and influence of dopants in the crystal lattice of the prepared samples. XRD patterns of the as-prepared strontium and barium digadolinium oxide phosphors were recorded and are shown in Fig. 9.2. The detailed structural information of SrGd$_2$O$_4$ phosphors was reported in our previous work (Singh et al. 2016). The diffraction pattern of phosphors show the dominance of the orthorhombic SrGd$_2$O$_4$ and BaGd$_2$O$_4$ phase (JCPDS Card No. 82-2320) with space group *Pnam* (62) along with the negligible presence of Gd$_2$O$_3$ phase. No diffraction peaks were observed due to the europium doping which indicates the successful incorporation of Eu^{3+} ions into Gd^{3+} lattice sites of the SrGd$_2$O$_4$ host matrix.

The microstructure of as-synthesized BaGd$_2$O$_4$:Eu^{3+} (4 mol%) phosphors were studied and are shown in Fig. 9.3. The FESEM images show agglomerated spherical morphology with particle sizes in the range 0.01-0.1 µm. However, the FESEM images of Eu^{3+} doped SrGd$_2$O$_4$ samples exhibit agglomerated rod-like structures with particle sizes in the range 0.3-3 µm

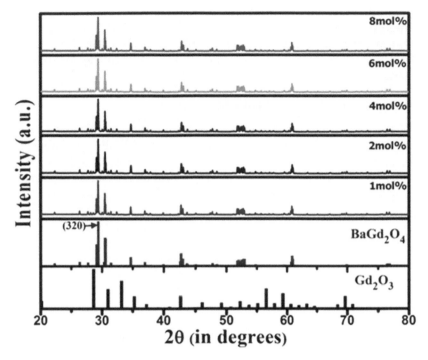

Fig. 9.2. XRD pattern of $BaGd_{2(1-x)}Eu_{2x}O_4$ (x = 1-8 mol%) phosphors.

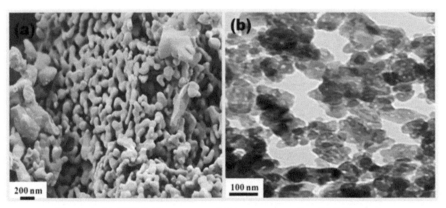

Fig. 9.3. (a) FESEM images and (b) corresponding TEM images of $BaGd_2O_4$: 4 mol% Eu^{3+} samples.

(Singh et al. 2016). The doping concentration did not show any impact on the morphology of the samples.

In order to investigate the effect of doping of Eu^{3+} ions in AGd_2O_4 (A = Ba or Sr) crystals more precisely, Transmission Electron Microscopy (TEM) of the $BaGd_2O_4$: Eu^{3+} (4 mol%) samples was also carried out and shown in Fig. 9.3b. TEM images reveal spherical agglomerates having average sizes

~0.01-0.1 μm. However, TEM images of $SrGd_2O_4$: Eu^{3+} (4 mol%) samples are presented in Figure [S1 (supplementary file)]. The TEM image of $SrGd_2O_4$: Eu^{3+} exhibits irregular agglomerates having average sizes ~0.3-3 μm. The TEM results of both the phosphors are consistent with the corresponding SEM data.

Luminescence Characterizations

Figure 9.4 shows the photoluminescence excitation (PLE) and emission spectra of Eu^{3+} doped $BaGd_{1.92}Eu_{0.08}O_4$ and $SrGd_{1.92}Eu_{0.08}O_4$ phosphors. The excitation spectra of prepared phosphors were recorded keeping the emission wavelength fixed at 615 nm. In general, the excitation spectrum of Eu^{3+} doped phosphor materials contains two different regions:

1. *Charge Transfer (CT) transition.*
2. *Intra 4f transitions.*

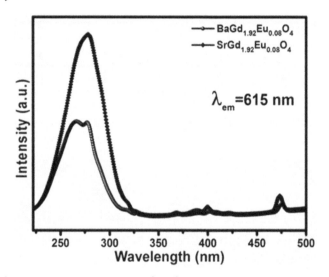

Fig. 9.4. Room temperature photoluminescence excitation spectra of $SrGd_{1.92}Eu_{0.08}O_4$ and $BaGd_{1.92}Eu_{0.08}O_4$ sample.

The broad band extending from 200-330 nm with band maximum at 276 nm in the higher energy region is associated with a Charge Transfer State (CTS), owing to an electron that is transferred from the oxygen 2p orbital to the empty 4f orbital of the europium.

It is worth noting that the f-f transition $^8S_{7/2} \rightarrow {^6I_{9/2}}$ at 276 nm of Gd^{3+} ions was overlapped with this strong CTS band in $SrGd_{1.92}Eu_{0.08}O_4$ samples. However, the f-f transition and CTS band can be distinctly seen in the excitation spectra of $BaGd_{1.92}Eu_{0.08}O_4$ samples. The sharp lines in the lower energy region correspond to direct excitation of the intra-4f forbidden transitions of Eu^{3+} ions: $^7F_0 \rightarrow {^5H_3}$ at 323 nm, $^7F_0 \rightarrow {^5D_4}$ at 362 nm, $^7F_0 \rightarrow {^5G_2}$ at 381 nm, $^7F_0 \rightarrow {^5L_6}$ at 394 nm, $^7F_0 \rightarrow {^5D_3}$ at 414 nm and $^7F_0 \rightarrow {^5D_2}$ at 465 nm respectively. Additionally, the typical excitation peak at 313 nm is assigned to f-f transitions of Gd^{3+} ($^8S_{7/2} \rightarrow {}^6P_{7/2}$) ions (Raju et al. 2014b, c; Zhou et al. 2007; Mari et al. 2011; Singh et al. 2016). The assignment of all the excitation and emission transitions is shown in Figure [S2 supplementary file)].

The effects of the Eu^{3+} contents on the emission behaviors in $BaGd_{2(1-x)}O_4:2xEu^{3+}$ and $SrGd_{2(1-x)}O_4:2xEu^{3+}$ phosphors are illustrated in Fig. 9.5 and Figure [S3 supplementary file)]. The intensity of the emission from the $^5D_{3,2,1}$ to the ^7F_J level decreases with increasing Eu^{3+} concentrations and the intensity of the emission from the $^5D_0 \rightarrow {}^7F_J$ transitions gradually increases. These results are owing to the higher energy level $^5D_{3,2,1}$ being quenched by a cross-relaxation mechanism (Raju et al. 2014b, c). This is a non-radiative process in which the excitation energy of the ion decaying from a higher excited state ($^5D_{3,2,1}$) of Eu^{3+} promotes a neighboring ion from the ground state to a metastable state by the following equations:

$$Eu^{3+} (^5D_1) + Eu^{3+} (^7F_0) \rightarrow Eu^{3+} (^5D_0) + Eu^{3+} (^7F_3)$$

$$Eu^{3+} (^5D_2) + Eu^{3+} (^7F_0) \rightarrow Eu^{3+} (^5D_1) + Eu^{3+} (^7F_4)$$

$$Eu^{3+} (^5D_3) + Eu^{3+} (^7F_0) \rightarrow Eu^{3+} (^5D_2) + Eu^{3+} (^7F_3)$$

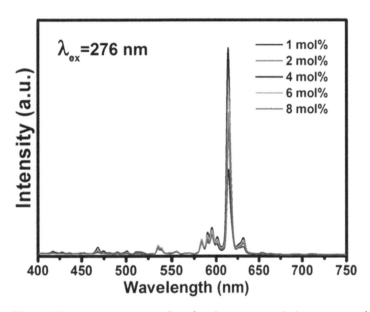

Fig. 9.5. Room temperature photoluminescence emission spectra of various $BaGd_{2(1-x)}Eu_{2x}O_4$ ($x = 1\text{-}8$ mol%) samples ($\lambda_{ex} = 276$ nm).

As the concentration of Eu^{3+} increases, a condition is met where the distance between two nearest Eu^{3+} ions becomes small enough to allow resonant energy transfer. Therefore, the energy can be easily transferred from one luminescent center to another. As a consequence, when the Eu^{3+} concentration is sufficiently high, the emission from the higher level ($^5D_{3,2,1}$)

can be easily quenched via cross-relaxation and the $^5D_0$ emission becomes dominant. The dominant emission peaks of europium at 594 and 615 nm are attributed to $^5D_0{\rightarrow}^7F_1$ and $^5D_0{\rightarrow}^7F_2$ transitions respectively.

All the emission bands are narrow in nature due to the characteristic electron shielding effect in trivalent rare-earth ions (Eu^{3+}) (Singh et al. 2016a, b). The intense emission peak observed at 615 nm is attributed to the electric dipole transition, which is hypersensitive in the host structure symmetry. The transitions, which are highly sensitive to the environment of the host matrix and become more intense, are generally called hypersensitive transitions. The parity allowed transitions at 594 nm is ascribed to the magnetic dipole transition of $^5D_0{\rightarrow}^7F_1$ and is insensitive to the site symmetry. It is worth noting that if the Eu^{3+} ions occupy an inversion symmetry center, only the magnetic dipole transition $^5D_0{\rightarrow}^7F_1$ is expected in the emission spectrum rather than electric dipole transition $^5D_0{\rightarrow}^7F_2$ (Raju et al. 2014b, c; Zhou et al. 2007; Mari et al. 2011). In the present case, the dominance of the $^5D_0{\rightarrow}^7F_2$ transitions indicates that the location of the Eu^{3+} ions deviates from the inversion symmetry i.e. at low symmetry sites (Chaker et al. 2003; Li et al. 2009; Karunadasa et al. 2005; Mari et al. 2011).

The photoluminescence intensity increases with increasing Eu^{3+} doping concentration. However, the PL emission intensity tends to decrease beyond a critical doping concentration ($x = 4$ mol%) in both the samples because of non-radiative energy transfer among Eu^{3+} ions (Fig. 9.6). This non-radiative energy transfer can occur generally due to exchange interaction, radiation reabsorption or multipole-multipole interaction.

Initially, the PL intensity increases monotonically with the increase in rare earth concentration. Then, a condition is encountered where the distance between two Eu^{3+} ions is such that the resonant energy transfer dominates,

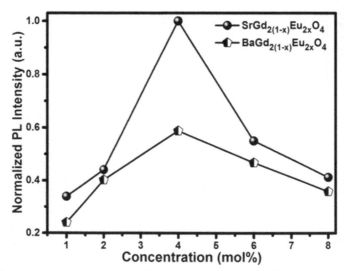

Fig. 9.6. Concentration quenching phenomenon of both the phosphors as a function of Eu^{3+} concentrations.

which will increase the non-radiative relaxation effectively resulting in the decreased PL intensity (Fig. 9.6). This effect is known as concentration quenching (Singh et al. 2016).

Comparison of Luminescence Characteristics of Eu^{3+} Doped $SrGd_2O_4$ and $BaGd_2O_4$ Phosphors

In order to understand the comparative luminescence behavior of phosphors, the normalized photoluminescence emission spectra of $SrGd_2O_4$:Eu^{3+} (4 mol%) and $BaGd_2O_4$:Eu^{3+} (4 mol%) were measured under the same excitation wavelength 276 nm. Although the two samples show the same characteristic emission transitions of Eu^{3+} ions from the excited state, $^5D_0$, to the ground states, ^7F_J (J = 1, 2, 3, 4) (Singh et al. 2016a, b), the relative intensity between each group of transitions in an individual sample is different. The interesting characteristics are as follows:

The PL emission spectra of $BaGd_2O_4$:Eu^{3+} sample shows the intense and dominant emission at 615 nm corresponding to $^5D_0 \rightarrow {}^7F_2$ transition. In addition, very weak emission bands were observed at 656 and 711 nm due to the transitions $^5D_0 \rightarrow {}^7F_3$ and $^5D_0 \rightarrow {}^7F_4$, respectively. It can be seen in Fig. 9.7 that the $^7F_1$ and $^7F_2$ energy levels of Eu^{3+} split into four (584, 586, 589 and 594 nm) and five (602.6, 611, 615, 626 and 631.7 nm) components, respectively, due to the completely lifted (2J+1) degeneracy. Generally, the $^7F_1$ energy levels show three splits, but in the present case the fourfold splitting in the $BaGd_2O_4$:Eu^{3+} phosphor confirms the fact that the Eu^{3+} emission comes from the occupation

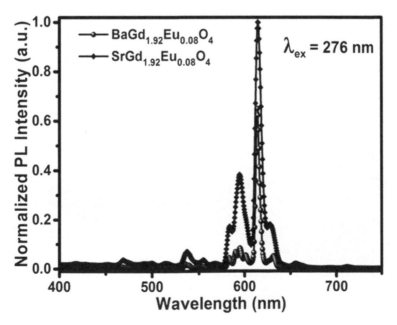

Fig. 9.7. Comparative PL emission spectra of $BaGd_{1.92}Eu_{0.08}O_4$ and $SrGd_{1.92}Eu_{0.08}O_4$ samples upon 276 nm UV excitation.

of Eu^{3+} into the two Gd^{3+} sites (one Gd site having C_s symmetry and other having non Cs symmetry). Similar results are reported by Raju et al. 2014 a, b. However, the PL emission spectra of $SrGd_2O_4:Eu^{3+}$ sample show only three (584, 594 and 598 nm (overlapped)) splits corresponding to $^7F_1$ energy levels which indicates that only one Gd^{3+} site may be occupied by Eu^{3+}, although, the position of the intense and dominant emission corresponding to $^5D_0{\rightarrow}^7F_2$ transition remains the same at 615 nm. In Fig. 9.7, it can be seen that the relative intensity of $^5D_0{\rightarrow}^7F_2$ transition, responsible for red emission in $SrGd_2O_4:Eu^{3+}$ sample is much higher in the $BaGd_2O_4:Eu^{3+}$ phosphor which indicates the more asymmetric environment around Eu^{3+} due to occupation of Gd^{3+} site having C_s symmetry in $SrGd_2O_4$ matrix than $BaGd_2O_4$ under 276 nm UV excitation.

Photometric Characterizations

Figure 9.8 displays the CIE (Commission Internationale de l'Eclairage) chromaticity diagram and corresponding digital images of $BaGd_{2(1-x)}Eu_{2x}O_4$ (x = 2, 4, 6 and 8 mol%) phosphors. The CIE parameters such as color coordinates (x, y) and color purity were calculated in order to know the photometric characteristics of the prepared samples. The color coordinates were obtained using the color calculator software (Som et al. 2014).

From the chromaticity diagram (Fig. 9.8 (a-b)), it can be seen that the color coordinates for both the samples traversed a wide range from

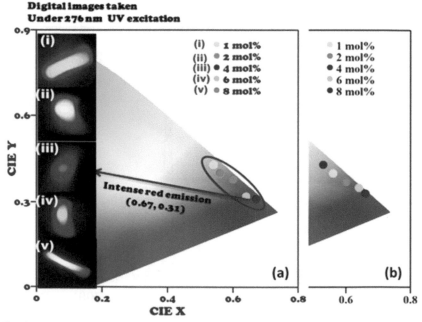

Fig. 9.8. Comparative CIE diagram for **(a)** $BaGd_{2(1-x)}Eu_{2x}O_4$ and **(b)** $SrGd_{2(1-x)}Eu_{2x}O_4$ (x = 1-8 mol%) samples (Inset shows the digital images of Eu^{3+} doped $BaGd_2O_4$ samples upon 276 nm UV excitation).

yellow to a deep red region on varying the Eu^{3+} concentration. The results demonstrate that the tuning of red color emission was possible by the alteration of Eu^{3+} concentrations. The intense red emission corresponding to the color coordinates (0.66, 0.33) and (0.67, 0.31) were observed for 4 mol% concentration of Eu^{3+} doped $SrGd_2O_4$ and $BaGd_2O_4$ respectively, which shows close approximation to the standard NTSC system (0.67, 0.33) and more saturated color coordinate values than the commercial red phosphor $Y_2O_2S:Eu^{3+}$ (0.65, 0.34) (Som et al. 2014). Afterwards, with the increase of Eu^{3+} concentration, the color coordinates shifted away from the deep red emission. The CIE parameters of Eu^{3+} doped phosphors are summarized in Table 9.1. It was reported that the variation in color coordinates occur due to the change of asymmetric ratio (R/O) which is defined as the ratio between the integrated intensity of the transitions $^5D_0{\rightarrow}^7F_2$ and $^5D_0{\rightarrow}^7F_1$ respectively (Som et al. 2014). The asymmetric ratio can be used as a local crystal field probe to measure the nature of the local Eu^{3+} surroundings. Thus this ratio indicates the degree of distortion from the inversion symmetry of the Eu^{3+} ion local environment. The emissions of Eu^{3+} will also contain the emission component from the upper excited levels of Eu^{3+}, but compared with the emissions of $^5D_0{\rightarrow}^7F_2$ and $^5D_0{\rightarrow}^7F_1$ of Eu^{3+} these contributions are very small and would not greatly affect the trends of R/O of Eu^{3+} (Bandi et al. 2012). The calculated symmetric ratios for both the samples are summarized in the Table 9.1. The asymmetric ratios of the samples were found to be the value more than 1, which is strong evidence that Eu^{3+} ions mainly occupy the lattice site without inversion symmetry (Xia et al. 2016; Tian et al. 2013; Luo et al. 2010; Lin et al. 2014; Kodaira et al. 2003; Judd 1962).

Normally, the light sources can be characterized by their dominant wavelength and color purity of emitted color. To investigate the effect on the color purity of the samples by the incorporation of Eu^{3+} ions, the color purity was calculated by using the following equation (Som et al. 2014; Dutta et al. 2015; Singh et al. 2016).

$$\text{Color purity} = \frac{\sqrt{(x_s - x_i)^2 + (y_s - y_i)^2}}{\sqrt{(x_d - x_i)^2 + (y_d - y_i)^2}} \times 100\%$$

The coordinates of a sample point are denoted by (x_s, y_s), (x_d, y_d) represent the coordinates of the dominant wavelength and the coordinates of illuminant point are denoted by (x_i, y_i). In the present study, (x_d, y_d) = (0.68, 0.32) and (x_i, y_i) = (0.3101, 0.3162) for the dominant emission wavelength at 615 nm (Singh et al. 2016a, b).

Judd-Ofelt Calculation

The Judd-Ofelt (J-O) theory, two identical formulations of a theory, was simultaneously and independently developed by Judd and Ofelt in 1962. The Judd-Ofelt theory is based on the static, free-ion and single configuration

Table 9.1. CIE coordinates and color purity of $SrGd_{2(1-x)}Eu_{2x}O_4$ phosphors

Samples	CIE coordinates		Color purity	Asymmetric ratio
	X	Y		
$SrGd_{2(1-x)}Eu_{2x}O_4$ (x = 1 mol%)	0.56	0.40	72%	2.90
$BaGd_{2(1-x)}Eu_{2x}O_4$ (x = 1 mol%)	0.54	0.43	48.1%	2.35
$SrGd_{2(1-x)}Eu_{2x}O_4$ (x = 2mol %)	0.60	0.37	80 %	2.85
$BaGd_{2(1-x)}Eu_{2x}O_4$ (x = 2mol %)	0.60	0.38	64.4 %	2.56
$SrGd_{2(1-x)}Eu_{2x}O_4$ (x = 4 mol%)	0.66	0.33	95%	2.59
$BaGd_{2(1-x)}Eu_{2x}O_4$ (x = 4 mol%)	0.67	0.31	94.68	2.98
$SrGd_{2(1-x)}Eu_{2x}O_4$ (x = 6 mol%)	0.64	0.35	89.6%	2.62
$BaGd_{2(1-x)}Eu_{2x}O_4$ (x = 6 mol%)	0.64	0.32	80%	2.75
$SrGd_{2(1-x)}Eu_{2x}O_4$ (x = 8 mol%)	0.53	0.43	67%	2.64
$BaGd_{2(1-x)}Eu_{2x}O_4$ (x = 8 mol%)	0.56	0.40	50.7%	2.48

approximations (Judd 1962; Ofelt 1962). Simply stated, the J-O theory significantly describes the spectral properties of Rare Earth ions (RE) in the solid states and is widely used for characterizing the optical transitions of trivalent RE ions (Judd 1962; Ofelt 1962; Marcantonatos 1986; Saraf et al. 2015). The detailed investigation of luminescence behavior and site symmetry of Eu^{3+} ions in $SrGd_2O_4$ and $BaGd_2O_4$ host was carried out by determining the Judd-Ofelt intensity parameters Ω_λ (λ = 2, 4, 6) (Judd 1962; Ofelt 1962). The intensity parameter Ω_2 determines the polarization and asymmetric behavior of the RE ligands (short range effect) whereas the other two parameters $\Omega_{4,6}$ depend on the long range effects. The radiative potential of RE ions in different hosts is generally revealed by the knowledge of these intensity parameters. These parameters are usually derived from absorption spectra, but Som et al. 2014 mentioned that it is tough to achieve an absorption spectrum of powder phosphors. However, owing to the special energy level structure of Eu^{3+} ions, the luminescence emission spectra can be used to calculate these intensity parameters (Zhang et al. 2012; Maekawa et al. 2007; Fu et al. 2006). Using the method described by Kodaira et al. 2003, these parameters can be calculated and used to estimate the spontaneous (radiative) decay rate of the emitting level $^5D_0$. According to this method, the radiative emission rates and integrated emission intensities are associated to each other by the relation (Judd 1962; Ofelt 1962),

$$\frac{A_{0-2,4}}{A_{0-1}} = \frac{I_{0-2,4}}{I_{0-1}} \frac{h\nu_{0-1}}{h\nu_{0-2,4}} \tag{1}$$

where I_{0-J} is the integrated emission intensity and $h\nu_{0-J}$ is the energy corresponding to the transitions $^5D_0 \rightarrow {}^7F_{1,2,4}$. As per Singh et al. 2016, the radiative emission rate A_{0-1} related to magnetic dipole transition $^5D_0 \rightarrow {}^7F_1$

has a value of ≈ 50 s^{-1}. Thus, the radiative emission rates related to forced electric dipole transitions can be obtained with the aid of A_{0-1} value. These radiative emission rates $A_{0-2,4}$ are also written as a function of the JO intensity parameters (Judd 1962; Ofelt 1962),

$$A_{0-J} = \frac{64\pi^4 \left(\nu_{0-2,4}\right)^3 e^2}{3hc^3} \frac{1}{4\pi\varepsilon_0} \chi \sum_{J=2,4,6} \Omega_J \left\langle {}^5D_0 \left| U^{(J)} \right| {}^7F_J \right\rangle^2 \tag{2}$$

where χ is the Lorentz local field correction factor as a function of the index of refraction (n) of the host $\chi = n(n^2+2)^2/9$, which convert the external electromagnetic field into an effective field at the location of the active center in the dielectric medium (Kodaira et al. 2003; Judd 1962; Ofelt 1962; Saraf et al. 2015). The value used as the refractive index of samples is 2.01 and 1.9 for SrGd$_2$O$_4$ and BaGd$_2$O$_4$, respectively calculated by using Singh et al. 2016. $\left\| U^{(\lambda)} \right\|$ is the squared reduced matrix elements whose values are independent of the chemical environment of the ion. For the case of Eu^{3+} ions, the non-zero square reduced matrix elements are solely $\left\langle {}^5D_0 \left| U^{(2)} \right\| {}^7F_2 \right\rangle^2 = 0.0032$ and $\left\langle {}^5D_0 \left| U^{(4)} \right| {}^7F_4 \right\rangle^2 = 0.0023$ (Judd 1962; Ofelt 1962). Thus, the values of $\Omega_{2,4}$ were obtained by using Eqs. (1) and (2). The value of Ω_6 could also be estimated by analyzing ${}^5D_0 \rightarrow {}^7F_6$ but in the present case this emission could not be detected due to instrumental limitations. From Eqn. (2), one can observe that each electric dipole ${}^5D_0 \rightarrow {}^7F_{2,4,6}$ transition depends only on square reduced matrix element. This unique feature of Eu^{3+} ion facilitates the determination of J-O parameters from the emission spectra. However, the remaining transitions from ${}^5D_0 \rightarrow {}^7F_{0,3,5}$ are forbidden both in magnetic and induced electric dipole schemes (since, the square reduced matrix element in both cases is zero).

The calculated J-O parameters have been used to derive some important radiative properties such as transition probabilities, radiative lifetime, branching ratios and lifetimes for the excited states of Eu^{3+} ions. The radiative transition probability of transition $\psi J \rightarrow \psi' J'$ can be calculated from the equation as $A_{rad} (\psi J, \psi' J') = A_{J-J'}$ (Judd et al. 1962; Ofelt et al. 1962; Saraf et al. 2015).

The total radiative transition probability (A_T) can be calculated by using the summation of radiative transition probability from a higher energy level to all the lower energy levels (Som et al. 2014):

$$A_T(\psi J) = \sum_{J'} A_{J-J'} \tag{3}$$

The radiative lifetime $\tau_{rad} (\psi J)$ of an excited state in terms of A_T, is given by (Judd 1962; Ofelt 1962):

$$\tau_{rad} (\psi J) = \frac{1}{A_T(\psi J)} \tag{4}$$

Table 9.2. Spectral intensity parameters of $SrGd_{2(1-x)}Eu_{2x}O_4$ and $BaGd_{2(1-x)}Eu_{2x}O_4$ phosphors

Samples	J-O intensity parameter	
	$\Omega_2 \, (pm^2)$	$\Omega_4 \, (pm^2)$
$SrGd_{2(1-x)}Eu_{2x}O_4$ (x = 1 mol%)	10.40	0.166
$BaGd_{2(1-x)}Eu_{2x}O_4$ (x = 1 mol%)	8.61	0.127
$SrGd_{2(1-x)}Eu_{2x}O_4$ (x = 2 mol%)	10.20	0.137
$BaGd_{2(1-x)}Eu_{2x}O_4$ (x = 2 mol%)	10.20	0.157
$SrGd_{2(1-x)}Eu_{2x}O_4$ (x = 4 mol%)	9.93	0.112
$BaGd_{2(1-x)}Eu_{2x}O_4$ (x = 4 mol%)	23.40	0.325
$SrGd_{2(1-x)}Eu_{2x}O_4$ (x = 6 mol%)	9.99	0.115
$BaGd_{2(1-x)}Eu_{2x}O_4$ (x = 6 mol%)	14.40	0.212
$SrGd_{2(1-x)}Eu_{2x}O_4$ (x = 8 mol%)	9.77	0.114
$BaGd_{2(1-x)}Eu_{2x}O_4$ (x = 8 mol%)	11.50	0.170

The branching ratio β (ψJ) corresponding to the emission from an excited level to its lowest lying levels is given by (Judd 1962; Ofelt 1962):

$$\beta(\psi J) = \frac{A_T \, (\psi J, \psi'J')}{A_T \, (\psi J)} \tag{5}$$

The Judd-Ofelt intensity parameters and the radiative parameters are presented in Tables 9.2 and 9.3. Since Ω_2 is most sensitive to the ligand environment, its value could reflect the asymmetry of the local environment at the Eu^{3+} ion site, while Ω_4 is related to the electron density on the surrounding ligands. The decrease in the Ω_4 parameter with Eu^{3+} concentration is also observed, which implies that the efficiency for the $^5D_0$ → $^7F_2$ transition increases. This indicates the enhancement of the red color emission (Som et al. 2014; Singh et al. 2016). In the $^5D_0$ → $^7F_2$ emission, the branching ratio (β) value is higher than the other transitions, indicating that the significant emission of Eu^{3+} belongs to $^5D_0$ → $^7F_2$ transition. This validates the pure red emission in the materials. The radiative lifetime for $^5D_0$ level is also calculated and summarized in this table. These results indicate that the Eu^{3+} doped $SrGd_2O_4$ and $BaGd_2O_4$ system can be effective luminescent materials.

Conclusion

Novel red-emitting Eu^{3+} doped $SrGd_2O_4$ and $BaGd_2O_4$ phosphors have been successfully synthesized by homogeneous precipitation method followed by a subsequent combustion process. All the samples revealed the orthorhombic phase formation and no phase change after doping was observed. The morphological studies of the samples revealed spherical and

Table 9.3. Spectral radiative parameters of $SrGd_{2(1-x)}Eu_{2x}O_4$ and $BaGd_{2(1-x)}Eu_{2x}O_4$ phosphors

Samples	$A_{0-2}(s^{-1})$	$A_{0-4}(s^{-1})$	$A_\tau(s^{-1})$	τ_{rad} (ms)	β_{0-1} (%)	β_{0-2} (%)	β_{0-4} (%)
$SrGd_{2(1-x)}Eu_{2x}O_4$ (x = 1 mol%)	82.49	0.853	133.3	7.49	37.49	61.86	0.639
$BaGd_{2(1-x)}Eu_{2x}O_4$ (x = 1 mol%)	56.65	0.544	107.2	9.32	46.63	52.86	0.507
$SrGd_{2(1-x)}Eu_{2x}O_4$ (x = 2 mol%)	80.90	0.704	131.6	7.59	37.99	61.47	0.535
$BaGd_{2(1-x)}Eu_{2x}O_4$ (x = 2 mol%)	67.26	0.667	117.9	8.48	42.40	57.03	0.565
$SrGd_{2(1-x)}Eu_{2x}O_4$ (x = 4 mol%)	78.73	0.577	129.3	7.73	38.66	60.88	0.446
$BaGd_{2(1-x)}Eu_{2x}O_4$ (x = 4 mol%)	154.2	1.384	205.6	4.86	24.31	75.00	0.673
$SrGd_{2(1-x)}Eu_{2x}O_4$ (x = 6 mol%)	79.19	0.593	129.7	7.70	38.52	61.01	0.457
$BaGd_{2(1-x)}Eu_{2x}O_4$ (x = 6 mol%)	94.50	0.905	145.4	6.87	34.38	64.99	0.622
$SrGd_{2(1-x)}Eu_{2x}O_4$ (x = 8 mol%)	77.45	0.588	128	7.81	39.04	60.49	0.459
$BaGd_{2(1-x)}Eu_{2x}O_4$ (x = 8 mol%)	75.87	0.727	126.6	7.89	39.49	60.00	0.574

rod-like agglomerates. The surface morphology was unaffected by Eu^{3+} doping. The photoluminescence spectra shows intense red emission for 4 mol% Eu^{3+} doped samples and after that concentration quenching occurs. The electric dipole transition $^5D_0 \rightarrow ^7F_2$ emerged as a dominant transition which is responsible for the intense red emission of the samples that can be attributed to the asymmetric surrounding of Eu^{3+} ions. Calculation of Judd-Ofelt intensity parameters Ω_2, Ω_4 and other spectral parameters confirmed the higher covalency and site symmetry of Eu^{3+} ion. The higher Ω_2 values suggested the dominance of $^5D_0 \rightarrow ^7F_2$ transition over magnetic dipole transition $^5D_0 \rightarrow ^7F_1$. Photometric characterization of the samples showed its applicability in lighting and display devices as the value of color coordinates are nearly equal to the standard red color (0.67, 0.33).

References

Antony, S.A., Nagaraja, K.S., Reddy, G.L.N., Sreedharan, O.M. 2001. A polymeric gel cum auto combustion method for the lower temperature synthesis of SrR_2O_4 (R = Y, La, Sm, Eu, Gd, Er or Yb). Mater. Lett. 51(5), 414-419.

Avci, N., Musschoot, J., Smet, P.F., Korthout, K., Avci, A., Detavernier, C., Poelman, D. 2009. Microencapsulation of moisture-sensitive CaS:Eu^{2+} particles with aluminum oxide. J. Electrochem. Soc. 156(11), J333-J337.

Bandi, V.R., Grandhe, B.K., Woo, H.J., Jang, K., Shin, D.S., Yi, S.S., Jeong, J.H. 2012. Luminescence and energy transfer of Eu^{3+} or/and Dy^{3+} co-doped in Sr$_3$AlO$_4$F phosphors with NUV excitation for WLEDs. J. Alloys Compd. 538, 85-90.

Barry, T.L., Roy, R. 1967. New rare earth-alkaline earth oxide compounds: Predicted compound formation and new families found. J. Inorg. Nucl. Chem. 29(5), 1243-1248.

Besara, T., Lundberg, M.S., Sun, J., Ramirez, D., Dong, L., Whalen, J.B., Siegrist, T. 2014. Single crystal synthesis and magnetism of the BaLn$_2$O$_4$ family (Ln = lanthanide). Prog. Solid State Chem. 42(3), 23-36.

Blasse, G. 1966. On the Eu^{3+} fluorescence of mixed metal oxides. IV. The photoluminescent efficiency of Eu^{3+}-activated oxides. J. Chem. Phys. 45(7), 2356-2360.

Bridot, J.L., Faure, A.C., Laurent, S., Riviere, C., Billotey, C., Hiba, B., Muller, R. 2007. Hybrid gadolinium oxide nanoparticles: Multimodal contrast agents for in vivo imaging. J. Am. Chem. Soc. 129(16), 5076-5084.

Carnall, W.T., Fields, P.R., Rajnak, K. 1968. Electronic energy levels in the trivalent lanthanide aquo ions. I. Pr^{3+}, Nd^{3+}, Pm^{3+}, Sm^{3+}, Dy^{3+}, Ho^{3+}, Er^{3+}, and Tm^{3+}. J. Chem. Phys. 49(10), 4424-4442.

Carnall, W.T., Fields, P.R., Rajnak, K. 1968. Spectral intensities of the trivalent lanthanides and actinides in solution. II. Pm^{3+}, Sm^{3+}, Eu^{3+}, Gd^{3+}, Tb^{3+}, Dy^{3+}, and Ho^{3+}. J. Chem. Phys. 49(10), 4412-4423.

Chaker, H., Kabadou, A., Toumi, M., Hassen, R.B. 2003. Rietveld refinement of the gadolinium strontium oxide SrGd$_2$O$_4$. Powder Diffr. 18(4), 288-292.

Das, S., Yang, C.Y., Lin, H.C., Lu, C.H. 2014. Structural and luminescence properties of tunable white-emitting Sr$_{0.5}$Ca$_{0.5}$Al$_2$O$_4$:Eu^{2+}, Dy^{3+} for UV-excited white-LEDs. RSC Adv. 4(110), 64956-64966.

De Vries, I.J.M., Lesterhuis, W.J., Barentsz, J.O., Verdijk, P., Van Krieken, J.H., Boerman, O.C., Bulte, J.W. 2005. Magnetic resonance tracking of dendritic cells in melanoma patients for monitoring of cellular therapy. Nature Biotechnol. 23(11), 1407.

Dubey, V., Kaur, J., Parganiha, Y., Suryanarayana, N.S., Murthy, K.V.R. 2016. Study of formation of deep trapping mechanism by UV, beta and gamma irradiated Eu^{3+} activated SrY$_2$O$_4$ and Y$_4$Al$_2$O$_9$ phosphors. Appl. Radiat. Isot. 110, 16-27.

Dutta, S., Som, S., Sharma, S.K. 2015. Excitation spectra and luminescence decay analysis of K$^+$ compensated Dy^{3+} doped CaMoO$_4$ phosphors. RSC Adv. 5(10), 7380-7387.

Fu, Z., Yang, H.K., Jeong, J.H., Zhang, S. 2008. Band structure calculations on orthorhombic bulk and nanocrystalline SrY$_2$O$_4$. J. Korean Phys. Soc. 52, 635.

Fu, Z., Zhou, S., Zhang, S. 2006. Preparation and optical properties of trivalent europium-doped bulk and nanocrystalline SrY$_2$O$_4$. J. Opt. Soc. Am. B 23(9), 1852-1858.

Judd, B.R. 1962. Optical absorption intensities of rare-earth ions. Phys. Rev. 127(3), 750.

Karunadasa, H., Huang, Q., Ueland, B.G., Lynn, J.W., Schiffer, P., Regan, K.A., Cava, R.J. 2005. Honeycombs of triangles and magnetic frustration in SrL$_2$O$_4$ (L = Gd, Dy, Ho, Er, Tm, and Yb). Phys. Rev. B 71(14), 144414.

Kikuchi, K. 2010. Design, synthesis and biological application of chemical probes for bio imaging. Chem. Soc. Rev. 39(6), 2048-2053.

Kim, T., Lee, N., Park, Y.I., Kim, J., Kim, J., Lee, E.Y., Na, H.B. 2014. Mesoporous silica-coated luminescent Eu^{3+} doped $GdVO_4$ nanoparticles for multimodal imaging and drug delivery. RSC Adv. 4(86), 45687-45695.

Kodaira, C.A., Brito, H.F., Felinto, M.C.F. 2003. Luminescence investigation of Eu^{3+} ion in the $RE_2(WO_4)_3$ matrix (RE = La and Gd) produced using the Pechini method. J. Solid State Chem. 171(1-2), 401-407.

Kolar, D., Skapin, S.D., Suvorov, D. 1999. Phase equilibria in the system $BaO-TiO_{~2}-Gd_{~2}O_{~3}$. Acta Chim. Slov. 46(2), 193-202.

Lakshminarasimhan, N., Varadaraju, U.V. 2008. Role of crystallite size on the photoluminescence properties of $SrIn_2O_4$:Eu^{3+} phosphor synthesized by different methods. J. Solid State Chem. 181(9), 2418-2423.

Li, H., Zhang, S., Zhou, S., Cao, X. 2009. Chemical bond characteristics, thermal expansion property and compressibility of AR_2O_4 (A = Ca, Sr, Ba; R = rare earths). Mater. Chem. Phys. 114(1), 451-455.

Lin, H.C., Yang, C.Y., Das, S., Lu, C.H. 2014. Red-emission improvement of Eu^{2+}–Mn^{2+} co-doped $Sr_2Si_5N_8$ phosphors for white light-emitting diodes. Ceram. Int. 40(8), 12139-12147.

Lojpur, V., Antić, Ž., Dramićanin, M.D. 2014. Temperature sensing from the emission rise times of Eu^{3+} in SrY_2O_4. Phys. Chem. Chem. Phys. 16(46), 25636-25641.

Luo, W., Liao, J., Li, R., Chen, X. 2010. Determination of Judd–Ofelt intensity parameters from the excitation spectra for rare-earth doped luminescent materials. Phys. Chem. Chem. Phys. 12(13), 3276-3282.

Maekawa, T., Kurosaki, K., Yamanaka, S. 2007. Thermophysical properties of BaY_2O_4: A new candidate material for thermal barrier coatings. Mater. Lett. 61(11-12), 2303-2306.

Marcantonatos, M.D. 1986. Multiphonon non-radiative relaxation rates and Judd–Ofelt parameters of lanthanide ions in various solid hosts. J. Chem. Soc. Faraday Trans. 2 82(3), 381-393.

Mari, B., Singh, K.C., Sahal, M., Khatkar, S.P., Taxak, V.B., Kumar, M. 2011. Characterization and photoluminescence properties of some MLn_2 (1 − x) O_4: $2xEu^{3+}$ or $2xTb^{3+}$ systems (M = Ba or Sr, Ln = Gd or La). J. Lumin. 131(4), 587-591.

Ofelt, G.S. 1962. Intensities of crystal spectra of rare-earth ions. J. Chem. Phys. 37(3), 511-520.

Park, S.J., Park, C.H., Yu, B.Y., Bae, H.S., Kim, C.H., Pyun, C.H. 1999. Structure and luminescence of SrY_2O_4:Eu. J. Electrochem. Soc. 146(10), 3903-3906.

Pavitra, E., Raju, G.S.R., Ko, Y.H., Yu, J.S. 2012. A novel strategy for controllable emissions from Eu^{3+} or Sm^{3+} ions co-doped SrY_2O_4:Tb^{3+} phosphors. Phys. Chem. Chem. Phys. 14(32), 11296-11307.

Poelman, D., Vercaemst, R., Van Meirhaeghe, R.L., Laflère, W.H., Cardon, F. 1993. Effect of moisture on performance of SrS:Ce thin film electroluminescent devices. Jpn. J. Appl. Phys. 32(8R), 3477.

Raju, G.S.R., Yu, J.S. 2014a. Novel orange and reddish-orange color emitting $BaGd_2O_4$:Sm^{3+} nanophosphors by solvothermal reaction for LED and FED applications. Spectrochim Acta A Mol. Biomol. Spectrosc. 124, 383-388.

Raju, G.S.R., Pavitra, E., Yu, J.S. 2014b. Cross-relaxation induced tunable emissions from the Tm^{3+}/Er^{3+}/Eu^{3+} ions activated $BaGd_2O_4$ nanoneedles. Dalton Trans. 43(25), 9766-9776.

Raju, G.S.R., Pavitra, E., Yu, J.S. 2014c. Pechini synthesis of lanthanide (Eu^{3+}/Tb^{3+} or Dy^{3+}) ions activated $BaGd_2O_4$ nanostructured phosphors: An approach for tunable emissions. Phys. Chem. Chem. Phys. 16(34), 18124-18140.

Reddy, A.A., Das, S., Ahmad, S., Babu, S.S., Ferreira, J.M., Prakash, G.V. 2012. Influence of the annealing temperatures on the photoluminescence of $KCaBO_3:Eu^{3+}$ phosphor. RSC Adv. 2(23), 8768-8776.

Saraf, R., Shivakumara, C., Behera, S., Nagabhushana, H., Dhananjaya, N. 2015. Photoluminescence, photocatalysis and Judd–Ofelt analysis of Eu^{3+}-activated layered BiOCl phosphors. RSC Adv. 5(6), 4109-4120.

Sasabe, H., Kido, J. 2013. Development of high performance OLEDs for general lighting. J. Mater. Chem. C 1(9), 1699-1707.

Sieber, K.D., Todd, L.B., Sever, B.R. 1995. U.S. Patent No. 5,391,884. Washington, DC: U.S. Patent and Trademark Office.

Singh, D., Tanwar, V., Bhagwan, S., Sheoran, S., Nishal, V., Samantilleke, A.P., ... Kadyan, P.S. 2016. Optoelectronic characterization of trivalent europium doped Gd_2O_3 and MGd_2O_4 (M = Ba or Sr) nanophosphors for display device applications. J. Nanoelectron. Optoelectron. 11(3), 305-310.

Singh, J., Manam, J. 2016a. Structural and spectroscopic behaviour of Eu^{3+}-doped $SrGd_2O_4$ modified by thermal treatments. J. Mater. Sci. 51(6), 2886-2901.

Singh, J., Manam, J. 2016b. Synthesis, crystal structure and temperature dependent luminescence of Eu^{3+} doped $SrGd_2O_4$ host: An approach towards tunable red emissions for display applications. Ceram. Int. 42(16), 18536-18546.

Smet, P.F., Moreels, I., Hens, Z., Poelman, D. 2010. Luminescence in sulfides: A rich history and a bright future. Materials 3(4), 2834-2883.

Smet, P.F., Parmentier, A.B., Poelman, D. 2011. Selecting conversion phosphors for white light-emitting diodes. J. Electrochem. Soc. 158(6), R37-R54.

Som, S., Das, S., Dutta, S., Visser, H.G., Pandey, M.K., Kumar, P., Sharma, S.K. 2015. Synthesis of strong red emitting $Y_2O_3:Eu^{3+}$ phosphor by potential chemical routes: Comparative investigations on the structural evolutions, photometric properties and Judd–Ofelt analysis. RSC Adv. 5(87), 70887-70898.

Som, S., Kunti, A.K., Kumar, V., Dutta, S., Chowdhury, M., ... Swart, H.C. 2014a. Defect correlated fluorescent quenching and electron phonon coupling in the spectral transition of Eu^{3+} in $CaTiO_3$ for red emission in display application. J. Appl. Phys. 115(19), 193101.

Som, S., Mitra, P., Kumar, V., Kumar, V., Terblans, J.J., Swart, H.C., Sharma, S.K. 2014b. The energy transfer phenomena and colour tunability in $Y_2O_2S:Eu^{3+}/Dy^{3+}$ micro-fibers for white emission in solid state lighting applications. Dalton Trans. 43(26), 9860-9871.

Srinivas, M., Rao, B.A. 2012. Luminescence studies of Eu^{3+} doped $BaGd_2O_4$ phosphor. Indian J. Sci. Technol. 5(7), 3022-3026.

Sun, X.Y., Liu, Y., Liu, X.L., Cao, R.P., Li, Y.N., Lin, L.W. 2014. Substitution of Y^{3+} for Gd^{3+} on the luminescent properties of $BaGd_2O_4:Eu^{3+}$ scintillating phosphors. Opt. Mater. 36(9), 1478-1483.

Sun, X.Y., Wang, W.F., Sun, S.Q., Lin, L.W., Li, D.Y., Zhou, L.P. 2013a. Synthesis and luminescent properties of novel $BaGd_2O_4:Eu^{3+}$ scintillating phosphor. Luminescence 28(3), 384-391.

Sun, X.Y., Zhou, Y.Z., Yu, X.G., Chen, H.H., Wang, H., Zhang, Z.J., ... Zhao, J.T. 2013b. Synthesis and luminescent properties of $BaGd_2O_4:Dy^{3+}$, an novel scintillating phosphor. Appl. Phys. B, 110(1), 27-34.

Tian, B., Chen, B., Tian, Y., Li, X., Zhang, J., Sun, J., ... Wang, Y. 2013. Excitation pathway and temperature dependent luminescence in color tunable $Ba_5Gd_8Zn_4O_{21}:Eu^{3+}$ phosphors. J. Mater. Chem. C 1(12), 2338-2344.

Tian, Y., Chen, B., Hua, R., Yu, N., Liu, B., Sun, J., ... Tian, B. 2012. Self-assembled 3D flower-shaped NaY (WO$_4$)$_2$:Eu^{3+} microarchitectures: Microwave-assisted hydrothermal synthesis, growth mechanism and luminescent properties. Cryst. Eng. Comm. 14(5), 1760-1769.

Wang, D., Wang, Y., Wang, L. 2007a. Photoluminescence properties of Sr (Y, Gd)$_2$O$_4$:Eu^{3+} under VUV excitation. J. Lumin. 126(1), 135-138.

Wang, H., Tian, L. 2011. Luminescence properties of SrIn$_2$O$_4$:Eu^{3+} incorporated with Gd^{3+} or Sm^{3+} ions. J. Alloys Compd. 509(6), 2659-2662.

Wang, W.N., Widiyastuti, W., Ogi, T., Lenggoro, I.W., Okuyama, K. 2007b. Correlations between crystallite/particle size and photoluminescence properties of submicrometer phosphors. Chem. Mater. 19(7), 1723-1730.

Xia, Z., Miao, S., Molokeev, M.S., Chen, M., Liu, Q. 2016. Structure and luminescence properties of Eu^{2+} doped Lu$_x$Sr$_{2-x}$SiN$_x$O$_{4-x}$ phosphors evolved from chemical unit cosubstitution. J. Mater. Chem. C 4(6), 1336-1344.

Zhang, J., Wang, Y., Guo, L., Huang, Y. 2012. Vacuum ultraviolet–ultraviolet, X-ray, and near-infrared excited luminescence properties of SrR$_2$O$_4$:RE^{3+} (R = Y and Gd; RE = Tb, Eu, Yb, Tm, Er, and Ho). J. Am. Ceram. Soc. 95(1), 243-249.

Zhang, Y., Geng, D., Shang, M., Zhang, X., Li, X., Cheng, Z., Lin, J. 2013. Soft-chemical synthesis and tunable luminescence of Tb^{3+}, Tm^{3+}/Dy^{3+}-doped SrY$_2$O$_4$ phosphors for field emission displays. Dalton Trans. 42(14), 4799-4808.

Zhou, J., Yu, M., Sun, Y., Zhang, X., Zhu, X., Wu, Z., Li, F. 2011. Fluorine-18-labeled Gd^{3+}/Yb^{3+}/Er^{3+} co-doped NaYF$_4$ nanophosphors for multimodality PET/MR/UCL imaging. Biomaterials 32(4), 1148-1156.

Zhou, L., Shi, J., Gong, M. 2007. Synthesis and luminescent properties of BaGd$_2$O$_4$:Eu^{3+} phosphor. J. Phys. Chem. Solids 68(8), 1471-1475.

Comparative Study on Synthesis of Gd$_2$O$_3$:Eu Phosphors by Various Synthesis Routes

Ruby Priya[1]*, O.P. Pandey[1] and Vikas Dubey[2]

[1] School of Physics & Materials Science, Thapar Institute of Engineering & Technology, Patiala - 147004, Punjab
[2] Department of Physics, Bhilai Institute of Technology Raipur, Kendri - 493661, India

Introduction

Nanoparticles have sparked intense interest in the field of material science due to their size dependent properties. The physical and chemical properties vary drastically with size variation and thus making them suitable for different applications (Siegel 1993; Siegel et al. 1998; Sun and Murray 2010). At nanometer size range, properties of crystallites are influenced by the presence of significant number of surface atoms. Due to a quantum confinement effect in these nanoparticles, the electronic states become more channelized which exhibit novel properties compared with their corresponding bulk phases (Kruis et al. 1998; Suryanarayana 1995; Wakefield et al. 1999).

Among the various nanoparticles, luminescent nanoparticles have attracted increasing technological and industrial interest. This interest has been mainly due to their novel optical properties that affect emission life time, luminescence quantum efficiency and concentration quenching. Luminescent materials, also called phosphors, are defined as the solids that can absorb and convert certain types of energy into radiation of light (Hase et al. 1990). In addition nanosized phosphors offer a number of potential advantages over traditional micro-scale phosphors in optical properties, such as high luminescence efficiency and high-resolution images, which contribute to their high surface to volume ratio and quantum size effects of the nanoparticles. Among all type of phosphors, Rare Earth (RE) materials

---
*Corresponding author: rubypriya1994@gmail.com

are of importance and attractive candidates. These usually exist in a trivalent state. With abundant f-orbital configurations, lanthanides (Ln) ion can exhibit sharp fluorescent emissions via intra-4f or 4f-5d transitions and thus are widely used as emitting species in various phosphors. Rare-earth oxide phosphors have been recognized to hold tremendous applications in the field of high performance luminescent devices, optoelectronic devices, sensors, catalysts, MRI contrast agents and other functional materials (Li et al. 2011; Perkins and Kaufmann 1935; Vrankic et al. 2015; Yang et al. 2018; Yanli et al. 2015).

Recently, the interest in rare earth doped oxide materials has increased due to their excellent luminescent efficiency, color purity, chemical, and thermal stability. Varieties of rare earth hydroxides and oxide compounds, such as $Y(OH)_3$, $Dy(OH)_3$, $Tb(OH)_3$, Y_2O_3, Eu_2O_3 and Gd_2O_3 have been synthesized. Among these rare-earth oxides, Gd_2O_3 has been identified as a promising host matrix for optical studies due to its good chemical, photo thermal and photo chemical stabilities as well as low phonon energy (Dhananjaya et al. 2010; Li et al. 2014; Liu et al. 2007; Maalej et al. 2015; Park et al. 2000; Tseng et al. 2010; Wang et al. 2014; Xing et al. 2014; Zhou et al. 2014). Moreover, Gd^{3+} is one of the very few RE^{3+} ions in the rare earth family, which acts as a luminescence sensitizer in the host. However, the research on Gd_2O_3 phosphors has been focused on the Eu^{3+} doped one dimensional nanostructures like, nanotubes, nanowires and nanofibres. These phosphors ($Gd_2O_3:Eu^{3+}$) exhibit a strong paramagnetic behavior (S = 7/2) as well as strong UV and cathode-ray excited luminescence, which are useful in a biological fluorescent label, contrast agent, and display applications. In addition, $Gd_2O_3:Eu$ is a very efficient X-ray and thermos-luminescent phosphor.

Europium ion in a trivalent state is one of the most studied rare earth elements because of the simplicity of its emission spectra and due to the wide application as red phosphor in color TV screens. Eu^{3+} *f-f* transitions are sensitive to its local environment. The monitoring of different concentrations of the Eu^{3+} content into a ceramic material is very interesting in understanding the nature of the lattice modifiers as well as the degree of order-disorder into its crystalline structure. The most intense *f-f* transition is the $^7D_0 \rightarrow F_2$ transition at 616 nm. When this ion is presented in a non-centrosymmetric site, it can be used as an activator ion with red emission which has been used in the most commercial red phosphor. Because of this, it is a promising candidate for biological sensors, phosphors, electroluminescent devices, optical amplifiers or lasers when it is used as a dopant in a variety of ceramic materials (Blasse 1994).

So far, several methods have been employed to fabricate phosphors of a small size and high quality $Gd_2O_3:Eu$ with reduced reaction temperature, such as hydrothermal methods (Lee et al. n.d.; Liu et al. 2007; Wang et al. 2014), precipitation methods (Priya and Pandey 2019), solution combustion methods (Dhananjaya et al. 2010; Li and Hong 2007), sol-gel methods (Jain and Hirata 2016) and spray pyrolysis methods (Iwako et al. 2010).

Literature Review

In order to gather information about the work done so far on the morphological and luminescent properties of Gd_2O_3:Eu, literature survey has been done in detail which is summarized below.

Pang et al. in 2003 synthesized Gd_2O_3:A (A= Eu^{3+}, Dy^{3+}, Sm^{3+}, Er^{3+}) phosphor films via Pechlini sol-gel soft lithography. XRD, AFM, SEM, Optical microscopy, UV/Vis transmission and photoluminescence (PL) spectra were used to characterize the films. It was observed that crystallinity increased with rise in temperature. The average size of the film was 70 nm and thickness 550 nm. Both the lifetime and PL intensity of rare earth ions increased with annealing temperature from 500 to 900 °C for the concentrations of Eu^{3+}, Dy^{3+}, Sm^{3+}, Er^{3+} 5, 0.25, 1 and 1.5% of Gd^{3+} in Gd_2O_3 films, respectively. Upon excitation into the host band at 230 nm, the rare earth ions Eu^{3+}, Dy^{3+}, Sm^{3+}, Er^{3+} show their characteristic red ($^5D_0$-$^7F_2$), yellow ($^4F_{9/2}$-$^6H_{13/2}$), orange ($^4G_{5/2}$-$^6H_{7/2}$) and green ($^4S_{3/2}$-$^4I_{15/2}$) emissions in crystalline Gd_2O_3 phosphor films.

Louis et al. in 2003 synthesized Gd_2O_3:Eu^{3+} phosphors by a sol-lyophilization technique. For temperatures lower than 1300K, highly crystalline samples with the cubic structure were obtained without concomitant growth of the particles (<50 nm). During the study, it was found that Eu^{3+} doping of 5% enabled an increase in the luminescent intensity as high as 3.5%. Here also, with an additional peak at about 609 nm at the vicinity of $^5D_0$-$^7F_{0...4}$ transitions was observed. This may be attributed to the size effect which disappeared with increase in size.

Bazzi et al. in 2004 synthesized luminescent nano-sized Eu_2O_3, Gd_2O_3:Eu and Y_2O_3:Eu particles by direct colloidal precipitation route at low temperature of 180 °C. The particles formed were ultrafine, equi-axed and monodisperse with mean grain diameter in the range 2-5 nm. The as-synthesized particles were characterized by HR-TEM, EDX, absorption spectroscopy and luminescent spectroscopy. High-resolution TEM showed that the nanocrystals formed were highly crystalline with cubic structure. The colloidal particles were well stabilized against agglomeration. Optical absorption spectra and luminescence spectra of Eu^{3+} based oxide was studied which corresponded to the known $^5D_0 \rightarrow {}^7F_n$. This indicated that luminescent centers were formed efficiently by this direct high-boiling alcohol precipitation. The luminescence spectra showed some important differences from those of the bulk materials, proving that reducing the scale to the nanometer range effectively leads to new optical properties.

Yang et al. in 2005 synthesized rare earth (Eu^{3+} - Er^{3+}) doped Gd_2O_3 nanocrystals using hot solution chemistry. Nanocrystals of Gd_2O_3 exhibited an anisotropic two-dimensional growth, leading to square-shaped nanoplates with an 11-16 nm edge length and 1.05 nm edge thickness. The crystals were characterized by XRD, TEM and PL. Gd_2O_3:Er^{3+} nanocrystals show characteristic peak near-infrared emissions at 1450-1650 nm due to $^4I_{13/2}$-$^4I_{15/2}$ transition. Considerably lower emission intensity of Gd_2O_3:Eu^{3+} nanocrystals were observed. This may be due to the fact that Eu^{3+} ions could possess a different, disordered environment compared to those in the center of bulk crystalline material, leading to inhomogeneous broadening of emission.

Lin and Li in 2006 synthesized one dimensional Gd$_2$O$_3$:Eu^{3+} phosphor nanowire by combining the sol-gel method and hydrothermal reactions followed by a sintering process at 1000 °C. In this silica material SBA-15 was used as the limiting reactor. Crystal structure was confirmed by XRD and morphology by TEM. The average size of the nanowires was found to be in 7-9 nm range. In comparison to bulk Gd$_2$O$_3$:Eu^{3+} materials, it was found that the photo-luminescent properties of the nanowires were different exhibiting size dependent properties. The main peaks were observed at 585, 597, 613 and 620 nm.

Engstrom et al. in 2006 synthesized Gd$_2$O$_3$ nanocrystals by the colloidal method. Diethylene glycol (DEG) was used as the capping agent. The nanocrystals formed were largely crystalline with sizes in the range of 5-10 nm. The oxidation state of the Gd$_2$O$_3$ was found by X-ray Photoelectron Spectroscopy (XPS). The ultra-small Gd$_2$O$_3$ particles were used to study the proton reflexivity. The pH dependency was investigated since the proton relaxivity of Gd chelates can be sensitive to pH. The higher relaxation rate at low Gd concentrations in Gd$_2$O$_3$-DEG nanoparticles solutions was found to be good that led to enhancements even at low doses.

Yang et al. in 2007 synthesized one dimensional Gd$_2$O$_3$:Eu nano and micro rods through a large scale and facile hydrothermal method followed by subsequent heat treatment process. X-Ray Diffraction (XRD), thermogravimetric analysis, differential scanning calorimetry (TGA-DSC), Scanning Electron Microscopy (SEM), Transmission Electron Microscopy (TEM), High Resolution Transmission Electron Microscopy (HRTEM), Selected Area Electron Diffraction (SAED), photoluminescence (PL) and Cathode-Luminescence (CL) spectra were used to characterize the samples. The size of the rods could be modulated from micro to nano size with an increase of pH using ammonia solution. Gd(OH)$_3$ rod-like crystals formed during the hydrothermal process can be used as good templates to synthesize 1D functional rare earth compounds such as Gd$_2$O$_2$S$_2$, by careful control of synthesis process. Due to the strong red emission corresponding to the $^5D_0$ -$^7F_2$ transition (610 nm) of Eu^{3+} under UV excitation (257 nm), low-voltage electron beams excitation (1-5 kV) and excellent dispersing properties of the obtained Gd$_2$O$_3$:Eu^{3+} rod-like phosphors, they are potentially applied in fluorescent lamps and field emission displays.

Lee et al. in 2008 synthesized various nanostructures of Gd(OH)$_3$:Eu (the molar ratio of Eu^{3+} = 0.04) at different pH by using the hydrothermal method. Subsequent dehydration at 500 °C led to the formation of Gd$_2$O$_3$:Eu. X-Ray Diffraction (XRD), Field-emission electron microscopy (FE-SEM), and Photoluminescent spectra were used to characterize the samples. The aspect ratios of phosphor particles are tunable by simply adjusting the pH of the initial solution during hydrothermal synthesis of Gd(OH)$_3$:Eu. It was observed that the selective control of Gd(OH)$_3$: Eu morphology provides a strategy for the selective control of one-dimensional oxide nano-phosphor Gd_2O_3:Eu.

Bai et al. in 2009 synthesized Gd_2O_3:Er^{3+}/Gd_2O_3:Yb^{3+} core-shell nanorods by the hydrothermal method. The average diameters of the NRs were approximately 20 nm and lengths were 150-200 nm. X-ray diffraction (XRD), Transmission Electron Microscopy (TEM), High Resolution Transmission Electron Microscopy (HRTEM) and up-conversion spectra were used to characterize the samples. The thickness of the Gd_2O_3:Yb^{3+} shells on the Gd_2O_3:Er^{3+} cores were 5 nm. Green emissions of $^2H_{11/2}$, $^4S_{3/2}$-$^4I_{15/2}$ and red emissions of $^4F_{9/2}$-$^4I_{15/2}$ were observed. Relative intensity from red to green and intensity ratio of $^2H_{11/2}$-$^4I_{15/2}$ to $^2H_{11/2}$-$^4I_{15/2}$ decreased in the core shell NRs compared to the bare NRs.

Jia et al. in 2009 synthesized highly uniform $Gd(OH)_3$ and Gd_2O_3:Eu^{3+} nanotubes by simple wet-chemical route at ambient pressure and low temperature. The diameter of the nanotubes was about 40 nm and lengths of 200-300 nm. The samples were characterized by XRD, EXD, TEM, SEM, SAED, PL, CL techniques. XRD techniques showed that crystallinity of the samples increased with a rise in temperature. SAED showed the hexagonal structure of $Gd(OH)_3$. Gd_2O_3:Eu^{3+} nanotubes exhibited strong red emission corresponding to the $^5D_0$-$^7F_2$ transition of the Eu^{3+} ions under UV light.

Jia et al. in 2010 synthesized highly uniform Gd_2O_3 hollow microspheres by template-directed synthesis. Melamine Formaldehyde (MF) was used as a template. The as-obtained Gd_2O_3 microspheres with a spherical shape and hollow structure having a uniform size distribution and the thickness of about 200 nm were obtained. The morphology, crystal structure and luminescent properties were characterized by XRD, FTIR, EDX, TGA-DSC, SEM, TEM and PL techniques. Gd_2O_3:Eu^{3+} hollow microspheres show strong red emission under UV light excitation. The Gd_2O_3:Er^{3+}, Gd_2O_3:Yb^{3+}/Er^{3+}, and Gd_2O_3:Yb^{3+}/Tm^{3+} samples exhibited bright green, red and blue colors under UV excitation, respectively. These hollow microspheres had multiple applications in drug delivery for biological applications.

Tian et al. in 2011 fabricated rare-earth doped Gd_2O_3 hollow spheres via the template-directed method using hydrothermal carbon spheres as sacrificed templates. SEM and TEM revealed that these hollow structure nanospheres have mesoporous shells that were composed of a large amount of uniform NPs. Crystal structure and other properties were characterized by XRD, EDX, FTIR and UV/Vis spectrometer. *In-vivo* UCL imaging and *in-vitro* relaxivity measurements demonstrated that the as-prepared Gd_2O_3:Yb/Er hollow spheres could serve as dual-imaging agent for optical properties. Drug release properties were investigated by using an IBU model.

Gai et al. in 2011 synthesized monodispersed and multicolour Gd_2O_3:Ln (Ln = Eu^{3+}, Tb^{3+}, Dy^{3+}, Sm^{3+}, Yb^{3+}/Er^{3+}, Yb^{3+}/Tm^{3+} and Yb^{3+}/Ho^{3+}) nanocrystals by homogeneous precipitation methods. XRD, SEM, TEM, FT-IR, TG-DTA, PL and CL were employed to characterize the samples. The crystals were of nanosize with a uniform diameter of 80 nm. The as-prepared Gd_2O_3:Ln phosphors demonstrated strong and multi-colored DC and UC emissions under ultraviolet and NIR excitations. Mesoporous silica coated Gd_2O_3:Eu^{3+} composites showed drug release properties suggesting the feasible application of luminescent NCs in biomedical fields.

Dhananjaya et al. in 2012 synthesized Bi^{3+} co-doped Gd_2O_3:Eu^{3+} nanophosphors by the combustion method. The nanophosphors were characterized by XRD, FTIR, TEM, SAED, EDX and UV-Vis. Spectroscopy and PL. XRD confirmed that the calcined products were in monoclinic with little cubic phases. The photoluminescence spectra of the synthesized phosphors excited with 230 nm showed emission peaks nearly at 590 nm, 612 and 625 nm. It was observed that a significant quenching was observed under 230 nm excitation when Bi^{3+} was co-doped. However, 5 mol% incorporation of the Bi^{3+} ions increased the luminescent intensity of the Gd_2O_3:Eu^{3+} nanophosphors. Using thermos-luminescence glow peaks, the trap parameters were evaluated and studied. The results revealed that the Gd_2O_3:Eu^{3+}, Bi^{3+} phosphors have promising applications in solid-state lightening.

Hazarika et al. in 2013 synthesized cubic-phase Gd_2O_3:Eu nanorods by the hydrothermal method. The as-synthesized nanorods were characterized by X-ray diffraction (XRD), High Resolution Transmission Electron Microscopy (HRTEM), optical absorption spectroscopy, photoluminescence (PL) spectroscopy, Raman spectroscopy and magnetic hysteresis measurements. The particle size was found out to be in the range of 5-6.5 nm. HRTEM micrographs revealed that the diameter of the nanorods prepared at pH = 13.3 (~7 nm) was much smaller than the rods prepared at pH = 10.8 (~19 nm). Furthermore, $M \sim H$ plot of the nanorod system (pH = 10.8) exhibited slight departure from the ideal superparamagnetic behavior, with low remanence and coercive field values.

Tamrakar et al. in 2014 synthesized Gd_2O_3 nanophosphors by combustion synthesis using gadolinium nitrate hexahydrate as a precursor and urea as fuel. Structural and surface morphology were studied by X-ray diffraction, transmission electron microscopy and scanning electron microscopy. Chemical analysis was done by FT-IR and optical properties were studied by photoluminescence techniques. The average size was observed to be 41 nm. In PL spectra, feeble emission at 490 nm (blue) and intense emission at approximately 545 nm (green) were observed after excitation at 300 nm.

Ren et al. in 2015 synthesized hollow Gd_2O_3:Eu nanospheres by the self-template method in which $Gd(OH)CO_3$:Eu acted as a template. The as-synthesized samples were characterized by XRD, TEM, FT-IR, PL and VSM. The effect of calcination temperature on the luminescence and magnetic properties were reported. The diameter of the prepared Gd_2O_3:Eu^{3+} nanospheres calcined at 600 °C was 200 nm and the thickness of Gd_2O_3:Eu^{3+} was about 20 nm. The luminescent intensity of hollow nanospheres was about 1.3 times compared to solid particles. Luminescent spectra showed that the I(610)/I(590) (asymmetry factor of luminescence) varied with the calcination temperature, indicating that the color purity and local environment of Eu^{3+} in Gd_2O_3 changes. The Magnetisation (M) of the samples also became stronger with increasing calcination temperature, which may be contributed to the rearrangement of Gd^{3+} and Eu^{3+}.

Jiang et al. in 2016 synthesized Gd_2O_3:Eu hollow microspheres by urea-assisted homogenous precipitation process using carbon spheres as a template followed by subsequent heat treatment. The as-synthesized particles were characterized by X-ray diffraction, Fourier transformed infrared spectroscopy, thermogravimetry, X-ray photoelectron spectroscopy, scanning electron microscopy, transmission electron microscopy and Brunauer-Emmett-Tellet surface area measurement. The results indicated that the final products can be indexed to a cubic Gd_2O_3 phase with high purity and have a uniform morphology at 500 nm in diameter and 20 nm in shell thickness. The as-synthesized Gd_2O_3 hollow microspheres exhibited a superior photo-oxidation activity to that of Gd_2O_3 powder and an effect similar to P25, significantly broadening the potential of Gd_2O_3 hollow microspheres.

Tamrakar et al. in 2017 synthesized Gd_2O_3:Eu by a solid state method. Structural properties of the as-synthesized particles were studied by transmission electron microscopy and X-ray diffraction analysis. The Photoluminescence (PL) technique was used to study luminescent properties of the material and its optical features were thoroughly investigated by calculating CIE chromaticity. The emission intensity for 613 nm, 584 nm and 467 nm were recorded as the function of dopant ion concentration. It was observed that the blue emission quenched at 1 mol% Eu^{3+} whereas red and green quenched after 2 mol% of Eu^{3+}.

Adam et al. in 2017 synthesized Y_2O_3:Eu and Gd_2O_3:Eu particles of various sizes using the wet chemistry route. The as-synthesized particles were characterized by XRD, SEM, TEM and PL spectroscopy. It was found that there is a size effect which is not dependent on the calcination temperature. These particles were characterized using three excitation methods: UV light at 250 nm wavelength, electron beam at 10 kV and X-rays generated at 100 kV. Regardless of the excitation source, it was found that with increasing particle diameter there is an increase in emitted light. Furthermore, dense particles emitted more light than porous particles. These results can be explained by considering the larger surface area to volume ratio of the smallest particles and increased internal surface area of the pores found in the large particles. For the small particles, the additional surface area hosts adsorbates that lead to non-radiative recombination, and in the porous particles, the pore walls can quench fluorescence. This trend was valid across calcination temperatures and was evident when comparing particles from the same calcination temperature.

Based on literature survey, the outcomes of different research papers have been summarized in Table 10.1 (Appendix 1).

Conclusion

Phosphor has been synthesized by a solid-state, hydrothermal, sol-gel, precipitation method. The hydrothermal synthesis route is morphology

Appendix 1

Table 10.1. Comprehensive study of literature

S. No.	Compound	Synthesis method	Morphology	Particle size	PL emission	PL excitation	Ref. no.
1.	Gd_2O_3:A (A = Eu^{3+}, Dy^{3+}, Sm^{3+}, Er^{3+}) Thin films	Pechlini sol-gel soft lithography	Uniform and cracked free non-patterned film	Size 70 nm, thickness 550 nm	572 nm, 610 nm	230 nm, 250 nm	Pang et al. 2003
2.	Gd_2O_3:Eu^{3+}	Sol-lyophilization technique.	Cylindrical	Diameter 15-20 nm, length 15-20 nm	611 nm	247 nm, 258 nm	Louis et al. 2003
3.	Eu_2O_3, Gd_2O_3:Eu, Y_2O_3:Eu	Precipitation method	-	2-5 nm	579 nm, 611 nm	250 nm	Bazzi et al. 2004
4.	Gd_2O_3:Er^{3+} / Er^{3+}	Hot solution chemistry	Nanoplates	Length 11-16 nm, Thickness 1.05 nm	612 nm (Eu), 1450-1650 nm (Er)	275 nm (Eu), 488 nm (Er)	Yang et al. 2005
5.	Gd_2O_3:Eu^{3+}	Sol-gel, Hydrothermal method	Nanowires	7-9 nm	585 nm, 597 nm, 611 nm, 620 nm	260 nm, 280 nm	Lin and Li 2006
6.	Gd_2O_3	Colloidal method	-	5-10 nm	-	-	Engstrom et al. 2006
7.	Gd_2O_3:Eu	Hydrothermal method	Nanorods, micro-rods	Nanorod (length 300-500 nm, diameter 20-120 nm)	617 nm	257 nm	Yang et al. 2007

(Contd.)

Table 10.1. (*Contd.*)

S. No.	Compound	Synthesis method	Morphology	Particle size	PL emission	PL excitation	Ref. no.
8.	$Gd(OH)_3$:Eu Gd_2O_3:Eu	Hydrothermal method	Nanowires, Nanotubes	Micro-rod (Length 8-15 μm, diameter 0.78-1.5 μm)	610 nm	257 nm	Lee et al. 2008
9.	Gd_2O_3:Er^{3+}/ Gd_2O_3:Yb^{3+}	Hydrothermal method	Core-shell nanorods	Diameter 20 nm, Length 150-200 nm	611 nm	254 nm	Bai et al. 2009
10.	$Gd(OH)_3$, Gd_2O_3:Eu^{3+}	Wet-chemical route	Nanotubes	Length 200-300 nm, Diameter 40 nm	610 nm	260 nm	Jia et al. 2009
11.	Gd_2O_3:Eu	Template directed method	Hollow spheres	Diameter 2-3μm, Thickness 200 nm	610 nm	254 nm	Jia et al. 2010
12.	Gd_2O_3:Yb/Er	Hydrothermal method	Hollow spheres	Thickness 20 nm, diameter 200-250 nm	625-700 nm	980 nm	Tian et Al. 2011
13.	Gd_2O_3:Ln (Ln = Eu^{3+}, Tb^{3+}, Dy^{3+}, Sm^{3+}, Yb^{3+}/Er^{3+}, Yb^{3+}/Tm^{3+} and Yb^{3+}/Ho^{3+})	Precipitation method	Hollow spheres	Diameter 80 nm	613 nm (Eu), 545 nm (Tb), 239 nm (Dy), 239 nm (Sm)	244 nm (Eu), 239 nm (Tb), 574 nm (Dy), 608 nm (Sm)	Gai et al. 2011

No.	Material	Method	Morphology	Size	Emission	Excitation	Reference
14.	Bi^{3+} co-doped Gd_2O_3:Eu^{3+}	Combustion method	Spherical particles	40-60 nm	590 nm, 612 nm, 625 nm	230 nm	Dhananjaya et al. 2012
15.	Gd_2O_3	Hydrothermal method	Nanorods	Length 50-130 nm, Diameter 5-20 nm	480 nm	300 nm	Hazarika et al. 2013
16.	Gd_2O_3	Combustion method	Agglomeration of different shape and size particles	41 nm	490 nm, 545 nm	300 nm	Tamrakar et al. 2014
17.	Gd_2O_3:Eu	Self-template method	Hollow spheres	Diameter 600 nm, Shell thickness 20 nm	610 nm	261 nm	Ren et al. 2015
18.	Gd_2O_3	Urea-assisted precipitation method	Hollow spheres	Diameter 500 nm, Shell thickness 20 nm	-	-	Jiang et al. 2016
19.	Gd_2O_3:Eu	Solid state method	-	Average size 76.78 nm	467 nm, 583 nm, 613 nm	237 nm, 254 nm, 275 nm	Tamrakar, et al. 2017
20.	Y_2O_3:Eu, Gd_2O_3:Eu	Wet chemistry route	-	Average size 35-40 nm	611 nm	250 nm	Adam et al. 2017

controlled among the various synthesis methods. In an attempt to produce various phosphors in the form of different nanostructures, the hydrothermal method has been widely adopted due to its simplicity, high efficiency and low cost. The hydrothermal treatments for colloidal precipitates of lanthanide hydroxides resulted in different nanostructures such as nanospheres, nano-rods, nanowires, nanoplates, nanotubes and nanobelts. The hydrothermal method is highly sensitive to the temperature, pH and ageing conditions. Thus, when the hydrothermal technique is applied to synthesize $Gd(OH)_3$:Eu as a precursor for the synthesis of Gd_2O_3:Eu nanophosphor, a strong pH dependence and temperature are observed in particle shapes. The combustion method is a low temperature method, but the agglomeration and non-uniform morphology is still an issue in this case. A solid-state method is a high temperature synthesis route and makes it insufficient to use from the economic point of view. The sol-gel method, Pechini method and precipitation methods are being efficiently used for nanosized particles. Various surfactants have been used for controlling the shape and size of the as-synthesized nanoparticles. Among the various nanostructures, one-dimensional structures owe tremendous applications over the others due to their shape specific properties. These one dimensional nanostructures also find various applications in the biomedical applications. Moreover, Gd_2O_3:Eu phosphors show dual behavior i.e. magnetic and luminescent behavior. Thus, these phosphors can be used for various applications such as light emitting diodes, bioimaging, crystal field displays and in various scintillator counters.

References

Adam, J., Metzger, W., Koch, M., Rogin, P., Coenen, T., Atchison, J.S., Konig, P. 2017. Light emission intensities of luminescent Y_2O_3:Eu and Gd_2O_3:Eu particles of various sizes. Nanomaterials 7, 26.

Bai, X., Song, H., Pan, G., Ren, X., Dong, B., Dai, Q., Fan, L. 2009. Improved upconversion luminescence properties of Gd_2O_3:Er^{3+}/Gd_2O_3:Yb^{3+} core-shell nanorods. J. Nanosci. Nanotechnol. 9, 2677-2681.

Bazzi, R., Flores, M.A., Louis, C., Lebbou, K., Zhang, W., Dujardin, C., Roux, S., Mercier, B., Ledoux, G., Bernstein, E., Perriat, P., Tillement, O. 2004. Synthesis and properties of europium-based phosphors on the nanometer scale: Eu_2O_3, Gd_2O_3:Eu, and Y_2O_3:Eu. J. Colloid Interface Sci. 273, 191-197.

Blasse, G. 1994. Reviews: Scintillator materials. Chem. Mater. 6, 1465-1475.

Dhananjaya, N., Nagabhushana, H., Nagabhushana, B.M., Chakradhar, R.P.S., Shivakumara, C., Rudraswamy, B. 2010. Synthesis, characterization and photoluminescence properties of Gd_2O_3:Eu^{3+} nanophosphors prepared by solution combustion method. Phys. B Condens. Matter 405, 3795-3799.

Dhananjaya, N., Nagabhushana, H., Nagabhushana, B.M., Sharma, S.C., Rudraswamy, B., Suriyamurthy, N., Shivakumara, C., Chakradhar, R.P.S. 2012. Synthesis, characterization, thermo- and photoluminescence properties of Bi^{3+} co-doped Gd_2O_3:Eu^{3+} nanophosphors. Appl. Phys. B 107, 503-511.

Engström, M., Klasson, A., Pedersen, H., Vahlberg, C., Käll, P.O., Uvdal, K. 2006. High proton relaxivity for gadolinium oxide nanoparticles. Magn. Reson. Mater. Physics, Biol. Med. 19, 180-186.

Gai, S., Yang, P., Wang, D., Li, C., Niu, N., He, F., Li, X. 2011. Monodisperse Gd₂O₃:Ln (Ln= Eu^{3+}, Tb^{3+}, Dy^{3+}, Sm^{3+}, Yb^{3+}/Er^{3+}, Yb^{3+}/Tm^{3+}, and Yb^{3+}/Ho^{3+}) nanocrystals with tunable size and multicolor luminescent properties. Cryst. Eng. Comm. 13, 5480-5487.

Hazarika, S., Paul, N., Mohanta, D. 2014. Rapid hydrothermal route to synthesize cubic-phase gadolinium oxide nanorods. Bull. Mater. Sci. 37, 789-796.

Iwako, Y., Akimoto, Y., Omiya, M., Ueda, T., Yokomori, T. 2010. Photoluminescence of cubic and monoclinic Gd₂O₃:Eu phosphors prepared by flame spray pyrolysis. J. Lumin. 130, 1470-1474.

Jain, A., Hirata, G.A. 2016. Photoluminescence , size and morphology of red-emitting Gd₂O₃:Eu³⁺ nanophosphor synthesized by various methods. Ceram. Int. 1-8.

Jia, G., Liu, K., Zheng, Y., Song, Y., Yang, M., You, H. 2009. Highly Uniform Gd(OH)₃ and Gd₂O₃:Eu³⁺ Nanotubes: Facile synthesis and luminescence properties. J. Phys. Chem. C 113, 6050-6055.

Jia, G., You, H., Liu, K., Zheng, Y., Guo, N., Zhang, H. 2010. Highly uniform Gd₂O₃ hollow microspheres: Template-directed synthesis and luminescence properties. Langmuir 26, 5122-5128.

Jiang, X., Yu, L., Yao, C., Zhang, F., Zhang, J., Li, C. 2016. Synthesis and characterization of Gd₂O₃ hollow microspheres using a template-directed method. Materials (Basel). 9, 323.

Kruis, F.E., Fissan, H.J., Peled, A. 1998. Synthesis of nanoparticles in the gas phase for electronic, optical and magnetic applications – A review. J. Aerosol Sci 8502, 511-535.

Lee, K., Bae, Y., Byeon, S. 2008. Nanostructures and photoluminescence properties of Gd₂O₃:Eu red-phosphor prepared via hydrothermal route. Bull. Korean Chem. Soc. 29, 2161-2168.

Lee, K., Bae, Y., Byeon, S., n.d. World's largest Science, Technology & Medicine Open Access Book Publisher pH Dependent Hydrothermal Synthesis and Photoluminescence of Gd₂O₃: Eu Nanostructures.

Li, G., Liang, Y., Zhang, M., Yu, D. 2014. Size-tunable synthesis and luminescent properties of Gd(OH)₃:Eu³⁺ and Gd₂O₃:Eu³⁺ hexagonal nano-/microprisms. Cryst. Eng. Comm. 16, 6670-6679.

Li, Q., Lin, J., Wu, J., Lan, Z., Wang, J., Wang, Y., Peng, F., Huang, M., Xiao, Y. 2011. Preparation of Gd₂O₃:Eu³⁺ downconversion luminescent material and its application in dye-sensitized solar cells. Chinese Sci. Bull. 56, 3114-3118.

Li, Y., Hong, G. 2007. Synthesis and luminescence properties of nanocrystalline Gd₂O₃:Eu³⁺ by combustion process. J. Lumin. 124, 297-301.

Lin, K.-M., Li, Y.-Y. 2006. Luminescent properties and characterization of one dimensional Gd₂O₃:Eu³⁺ phosphor nano-wire for field emission application. Nanotechnology 17, 4048-4052.

Liu, G., Hong, G., Wang, J., Dong, X. 2007. Hydrothermal synthesis of spherical and hollow Gd₂O₃:Eu³⁺ phosphors. J. Alloys Compd. 432, 200-204.

Louis, C., Bazzi, R., Flores, M.A., Zheng, W., Lebbou, K., Tillement, O., Mercier, B., Dujardin, C., Perriatb, P. 2003. Synthesis and characterization of Gd₂O₃:Eu³⁺ phosphor nanoparticles by a sol-lyophilization technique. J. Solid State Chem. 173, 335.

Maalej, N.M., Qurashi, A., Assadi, A.A., Maalej, R., Shaikh, M.N., Ilyas, M., Gondal, M.A. 2015. Synthesis of Gd₂O₃:Eu nanoplatelets for MRI and fluorescence imaging. Nanoscale Res. Lett. 10, 215.

Pang, M.L., Lin, J., Fu, J., Xing, R.B., Luo, C.X., Han, Y.C. 2003. Preparation, patterning and luminescent properties of nanocrystalline Gd_2O_3:A (A=Eu^{3+}, Dy^{3+}, Sm^{3+}, Er^{3+}) phosphor films via Pechini sol-gel soft lithography. Opt. Mater. (Amst). 23, 547-558.

Park, J.C., Moon, H.K., Kim, D.K., Byeon, S.H., Kim, B.C., Suh, K.S. 2000. Morphology and cathodoluminescence of Li-doped Gd_2O_3:Eu^{3+}, a red phosphor operating at low voltages. Appl. Phys. Lett. 77, 2162-2164.

Perkins, T.B., Kaufmann, H.W. 1935. Luminescent materials for cathode-ray tubes. Proc. Institure Radio Eng. 23, 1324-1333.

Priya, R., Pandey, O.P. 2019. Photoluminescent enhancement with co-doped alkali metals in Gd_2O_3:Eu synthesized by co-precipitation method and Judd Ofelt analysis. J. Lumin. 212, 342-353.

Ren, X., Zhang, P., Han, Y., Yang, X., Yang, H. 2015. The studies of Gd_2O_3:Eu^{3+} hollow nanospheres with magnetic and luminescent properties. Mater. Res. Bull. 72, 280-285.

Siegel, R.W. 1993. Nanostructured materials-mind over matter. Nano Struct. Mater. 3, 1-18.

Siegel, R.W., Hu, E., Roco, M.C. 1998. Nanostructure Science and Technology. Springer.

Sun, S., Murray, C.B., 2010. Synthesis of monodisperse cobalt nanocrystals and their assembly into magnetic superlattices. J. Appl. Phys. 85, 4325-4330.

Suryanarayana, C. 1995. Nanocrystalline materials. Int. Mater. Rev. 40, 41-64.

Tamrakar, R.K., Upadhyay, K., Sahu, I.P., Bisen, D.P. 2017. Tuning of photoluminescence emission properties of Eu^{3+} doped Gd_2O_3 by different excitations. Optik (Stuttg). 135, 281-289.

Tian, G., Gu, Z., Liu, X., Zhou, L., Yin, W., Yan, L., Jin, S., Ren, W., Xing, G., Li, S., Zhao, Y. 2011. Facile fabrication of rare-earth-doped Gd_2O_3 hollow spheres with upconversion luminescence, magnetic resonance, and drug delivery properties. J. Pysical Chem. C 115, 23790-23796.

Tseng, T.K., Choi, J., Jacobsohn, L.G., Yukihara, E., Davidson, M., Holloway, P.H. 2010. Synthesis and luminescent characteristics of one-dimensional europium doped Gd_2O_3 phosphors. Appl. Phys. A Mater. Sci. Process. 100, 1137-1142.

Vrankic, M., Grz, B., Lu, D., Bosnar, S., Ankica, S. 2015. Chromium environment within Cr-doped $BaAl_2O_4$: Correlation of X-ray di ff raction and X-ray absorption spectroscopy investigations. Inorg. Chem. 54, 11127-11135.

Wakefield, G., Keron, H.A., Dobson, P.J., Hutchison, J.L. 1999. Synthesis and properties of sub-50-nm europium oxide nanoparticles. J. Colloid Interface Sci. 182, 179-182.

Wang, Z., Wang, P., Zhong, J., Liang, H., Wang, J. 2014. Phase transformation and spectroscopic adjustment of Gd_2O_3:Eu^{3+} synthesized by hydrothermal method. J. Lumin. 152, 172-175.

Xing, G., Guo, Q., Liu, Q., Li, Y., Wang, Y., Wu, Z., Wu, G. 2014. Highly uniform $Gd(OH)_3$ and Gd_2O_3:Eu^{3+} hexagram-like microcrystals: Glucose-assisted hydrothermal synthesis, growth mechanism and luminescence property. Ceram. Int. 40, 6569-6577.

Yang, J., Li, C., Cheng, Z., Zhang, X., Quan, Z., Zhang, C., Lin, J. 2007. Size-tailored synthesis and luminescent properties of one-dimensional Gd_2O_3:Eu^{3+} nanorods and microrods. J. Phys. Chem. C 111, 18148-18154.

Yang, J., Wang, X., Song, L., Luo, N., Dong, J., Gan, S., Zou, L. 2018. Tunable luminescence and energy transfer properties of $GdPO_4$:Tb^{3+}, Eu^{3+} nanocrystals for warm-white LEDs. Optical Materials 85, 71-78.

Yanli, W.U., Xianzhu, X.U., Qianlan, L.I., Ruchun, Y. 2015. Synthesis of bifunctional Gd$_2$O$_3$ Eu^{3+} nanocrystals and their applications in biomedical imaging. J. Rare Earths 33, 529-534.

Zhou, C., Wu, H., Huang, C., Wang, M., Jia, N. 2014. Facile synthesis of single-phase mesoporous Gd$_2$O$_3$:Eu nanorods and their application for drug delivery and multimodal imaging. Part. Syst. Charact. 31, 675-684.

Significance of TL Radiation Dosimetry of Carbon Ion Beam in Radiotherapy

Karan Kumar Gupta[1,2], N.S. Dhoble[3], Vijay Singh[4] and S.J. Dhoble[2]*

[1] Department of Chemical Engineering, National Taiwan University, Taipei, Taiwan, ROC
[2] Department of Physics, R.T.M. Nagpur University, Nagpur - 440033, India
[3] Department of Chemistry, Sevadal Mahila Mahavidyalaya, Nagpur - 440009, India
[4] Department of Chemical Engineering, Konkuk University, Seoul, 143701, Korea

Introduction

As it is known that radiation is a type of electromagnetic wave which can travel through vacuum or most of the matter having a different media. There are mainly two types of radiation: (1) ionizing and (2) non-ionizing. Here we are concerned only with ionizing radiations. When the ionizing radiations pass through any medium the medium gets ionized. In Radiation Therapy (RT) these high ionizing radiations are used to kill malignant tissues in the treatment of cancerous tumors. The discovery of X-ray by Roentgen in the year 1895 led to the use of photons in external beam radiotherapy (EBRT), which is a major type of treatment modality in radiotherapy. In the year 1946, Dr. Robert R. Wilson was the first person who worked on developing a particle accelerator and proposed the medical use of protons in the treatment of cancer (Wilson 1946). Proton radiotherapy is associated with some main advantages over photon radiotherapy such as deposition of a very low entrance dose and a maximum dose deposition when it reaches the target and stops, thereby eliminating an exit dose (Parker 1985). In spite of better dose distribution, the Relative Biological Effectiveness (RBE) of proton beams is slightly greater than photon beams (Eickhoff et al. 1999).Whereas the physical and biological properties of proton and heavy charged particles are quite different from each other due to high LET value of heavy charged particles. Thus, particle therapy can be further subdivided in two categories: one with proton having low LET value and another with heavy charged

*Corresponding author: sjdhoble@rediffmail.com

particle having high LET value (Ertner et al. 2006). Since a carbon ion beam is a type of heavy charged particle beam having LET higher than a proton, and possesses significantly increased RBE value particularly in the Bragg peak region, it becomes very popular and advantageous in the treatment of malignant tissues which lead to the double strand DNA break during its single hit to the targeted tumor (Ertner et al. 2006; Kraft 2000; Tsuji and Kamada 2012). In the treatment of cancerous tumors and other malignant tissues with ionizing radiation, it is important to measure the amount of dose (absorbed dose) delivered to the targeted area, within an uncertainty of 5% at 2σ (two standard deviation) level, to reduce the risk of normal tissue complication (Pour et al. 2018). Thus with the help of a suitable dosimetry (which helps to determine the absorbed dose), we can control the radiation exposure to targeted or other areas within the patient's body.

Instead of several electronics and other dosimeters, TL dosimeter has an advantage of a small size and does not need high voltage supply, wires or cables i.e. easy to handle characteristics, which makes thermoluminescence dosimeters (TLDs) as one of the most preferred choices of detectors as a measurement of ionizing radiation in environmental and medical applications over the decades. For this purpose several TL materials (LiF:Mg, Ti; LiF:Mg, Cu, P; $CaSO_4$:Dy etc.) are investigated (Samuel et al. 2017).

After the development of several particle accelerators, the efficacy of heavy charged particles in clinical use has been proved in radiotherapy, which leads to the investigation of proper dosimetry for HCPs (Karger et al. 2010; Bert et al. 2010). As mentioned earlier, the carbon ion beam is a type of heavy charged particle beam and these days it is mostly used in external beam radiotherapy which makes the necessity of the suitable carbon ion beam dosimetry to monitor the desired amount of dose which should be delivered to the patient. In this chapter we deal with some possible aspects that can be obtained using thermoluminescence dosimetry to measure the dose delivered to a patient in carbon radiotherapy and for this purpose we tried to develop the idea of carbon RT and its advantages over other low ionizing radiation with reference to Linear Energy Transfer (LET) and Relative Biological Effectiveness (RBE) of the carbon ion beam. To detect ionizing radiation, especially ion beam radiation with the help of the TL technique, it is important to know how ion beam radiation interacts with the TL phosphor and leads to the creation of defects inside the phosphor. Since, the TL phenomenon completely depends on the type and position of defect formation, which acts as a trapping center for electron and holes inside the TL materials, in this chapter we have tried to make a clear understanding of HCP interaction with TL materials and defect formation by the same and also the TL response of these materials with respect to another reference radiation like ^{60}Co or ^{137}Cs. However, to understand the effect of ionization density on TL response of various TL materials and relative HCP or photon TL efficiencies, we have taken help of three models/theories in this chapter as Track Structure Theory (TST), Modified Track Structure Theory (MTST) and Micro Dosimetric Target Theory (MTT). The actual problem on ion beam irradiation of TLD materials arises due to saturation effect of TL response at high fluences of the ion beam (Horowitch et al. 2001; Brandan et al. 2002; Salah 2011). Track Interaction Model (TIM), Extended Track Interaction Model (ETIM) and Unified Interaction Model (UIM) have been successfully described for explaining this saturation effect and to understand the incorporation of the trapping center, luminescence center and competitive center along the HCP track, which is responsible for the TL. To show how we can measure the carbon ion beam radiation dose with the help of the TLDs, some of the TL phosphor irradiated to different energies and at different fluences of carbon beam at ion per cm^2 within a certain range have been included in this chapter.

Radiotherapy

Radiotherapy is a type of medical treatment which uses high ionizing radiation to control or kill malignant cells. Thus, radiotherapy is also known as radiation therapy. Radiation therapy is commonly applied or may be curative in various types of cancerous tumors if they are localized in the body. In radiotherapy high energy beams are delivered to damage the DNA of the cancerous tissue leading to cellular death. This treatment can be used both in terms of a cure as well as to provide pain control. Radiotherapy may also be used as a part of the adjuvant therapy to reduce the risk of tumor recurrence on the area around the original cancer after surgery (Overman et al. 2010).

In radiotherapy high energy beams can also affect both cancer cells and normal tissue (such as skin or organs through which radiation must pass through to treat the cancer). To prevent normal tissue from being excessively damaged, shaped radiation beams are aimed from several angles of exposure to match the shape of the tumor located inside the patient's body, providing a much larger absorbed dose at the tumor site instead of the surroundings. Radiotherapy may be combined together with surgery, chemotherapy, immunotherapy, hormone therapy or a combination of all these four therapies. During bone marrow transplant many patients go through radiotherapy treatment which is also known as Total Body Irradiation (TBI). Total body irradiation is a technique which makes sure that body of a patient is ready to receive bone marrow or stem cell transplant. Radiation oncologists also used TBI together with high dose chemotherapy treatment to damage the cancer cells in areas (skin, bones, nervous system, and testes in men) not affected after chemotherapy. Radiotherapy can be further subdivided into two main categories known as external radiotherapy and internal radiotherapy. In external radiotherapy, the radiation source is placed outside the patient's body whereas in internal radiotherapy such as in Brachytherapy, the radiation source is sent inside the patient's body via a protective capsule or wire to the area requiring treatment (Skowronek 2017).

Radiotherapy finds its application in malignant and several non-malignant conditions such as in the treatment of acoustic neuromas or neurolemmomas (hearing loss), trigeminal neuralgia or prosopalgia (pain in the face), severe thyroid eye disease or thyroidopthalmopathy (starey eyes), pigmented villonodular synovitis etc. (Bagheri et al. 2004; Behbehani et al. 2004). It can also be used in the hindrance of keloid scar growth, heterotopic ossification (three types - Myositis ossification progressive, Traumatic myositis ossification and Neurogenic heterotopic) and vascular restenosis. However, the uses of radiation therapy in non-malignant treatment have some side effects like the risk of radiation persuaded cancer. In radiotherapy treatment, at low doses there is a minimum side effect whereas high doses can cause different types of side effects. These side effects are mainly: (1) Acute side effects like nausea and vomiting, swelling, infertility, mouth, throat and stomach sores (Eric 2000), damage to the epithelial surfaces etc. (2) Late side effects like cancer, heart disease (Taylor et al. 1990), dryness etc. (3) Cumulative side effects (Nieder et al. 2000). To overcome the side effects of high doses, different amounts of ionizing radiation in different intervals of time is used (also called Fractionation) in the course of treatment.

Carbon Ion Beam Radiotherapy (In Comparison to Photon and Proton RT)

After the discovery of X-rays in the year 1895 by Roentgen, X-rays, electron beams and gamma rays, also known as photons, have been mainly used in the external beam radiotherapy in the treatment of malignant cancerous tumors as conventional radiotherapy. In the year 1946 Dr. Robert R. Wilson was the first person who discovered the medical usage of protons for cancer treatment (Wilson 1946). However, just within the next 10 years of its discovery protons were used in radiotherapy to treat different types of cancerous patients and first patient was treated with the proton beam in the year 1954 at Lawrence Berkeley National Laboratory (LBNL) in USA (Ertner et al. 2006). In the years between 1977 and 1992 the first clinical trial with Helium and Neon ions with promising results also took place at LBNL (Castro et al. 1994). The first clinical experience in the treatment of malignant tissues with the help of carbon ion beam was launched at the National Institute of Radiological Science (NIRS) in the year 1994 using Heavy Ion Medical Accelerator in Chiba (HIMAC) (Torikoshi et al. 1995). Currently there are five carbon ion beam radiotherapy centers established all over the world, three in Japan, one in China and other one in Germany. In addition, right now seven new carbon ion RT centers are under construction, Italy - one, Germany - two, Austria - one, China - one and Japan - two (Bhatt 2013).

Radiotherapy works by damaging the gene (DNA) of the cancerous cell to prevent its growth and division. However irreparable damage of DNA of the cell can occur with the help of double strand DNA break instead of single strand DNA break. Radiation treatment with conventional photons leads only to a single strand DNA break and repeating the treatment with this photon radiation leads to cellular death. However, treating the lesion with ionizing radiation several times can harm the normal tissues around the lesion and also damage the other parts of the body which may lead to complications with normal tissues. This type of normal tissue complication problem can be overcome with the help of high ionizing radiation having high Linear Energy Transfer (LET) values. Carbon ion beam provides high ionizing radiation having greater energy than photons leading to double strand DNA break during its single hit to targeted lesion due to its maximum dose deposition at the Bragg peak region (Kanematsu et al. 2019; Ishikawa et al. 2019). Since, the relative biological effectiveness of carbon ion beam, due to its high LET value, is greater than that of proton and photon, it is extensively used nowadays in external beam radiotherapy.

Carbon Ion Beam or Heavy Charged Particle Beam

Particle Nature

Carbon ion beam is a type of heavy ion beam or heavy charged particle beam having masses several times more than that of an electron. Here heavy ion stands for the ion larger than the proton. Heavy charged particles like alpha particles, carbon, oxygen, argon, neon etc. can be generated with the help of a particle accelerator (either electrostatic or oscillating field accelerator). These heavy charged particles have energies higher than that of electrons and photons such that it can ionize the medium deeper than electrons and photons.

Beam Features

The depth dose distribution of HCP beams, when the HCP beam incident on any medium is very low in the entrance region, shows a straight trajectory, and prominent dose (as a peak) in the stopping region, also called Bragg peak, which results in the irradiation of very small confined tumors or other malignant cells within the patient's body. HCP passing through any medium loses its maximum energy due to interaction with atomic electrons of the medium (Tobias et al. 1952). It also undergoes elastic or inelastic collisions with the nucleus. The main difference of the electron beam and heavy charged particle beam is due to the rest mass of the electron which is less than the HCP mass. Thus, the electron beam loses its maximum energy during its first interaction with the matter but the scattering angle of HCP is very low, resulting in a sharper lateral dose distribution and traversing a long distance through the medium along with producing secondary electron and higher order electron.

Beam Superiority

The biological effect of any beam completely depends upon the beam quality i.e. upon LET which measures the energy loss distribution of a particular beam along its tracks. It is observed that heavy ion passing through any medium slows down and makes the interaction with nuclei resulting in the disintegration of the incident ion. The above destruction of the incident ion depends on the ion type, ion energy and also on the surroundings of the medium. According to Bichsel the variation in energy loss directly depends on the square of the nuclear charge, which results in a large discrepancy in the energy loss and their ranges in a medium for different type of particles (Bichsel 1972). In-elastic collision with the absorber nucleus leads to fragmentation of the absorber nucleus. The energy of target nucleus fragments has less energy than the projectile fragments and therefore the dose due to target nucleus fragments is locally deposited.

Thermoluminescence Dosimetry

TL dosimetry is used as a dosimetry of ionizing radiation since the last several decades and can be used in the measurement of ionizing radiation in radiation affected areas where the absorbed dose is in the range of micro grays and also in radiotherapy where doses are in the range of several grays. TL material also found its application in measurement of nuclear radiation doses from the early 1950s, where the patient swallowed the crystal when he/she was injected with radioactive isotopes and after his recovery in few days, the accrued dose was measured with the help of the intensity of TL signal emitted by the crystal in a particular unit (Su et al.1985).Thus for this purpose several types of TL materials are developed nowadays. Some of the materials that have been investigated earlier are $CaSO_4$: Dy, CaF_2: Dy, LiF: Mg, Ti and Al_2O_3: C. Instead of several electronics and other dosimeters, TLDs are useful in point dose measurement *in vivo* dosimetry due to their small size and need no requirement of high voltage supplies.

TL is the phenomenon of emitting light during heating of some special type of crystalline materials, after irradiation with different types of electromagnetic or other ionizing radiation which may be gamma rays, alpha rays, electron beam or may be different types of heavy charged particle beams. This phenomenon was first observed by Robert R. Boyle in 1663 after observing a glimmering light from diamond on heating in a dark room (Boyle 1664). The use of TL as a radiation detector has a long history from early 1895, when this process was used as a measurement of radiation due to electrical discharge with the help of artificially prepared $CaSO_4$: Mn phosphor by Wiedmann and the same method was used again by Lyman in the year 1935 for far-ultraviolet range (Wiedman and Schmidt 1895; Lyman 1935). However the word, TL was first coined in literature by E. Weidman and Schmidt in the year 1895 (Wiedman and Schmidt 1895). Randall and Wilkins along with Garlick and Gibson in the year 1945 and 1949 were able to provide some understanding of certain features of the TL process. Whereas, this phenomenon became a subject of inclusive interest from the year 1953 when its effectiveness in radiation dosimetry was established by Daniels et al. (Daniels et al. 1953), on the other hand Cameron et al. were successful in utilizing the property of thermoluminescence of LiF as a radiation dosimeter for the measurement of X-rays, gamma rays, electron and beta rays and thermal neutrons (Cameron et al. 1964). Since, the use of TLDs as a dosimetry of ionizing radiation has many advantages over the other dosimeteric systems viz. small size, roughness, range, effortlessness in reading and sensitivity and also due to its low cost it has attracted a lot of attention from investigators. During the 60s and 70s a lot of work was done in this field and the technique is still a part of recent investigations.

Heavy Charged Particle (HCP) Interaction with TL Materials

Carbon ion beam is a type of heavy charged particle beam and, as we know, when the heavy charged particle beam interacts with matter the ion beam loses its energy by a number of mechanisms, such as (i) Ionization and excitation, (ii) Nuclear collisions, (iii) Photon generation and (iv) Nuclear reactions. When TL materials are irradiated with HCP, total HCP stopping power is due to the contribution of the first two mechanisms electronic and nuclear stopping powers. Total electronic energy loss can further be sub-divided into three main energy regions. They are (i) high energy region, (ii) intermediate energy region and (iii) low energy region. In the high energy region the velocity of a heavy charged particle is too high given by the formula $v > v_0 Z_1^{2/3}$ (where v_0 indicates $c/137$, here c is the velocity of the light) i.e. incident ion velocity is faster than the atomic electrons of the medium. According to the corrected Bethe formula electronic energy loss related to the medium through which HCPs travels and also on the velocity of the projectiles (Bethe 1930). This Bethe formula is not appropriate in the case of low ion velocities. In the low energy region the ion's velocity is very low given by the formula $v << v_0 Z_1^{2/3}$ and also the low energy region practically shows the neutrality to both target and projectile. In this region total stopping power is given with the help of the theory of Lindhard-Scharff (Lindhard and Scharff 1961). But in the intermediate energy region the total electronic energy loss can be predicted by the semi-empirical formula of Biersack and Haggmark (Biersack and Haggmark 1980) which is actually the combination of both Bethe and Lindhard-Schraff formulation.

In the high and intermediate energy region the contribution of nuclear stopping power is very little but is predominant in the low energy region. The total nuclear stopping power can be calculated in three different steps: in first step we can calculate the abundant projectile target scattering potential (Berger and Seltzer 1964); in second step we can determine the elastic scattering cross section with the help of classical mechanics and finally the last step consists of the total transferred energy calculation of the target atom through elastic scattering. In passing through a medium, HCPs cover a distance before coming to rest which is also called range of the HCPs. The stopping power of charged particle can be defined as the loss of particle energy when these HCPs pass through any medium; its inverse provides the distance covered per unit energy loss. For comparison purposes the track of both HCP and low charge particle is shown in effect and defect creation by HCP.

The defect structures present in the TL materials are responsible for the TL signal, and also represent the dosimetric properties of TL materials. Irradiation of TL sample with high energy particle introduces defects in TL materials through the creation of electrons, holes and exciton (Horowitz et al. 2001). These electrons and holes may recombine or may diffuse through the materials and can be trapped in the forbidden energy gap of the materials while these trapped charges can again be excited by the low energy radiation of HCP and results in the trapping of these charge carriers at the deep traps. This may lead to permanent damage of the crystal, also known as ionization damage. Instead of focusing attention on the defect creation mechanism inside the TL materials after irradiation it is important to know the type of defects present inside the TL materials at the thermodynamic equilibrium. The vacancy defect (simplest point defect) also called Schottky defect and

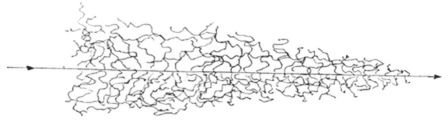

Fig. 11.1a. Diagram of highly localized HCP track follow almost straight trajectory inside the matter. (Reprinted with permission from Y.S. Horowitz, "Theory of heavy charged particle response (efficiency and supralinearity) in TL materials", *Nuclear Instruments and Methods in Physics Research B*, 184 (2001): 85-112)

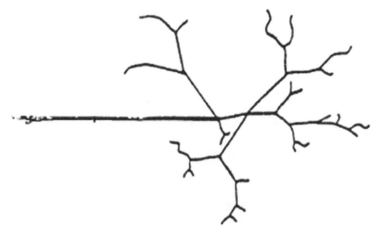

Fig. 11.1b. Diagram of low energy particle track which shows twisted path inside the matter. (Reprinted with permission from Y.S. Horowitz, "Theory of heavy charged particle response (efficiency and supralinearity) in TL materials", *Nuclear Instruments and Methods in Physics Research B*, 184 (2001): 85-112)

interstitial defect or the pair of vacancy and interstitial known as Frenkel defects is found to be present in the most of TL materials, whereas Frenkel defects are mainly found at a higher temperature. Moreover, a number of point defects can be created inside the crystal after its irradiation with high energy particles in which the incident particle displaces the atom from its lattice points and results in the creation of new point defects (Doan and Martin 2003; Yu et al. 2019). This type of radiation damage inside the materials is known as displacement damage. Here point defect creation is the complete function of the incident particle and the nature of the TL materials or crystals and independent of temperature (Yanwen et al. 2014). Since the TL materials are mainly insulators or semiconductors having very less number of free charge carriers on irradiation to very low doses can lead to the modification in electrical and optical properties of any material. However, on the basis of some experimental results it has been observed that the nanocrystalline materials show a prominent resistance against the irradiation, while few experiments show structural disorder in the nanomaterials as compared to bulk materials after irradiation (Zhou et al. 2018). In spite of thermal growth mechanism of nanomaterials, the irradiation induced growth mechanism of various nanomaterials has also been reported by many investigators while nanocrystalline materials become amorphous in nature at low doses of ionizing radiation. The irradiation of nanomaterials, however, can be used to modify the grain size, phase structure, optical properties and physical properties of various materials (Andrievski 2011; Krasheninnikov and Nordlund 2010).

Radial Dose Distribution along the HCP Track

Total amount of energy transported to the medium through HCP irradiation is known by radial dose distribution and is also helpful in describing the radiation preoccupation stage of HCP encouraged TL processes. The understanding of radial dose distribution around the HCP can help us in theoretical evaluation of effective HCP TL efficiencies. Dose distribution of the HCPs is inhomogeneous and highly localized within the medium due to straight trajectories of the HCP within the medium. Secondary electrons produced by the HCPs due to transfer of their energy to the atomic electrons, passing through the medium, yield a radially identical spreading of dose around the HCP. In TL materials, it is not possible to directly calculate the radial dose distribution of HCP because energy deposition is of the order of few MeV per nucleon within 1000 Å from the HCP path. In Fig. 11.2 it can be seen that near the ion track radial dose distribution of He ions (in LiF TLD material) having initial energy 4 MeV increases as the ion energy decreases which is in good agreement with the stopping power behavior. Secondary electron loses its energy and this energy accumulates in the form of concentric cylindrical shells (which is divided in to six transverse slices) around the beam direction z as shown in Fig. 11.2 (Horowitz et al. 2001). The total amount of radial dose distribution can be calculated by dividing the

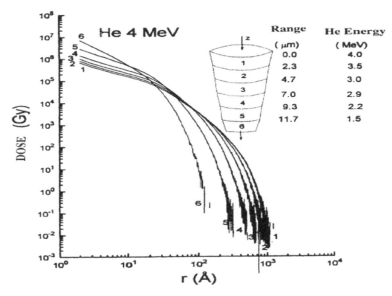

Fig. 11.2. Radial dose distribution along the track of He ion beam having initial energy 4 MeV in solid state LiF. The whole track is separated into six slices and the energy of the beam arriving each slice has been shown as a function of depth in material. (Reprinted with permission from Y.S. Horowitz, "Theory of heavy charged particle response (efficiency and supralinearity) in TL materials", *Nuclear Instruments and Methods in Physics Research B*, 184 (2001): 85-112)

amount of energy stored in volume of the shell per unit material (Horowitz et al. 2001).

The radial dose spreading is prolonged to higher radial distance for the initial slices where secondary electrons have maximum energy. The radial extension of the track is less affected as the charge (Z) and mass (M) of the heavy charged particle increases but dose distribution increases as shown in Fig. 11.3 for H, He and C ions with initial energy 2.5 MeV per nucleon (Horowitz et al. 2001).

HCP Thermoluminescence Efficiency

Thermoluminescence efficiency α is given as the ratio of mean energy $\bar{E}_0$ released as a TL signal to the mean energy $\bar{E}$ given to TL material, after irradiation of material (Ezra and Horowitz 1982; McKeever and Horowitz 1990). Thus TL efficiency can be written as:

$$\alpha = \frac{\bar{E}_0}{\bar{E}} \tag{1a}$$

where α = TL efficiency,

$\bar{E}_0$ = mean energy emitted as TL light, and

$\bar{E}$ = mean energy given to the TL material during irradiation.

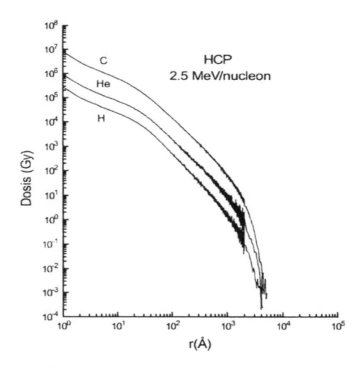

Fig. 11.3. Radial dose distribution along the track of H, He, and C ion beam having initial energy 2.5 MeV. (Reprinted with permission from Y.S. Horowitz, "Theory of heavy charged particle response (efficiency and supralinearity) in TL materials", *Nuclear Instruments and Methods in Physics Research B*, 184 (2001): 85-112)

Since the TL efficiency of various types of TL materials can be influenced by different types of ionizing radiation, it is important to calculate relative TL efficiency of HCP with respect to another reference radiation, typically ^{60}Co or ^{137}Cs gamma radiation. Thus, relative TL efficiency can be defined as the ratio between TL responses per unit mass per unit dose by the type of radiation under study with respect to TL response per unit mass per unit dose generated by reference radiation.

$$(R_e)_{i,ref} = \frac{R_i/D_i}{R_{ref}/D_{ref}} \tag{1b}$$

where R_e = Relative efficiency,

R_i = TL response per unit irradiated mass by radiation under study,

D_i = Dose distributed by the radiation under study,

R_{ref} = TL response per unit irradiated mass for reference radiation, and

D_{ref} = Dose distributed by the reference radiation.

Track Interaction Model (TIM), Unified Interaction Model (UNIM) and modified track structure theory are capable of explaining the relative TL efficiencies of HCP as a function of ionization density, energy and particle type. It was observed that as the LET of the particle increases, the relative TL efficiencies decreases. The relative TL efficiencies are also persuaded by the number of experimental aspects such as the material temperature, read out parameters and annealing parameters.

Relative TL Efficiency of Carbon Ion Beam and other HCPs

The dependence of $(R_e)_{i,ref}$ on LET of heavy charge particle per unit density of TL materials has been investigated by many researchers and also by the Tanaka and Furuta on the basis of the assumption that the relative TL response can only be influenced by the stopping power of HCPs (Tanaka and Furuta 1974). Additionally many TL materials such as LiF, CaF_2:Mn, $CaSO_4$:Tm etc. have only LET per unit density dependence. A clear lack of correlation has been observed between $(Re)_{i,ref}$ and LET per unit density for some materials (e.g. CaF_2: Mn, $Li_2B_4O_7$ Mn and LiF TLDs) with the help of data compiled by Horowitz (Horowitz 1984). According to Geiss et al. (Geiss et al. 1998), Schmidt (Schmidt et al. 1990) and the group of IFUNAM, Mexico, it is generally observed that there is a decrease in the value of $(Re)_{i,ref}$ at LET per unit density of TL materials more than 10^2-10^3 MeV g^{-1} cm^2 but a greater spread in values for LET per unit density is also observed for values larger than 10^3 MeV g^{-1} cm^2 (Horowitz et al. 2001). In the case of different types of particles having same LET value it is also observed that the value of Re (Relative efficiency) is different and even for the similar particle types. In Fig. 11.4 the relative TL efficiency of LiF: Mg, Ti TLD material to ^{12}C along with ^{1}H, ^{3}He, ^{16}O and ^{20}Ne ion beam with respect to ^{60}Co as a reference radiation has been shown and it has been observed that as the LET increases to a particular value for each ion beam radiation the relative efficiency also decreases (Massillon et al. 2006).

A large spread in *Re* values has been recognized due to the various experimental factors, such as (1) doping concentration in host of TL materials, (2) different types of experimental techniques used in the synthesis of TL materials, (3) annealing of the sample before and after irradiation, and (4) temperature interval used during the reading of TL signal. Thus, it is important to perform the whole experiment in the same laboratory with a prudently demarcated protocol having same concentration doping for a specified dosimetry with same annealing and read out procedures.

Theoretical Models

The relative TL response of various types of materials can be described with the help of three main models/theories, which have been developed from many years. They are: (1) Track Structure Theory (TST) developed by Waligroski and Katz in the year 1980 (Waligroski and Katz 1980). (2) Modified

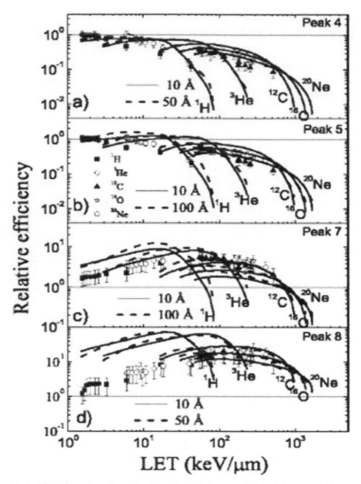

Fig. 11.4. Plot showing the relative TL efficiency of LiF: Mg, Ti TLD materials using ^{12}C, ^{1}H, ^{3}He, ^{16}O and ^{20}Ne ion beam having initial incident energies 1, 3, 15, 25 and 40 MeV u^{-1} respectively with respect to Co^{60} gamma radiation. (Reprinted with permission from G. Massillon-JL, "Observation of enhanced efficiency in the excitation of ion-induced LiF:Mg,Ti thermoluminescent peaks", *Journal of Applied Physics*, 100 (2006): 103521-6)

Track Structure Theory (MTST) introduced by Kalef-Ezra and Horowitz in the year 1982 (Ezra and Horowitz 1982). (3) Microdosimetry target theory (MIT) introduced by Olko et al. in 2002 (Olko et al. 2002).

Track Structure Theory (TST)

According to the track structure theory, the response of all physical, biological and chemical medium to HCP irradiation is due to the creations of secondary electrons and higher order electrons which are generated when HCP pass through the material, and slows down. Thus, track structure theory helps us to provide the information about the general behavior of all solid-state matters to HCP radiation. The recombination mechanism plays an important role in TL, and the dose dependent effectiveness of competing centers present in TL materials which do not contribute to the generation of TL light necessitates an erudite approach. Distribution of the dose due to the secondary electron generated by weakly ionizing radiation is uniform in the medium and leads to rather uniform occupation probabilities of different centers (e.g. trapping center, competing center and luminescent center) in the medium responsible for TL processes (Horowitz 2001). Whereas in the case of TL materials irradiated with HCP it follows almost a straight path in the medium resulting in a very inhomogeneous and highly localized spatial dose distribution. Thus, highly non-uniform occupation probability of different centers as mentioned in the case of weakly ionizing radiation is observed as 100% near the track axis and 0% apart from the track axis in the range of 1000 Å. TL efficiency of the materials to HCP irradiation depends also on the spatial non-uniform occupation probability of different centers such as trapping center, competing center and luminescence center. Track Structure Theory (TST) is generally based on the spatial difference of the dose to measure the HCP response. The radiation end effect of HCP and gamma/electron radiation is due to the secondary electron produced by both. Thus, different HCP radiation response can be measured with the help of the information of the dose response of any type/energy of electron or gamma radiation.

Modified Track Structure Theory (MTST)

According to Kalef-Ezra and Horowitz (Ezra and Horowitz 1982) who developed the principle of Modified Track Structure Theory (MTST), focused their attention on the significance of toning the energy spectra of the secondary electrons or delta rays produced by the reference gamma radiation and HCPs. The "dose response function" introduced in MTST measured with respect to reference radiation is denoted by $f_\delta(D)$ and is also known as TL response function of reference radiation under test, which simulates as much as possible the energy spectra and irradiated volume due to secondary electron generated by desired HCP radiation. Thus in MTST Re HCP to gamma can be written as

$$(R_e)_{HCP,\gamma} = (R_e)_{HCP,\gamma} = \frac{W_\gamma \int^{R_{max}} \int^{r_{max}} f\delta(D)D(r,l)2\pi r dr dl}{W_{HCP} \int^{R_{max}} \int^{r_{max}} D(r,l,E)2\pi r dr dl} \quad (2)$$

Here $(Re)_{\delta,\gamma}$ is the relative efficiency of delta ray or secondary electron generated by HCP with respect to reference gamma radiation. W_γ and W_{HCP} indicate the amount of energy required for the production of electron hole pair by the radiation of gamma and HCP respectively. The infinitesimal radial dose spreading nearby the HCP path is denoted by $D(r, l, E)$. During the radiation absorption stage of materials HCP passing through the medium slows down by emitting the charge particles which travels axial and radial distance, thus maximum axial and radial distance travelled by this charge particles are denoted by R_{max} and $r_{max} f_\delta(D)$ indicates experimentally measured TL dose response function of reference radiation and can be expressed as

$$f_\delta(D) = \frac{F(D)/D}{F(D_1)/D_1} \tag{3}$$

Here $F(D)$ and $F(D_1)$ represent the TL signal obtained at a particular dose D and low dose D_1 respectively. At low doses TL dose response is found to be linear. With the help of $f_\delta(D)$ we can measure the TL efficiency as a function of macroscopic dose. It assigns a different value in linear and supra-linear regions as unity and more than unity and possesses the value of less than unity in sublinear regions following saturation. To determine relative TL efficiency of alpha particles in LiF materials Horowitz et al successfully applied MTST by using tritium beta particle as a test radiation (Ezra and Horowitz 1982; Horowitz 1984). The material BeO has some interesting properties like HCP relative TL efficiency is greater than unity because of the highly supralinear macroscopic dose response. Thus, MTST was also applied to the BeO material for the determination of the relative TL efficiency of C, Ar and Ne ion irradiation using 15 kV X-rays. In MTST the region near the ion path for which dose distribution is greater than the higher dose distribution value and for which also $f_\delta(D)$ is also zero is defined by "efficiency saturation radius". Thus, deposition of energy in this region gives rise only slightly to the efficiency because of the existence of very high dose levels in the core which results in the wastage of the dose due to very high existence of the electron and hole trapping center and luminescent center in this region (Avila and Brandan 2004).

Since in MTST the reference test radiation is chosen on the basis of toning the secondary electron spectra or energy spectra of delta ray with that produced by HCPs, it can be briefly said that MTST can be successfully applied to those materials in which macroscopic dose response is independent of the gamma ray energy or reference test radiation.

Microdosimetric Target Theory (MTT)

The microdosimetric model is based on the physical and biological response of any microscopic volume due to the energy deposition by ionizing radiation after the irradiation of that particular volume (Kellerer and Rossi 1972, Zaider 1990). However, the distribution of this deposited energy of ionizing radiation is not uniform in the microscopic volume and transferred to the matter in discrete amounts, which results in ionization and excitation. In general the microdosimetric target theory is based on the assumption that with the help of integral convolution of two distinct functions like $f(z; D)$ and $r(z)$, the dose effect relationship can be expressed as (Horowitz and Olko 2004)

$$R_k = \int f(z; D) \, r(z) dz \tag{4}$$

Here $f(z; D)$ is known as the microdosimetric distribution function and is used to describe the fluctuation of ionization incident and $r(z)$ is known as the response function used to express the probability of effect occurring after the energy deposition incident z. Here z is used to denote the specific energy (total quantity of dose deposited to a microscopic volume) while its value increases as the target volume decreases. With the help of one hit detector model the saturation effect occurring in dose response of LiF:Mg, Cu, P without supralinearity at high doses can be explained. The one hit detector model is based on the assumption that a detector has a large number of microscopic target volume and this target volume results in the emission of TL light (response) during the single energy deposition incident of ionizing radiation. Similarly, in multi hit target theory the target volume is able to tolerate $n - 1$ hit but as the number of hits increases to 'n' or more than 'n' the target is stimulated and results in the emission of TL light. Thus, the rapid saturation effect of the response around the tracks of densely ionizing radiation arises due to the saturation of the available non-hit target volume. However, with the help of microdosimetric theory the relative TL efficiency of various TL detectors to HCP and photon energies with respect to reference ionizing radiation energy has been studied by many investigators.

Track Interaction Model (TIM)

LiF has been used since the 1960s as one of the passive radiation dosimetric materials for the measurement of the dose of ionizing radiation. It is necessary to explain the behavior related to saturation of HCP TL fluence response and ionization density dependence of supralinearity for the dose response of peak 5 in LiF:Mg,Ti (TLD-100). Therefore a number of phenomena (basis of track interaction model) related to ionization density dependence of TL signal emitted by TLD-100 was conceived during the 1970s (Attix 1975; Claffy et al. 1968). A mathematical model to explain both behaviors as mentioned above to ionization density or different doses of radiation in LiF (TLD-100) is well described in the framework of track interaction model (Horowitz et al. 1996; Moscovitch and Horowitz 1988; Rosenkrantz and Horowitz 1993). Whereas the first track interaction model was proposed by Claffy and Atix (Attix 1975; Claffy et al. 1968). According to TIM when any TL material gets irradiated by ionizing radiation of low doses/fluence, an electron and hole pair is created, which acts as a localized entity (densely ionized region), gets trapped around the HCP track, and results in emission of TL light by heating of the materials due to radiative recombination of the localized entity. In the case of low dose/fluence the charges recombine in their own track and also if they are not able to reach the recombination center, they get destroyed non-radiatively. Thus, in the case of linear region of HCP TL fluence response the role of competitive processes gets suppressed. Whereas in the case of high dose/fluence (supralinear region), the total number of tracks get increased and also the distance between these tracks is reduced leading to intersection between the tracks, which results in the increase in number of localized entity providing enhanced TL efficiency by radiative recombination of trapped charges due to increase in recombination probability. The saturation of HCP TL fluence response was explained with the help of theoretical consideration of HCP radial dose profile which shows that due to overlap in tracks at radiation absorption stage on high dose/fluence it leads to saturation in dose response (Horowitz et al. 1982). Thus, to show the complete mechanism involved in dose response of TL materials, one has to consider both the absorption and recombination stage.

In the model developed by Horowitz (Horowitz et al. 2002) it has been clearly shown that HCP passing through TL materials ionize the electron and hole across their track, and this ionized particle moves outward, gets localized surrounding the track results in the highly occupation probability of Charge Carriers (CCs), Trapping Centers (TCs) and luminescence recombination centers (LCs) in the form of a cone having a circular cross section. Here R is the distance between the two-track axes and r_{eff} is the radius of the circular cone. As the HCP passing through the matter destroys the competitive center across the track by capturing the electron in the radiation absorption stage. These competitive centers are still dominant in the inter track region at low doses. Whereas in the case of high doses the inter track region gets reduced or it can be said that the electrons escaping from the parent track reach the populated luminescence center of the other track near the parent track and results in the production of TL light.

Unified Interaction Model (UNIM)

The unified interaction model, developed by BGUG (Ben Gurion University Group) to explain the theory of dose response, is capable of explaining the phenomenon like ionization density dependence of supra-linearity and sensitization of several glow curves in LiF-TLD (100) incorporated with the physical ideas of defect interaction model (DIM) (Mische and McKeever 1989) for gamma rays and electrons (uniform ionizing radiation) along with all the important features of the extended track interaction model for densely ionizing radiation or Heavy Charged Particles (HCPs), into an integrated mathematical background. Whereas the basic assumption of UNIM is based on the recombination processes involved in localized and delocalized entity which give rise to the luminescence process, and is also able to explain the dose dependence of supra-linearity and sensitization behavior of LiFTLD(100) to both gamma ray and HCP. According to this model, materials on irradiation with gamma and electron at low doses give rise to a localized entity which is mainly TCs and LCs and are spatially coupled together with a long range interaction (Fig. 11.5), and make recombination results in the linear region of the dose response curve where migration of charged carrier to conduction

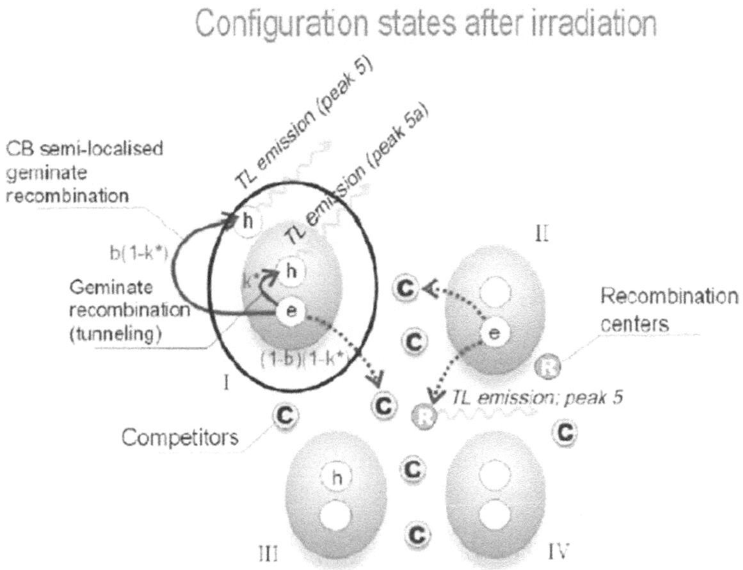

Fig. 11.5. Schematic diagram of the configuration states of spatially coupled TC/LC following irradiation: (i) locally trapped e-h pair into TC/LC complex, (ii) singly trapped electron in TC/LC complexes, (iii) singly trapped hole in TC/LC complexes and (iv) both TC/LC complex are emptied. Here k is used to denote the relative geminate recombination probability within the TC/LC complex. (Reprinted with permission from Y. Horowitz, "Thermoluminescence theory and analysis: Advances and impact on applications", *Encyclopedia of Spectroscopy and Spectrometry*, Third Edition: 444–451)

band is not involved and hence competitive processes are suppressed (Horowitz et al. 2017).

Extended Track Interaction Model (ETIM)

In the frame work of the extended track interaction model, special emphasis is given to the role of the individual track to understand the track interaction effects in HCP TL response. The normalized HCP TL response function is denoted by $f(n)$ and can be written in mathematical form as:

$$f(n) = [F(n)/n]/[F(n^*)/(n^*)] \qquad (5)$$

Here $F(n)$ is used to denote the TL signal at a fluence n per cm^2 and $F(n^*)$ is used to indicate the TL response at a fluence of n^* somewhere in the linear region of the response. The HCP TL response takes the value of unity in the linear region and greater than unity in supralinear region with less unity in the region where the HCP TL fluence response shows saturation. Moreover, in the extended track interaction model it has been explained that the saturation effect not only arises in the region where the value of $f(n)$ is less than unity but also effective at lower level of fluences and it may be in cooperation with the things giving rise to supralinearity over a significant part of the range of fluence where supralinearity is detected. Thus ETIM is able to theoretically explain both supralinearity and saturation effects in this way. Therefore, in this model HCP TL fluence response is approximated by following equation

$$f(n) = [1 + fsup(n)][fsat(n)] \qquad (6)$$

In the above Equation (6) the TL signal arising from the contribution of single track. '$fsup(n)$' corresponds to supralinear contribution and $fsat(n)$ corresponds to saturation effects due to growing probability of overlapping of tracks at high fluences. Implicit in the above equation based on the postulation that the both linear and supralinear behavior are affected in almost the same manner due to the non-obtainability of extra trapping and luminescence centers (TC and LC) in the track core overlap region. Here $fsup(n)$ discussed briefly according to the expression shown below.

$$fsup(n) = [Ne / Nw] \sum_{i=1}^{4} \int_{r_0} g(r_h, R_i) \exp(-R / \lambda_0) P_i(n, R_i) dR \qquad (7)$$

N_e and N_w are used to denote the total number of electrons that leave or escape from the parent track and number of electrons that recombine within the parent track results in the emission of TL signal respectively. r_0 denotes the maximum distance from the track axis at which electrons have a major probability of being escaped from its parent track and λ_0 is assigned for the mean free path for the diffusion of the charge carrier within the inter track region. Here r_h and R_i are assigned for the actual recombination radius for the adjacent tracks and distance between neighboring tracks respectively. The $g(r_h, R_i)$ is called solid angle factor between adjacent tracks and can also be written as $r_h/(2R)$ while $p_i(n, R_i)$ is the ith nearest neighbor probability distribution function and has already been calculated by many investigators up to the 4th nearest neighbor contribution (Fuks et al. 2011; Horowitz et al. 2003; Moscovitch and Horowitz 1988).

Carbon Beam Irradiated TLD Materials

It is known that TL materials are mainly insulators or semiconductors on which irradiation with different exposures to ionizing radiation like gamma rays, electron, alpha particles and heavy charged particles exhibit the characteristics of the cold emission of light on heating the materials. This above technique is known as thermoluminescence and is widely used these days in the field of dosimetry of ionizing radiation. However, since the age of the discovery of TL technique to the modern age many TL materials have been investigated and they are preferably used in the radiation dosimetry areas of low ionizing radiation (Fuks et al. 2011; Horowitz et al. 2003). Some of the materials in microcrystalline forms are LiF:Mg, Ti; LiF:Cu, Mg, Ti; CaSO$_4$:Dy, LiNaSO$_4$: Eu and K$_3$Na(SO$_4$):Dy etc. In the case of heavy charged particle irradiation of the above materials in their original microcrystalline form, however, does not provide a better TL dose response even at low fluences (Satinger 2000) but with the increase of the dose a significant variation in the glow curve structure is observed which is unwanted for the dosimetry purposes. The poor TL dose fluence response to HCP is mainly due to the saturation effect which arises due to the overlapping of tracks at higher doses as explained earlier in this chapter in the section of track interaction model. However, it has been experimentally proved that the same materials in their nanocrystalline form show a better TL dose fluence response even at the higher doses. At present a large number of materials have been studied for γ ray dosimetry but there are very few studies for C^{5+} ion irradiation. Here we deal with the study of some TLD materials like $CaSO_4$:Dy, $CaMg(SO_4)_4$:Dy, $LiCaBO_3$:Dy or Ce and $Li_2BaP_2O_7$:Dy which have been irradiated with carbon ion beam (Salah et al. 2006; Salah 2008; Kore et al. 2014; Oza et al. 2015; Wani et al. 2015).

$CaSO_4$:Dy nano crystal with cubic/rectangular shape

As it is known sulfate-based materials are widely used in radiation dosimetry and have gained popularity due to their high sensitive TL response. In the year 1895 calcium sulfate ($CaSO_4$) became the first material which was used to measure ionizing radiation dose, and thereafter a lot of investigators showed interest to synthesis this material with various dopants to improve its TL response. However, $CaSO_4$ doped with Dy gained a lot of attraction of investigators and the first was prepared by the Yamshita et al. in 1971 (Yamshita et al. 1971), which has been used by many countries for dosimetry purposes to ionizing radiation like gamma rays, X rays, electron beam and also for alpha particles, due to its high sensitivity, simple glow curve structure, minimum fading, easy method of preparation and excellent reusability. Since the TL saturation effect arises in its microcrystalline form at very high doses of gamma rays, it has been observed that the same material shows enhanced TL response and sensitivity in its nanocrystalline form. Thus the nanocrystalline form of $CaSO_4$:Dy materials can be used to investigate its TL response to HCP. The nanocrystalline $CaSO_4$:Dy phosphor was prepared for the first time by Salah et al. with the help of chemical co-precipitation method (Salah et al. 2006).

The pellet size of 6 mm thickness and 1 cm diameter was prepared for both the nano and microcrystalline sample by using 50 mg and 100 mg of $CaSO_4$:Dy sample with 1 mg of Teflon powder respectively and then the pellets were irradiated to C^{6+} ion beam at an energy of 75 MeV at room temperature at different ion fluences within the range of 1×10^9 to 1×10^{13} ion cm^{-2} with the help of a 16 MV Tandem Van De-Graff type electrostatic accelerator (15 UD Pelletron) at Inter University Accelerator Center (IUAC) New Delhi, India (Kanjilal et al. 1993).

According to Fig. 11.6 it can be clearly seen that the TL glow curves for both micro and nanocrystalline samples are nearly same in their shape and two peaks are observed for both at 160 °C and 210 °C with a slight difference (due to the change in total number of luminescent and trapping centers as a result of high energetic ion beam irradiation) in their TL response. In the case of the microcrystalline sample the second peak at 210 °C was found to be dominant at low fluences while in nanoparticles the first peak at 160° C was dominant even at higher fluences. From the plot of peak intensity ratio of

a

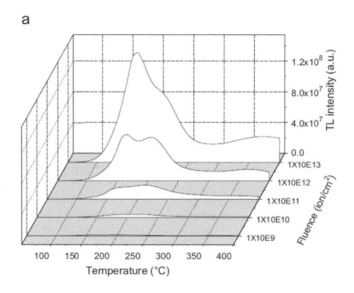

b

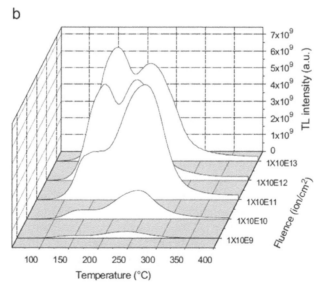

Fig. 11.6. (a) TL glow curve of nanocrystalline $CaSO_4$: Dy at different ion fluences. (b) TL glow curve of microcrystalline $CaSO_4$:Dy. (Reprinted with permission from Numan Salah, "Nanocrystalline materials for the dosimetry of heavy charged particles: A review", *Radiation Physics and Chemistry*, 80 (2011): 1–10)

two glow peaks at 160 and 210° C as a function of ion beam fluence for both micro and nanoparticles it was observed that the TL intensity ratio varies little at low fluences while at higher fluences the change was large.

The total amount of the dose absorbed by the micro and nano particles after C^{6+} beam irradiation was calculated with the help of the relation (Geiß et al. 1998):

$$D = 1.602 \times 10^{-10} \times 1/\rho \times de/dx \times \varphi \tag{8}$$

Here the absorbed dose D is expressed in the unit of Gy, de/dx is mean energy loss in MeVcm^{-1}, φ is the particle fluence in ion cm^{-2} and finally ρ is the density of the material in gmcm^{-3}. The total energy loss de/dx [electronic energy loss $(de/dx)_{elc}$ and nuclear energy loss $(de/dx)_{nuc}$] was calculated using a monte carlo simulation based on the trim code and tabulated in Table 11.1, while both nano and microcrystalline materials absorbed the same amount of dose 0.299 to 2990 Gy as the fluence goes to 1×10^9 to 1×10^{13} ion cm^{-2}.

Table 11.1. TRIM calculation on nano and microcrystalline CaSO$_4$:Dy phosphor irradiated with 75 Mev C^{6+} ion beam

Sample	Density (ρ) (gm/cm^3)	$(de/dx)_{elc}$ (MeV/mm)	$(de/dx)_{nuc}$ (MeV/mm)	Range (μm)
Nanocrystalline	1.06	198	1.064	244.24
Microcrystalline	2.12	396	1.129	122.12

Source: Data from Numan Salah, "Nanocrystalline materials for the dosimetry of heavy charged particles: A review", *Radiation Physics and Chemistry*, 80 (2011): 1–10.

Disk Shaped CaMg$_3$(SO$_4$)$_4$:Dy Phosphor

As mentioned earlier the sulfate based phosphors are most popular for their use in the measurement of ionizing radiation using the TL technique. Some examples of sulfate based materials are CaSO$_4$:Dy, Li$_2$SO$_4$:Dy,Eu and K$_2$Ca$_2$(SO$_4$)$_3$:Eu, etc. and halo sulfates like K$_3$Ca$_2$(SO$_4$)$_3$Cl: Dy and Na$_{21}$Mg(SO$_4$)$_{10}$Cl$_3$ etc. Sulfate based phosphor have been investigated by many researchers due to their outstanding TL properties and became part of recent investigation, to enhance the TL properties of these known materials or either to develop new high sensitive materials with idyllic TL characteristics. Thus CaMg$_3$(SO$_4$)$_4$ also called CMS phosphor was first of all developed by Kore et al. (Kore et al. 2014) by using an acid distillation method to measure the gamma ray dose and ion beam radiation of C^{5+}.

The phosphor in the form of pellets and having a thickness 0.075 cm were irradiated by C^{5+} ion beam at energies of 50 MeV and 75 MeV at room temperature at different ion fluences within the range of 15×10^{10} to 30×10^{12} ion cm^{-2} at Inter University Accelerator Center (IUAC), New Delhi. The TL analysis of CMS phosphor was performed for different ionizing radiation by Kore et al. with gamma rays from ^{60}Co source at 15 Gy dose, ^{137}Cs source at 2300 mRad and C^{5+} ion beam at 50 and 75 MeV with 216 KGy (Kore et al. 2014). Here $CaSO_4$:Dy phosphor was also used for a comparative study. As shown in Fig. 11.7 it can be seen that the TL glow curve shape for both ^{60}Co and ^{137}Cs was nearly same and it also resembles the glow curve shape of CMS phosphor on C^{5+} ion beam (50 and 75 MeV) irradiation with a slight change in the glow peak position and TL intensity. This may occur due to the creation of new trapping and luminescent centers. However, the glow curve shape (as shown in Fig. 11.7d) of 50 MeV C^{5+} ion beam and 75 MeV ion beam irradiation were found to be same with a little change in the peak position and TL intensity. The TL glow curve of CMS phosphor for gamma ray irradiation with ^{60}Co (as shown in Fig. 11.7) indicates two glow peak at 260° C and 150° C with prominent peak at 260° C.

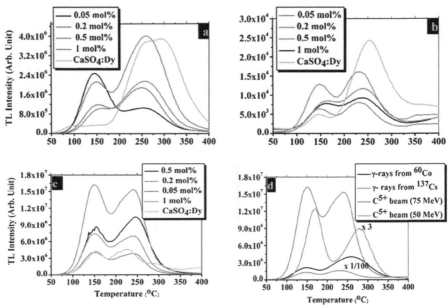

Fig 11.7. TL study of CMS phosphor: (a) TL response of ^{60}Co γ ray irradiated CMS phosphor having total dose of 15 Gy. (b) TL response of ^{137}Cs γ ray irradiated CMS phosphor having total dose of 2300 mRad. (c) TL response of CMS phosphor with C^{5+} ion beam having 75 MeV energy with dose of 216 KGy. (d) Comparison of different TL responses irradiated by ^{60}Co, ^{137}Cs and C^{5+} ion beam. (Reprinted with permission from B.P. Kore, "A new highly sensitive phosphor for carbon iondosimetry", *RSC Advances*, 4 (2014): 49979-49986)

Borate and Phosphate-based TL Phosphors

Borate based compounds are well known for their larger band gap and can preferably be used in the field of luminescence. The microcrystalline powder of $LiCaBO_3$ doped with Dy or Ce was synthesized by Oza et al., using a modified solid-state synthesis method (Oza et al. 2015). The as-synthesized phosphor was irradiated in the form of pellets by using 75 MeV C^{5+} ion beam radiation at different fluences within the range of 3.75×10^{12} to 7.5×10^{13} ion cm^{-2}. However, the maximum TL intensity for Dy doped phosphor was observed at 1.875×10^{13} ion cm^{-2} and for Ce doped phosphor the maximum TL intensity was observed at 3.75×10^{12} ion cm^{-2}. As the fluence increases beyond a certain limit, the TL intensity was found to decrease for both the phosphors. At different fluences the shape of all TL glow curves was found to be same with a little variation in the glow peak temperature. The glow peak temperature at different fluences for both Dy and Ce dopant are shown in Table 11.2. The saturation at higher fluences can be explained with the help of TIM as discussed earlier in this chapter.

Table 11.2. Glow peak temperature of $LiCaBO_3$:Dy or Ce phosphor at different fluences (Oza et al. 2015)

Fluence (ion cm⁻²)	Glow peak temperature (°C)	
	LiCaBO₃:Dy (0.5 m%)	LiCaBO₃:Ce (1 m%)
3.75×10^{12} (1 min)	115, 312	100, 154, 242
1.875×10^{13} (5 min)	158, 253	102, 164, 234
3.75×10^{13} (10 min)	124, 312	110,228
5.625×10^{13} (15 min)	111, 312	106,235
7.5×10^{13} (20 min)	113, 166, 312	106, 227

In 2002, Kovacheva et al. synthesized $Li_2BaP_2O_7$ material for the first time and determined its crystal structure (Kovacheva et al. 2001). $Li_2BaP_2O_7$ doped with Dy was synthesized by Wani et al. using solid state reaction method, to study its luminescence property (Wani et al. 2015). To study the thermoluminescence property, $Li_2BaP_2O_7$ doped with Dy sample was irradiated by 75 MeV C^{5+} ion beam at Inter University Accelerator Center, New Delhi. The prepared sample was irradiated at different fluences within the range of 15×10^{10} to 30×10^{10} ion cm^{-2}. The maximum TL intensity of $LiBaP_2O_7$:Dy phosphor was found to be maximum at 407K for 0.5 mol% of Dy at a fluence of 15×10^{10} ion cm^{-2} after 75 MeV C^{5+} ion beam irradiation (Fig. 11.8).

Conclusion

Due to the increasing interest of carbon ion beam in the field of radiotherapy to treat different types of cancerous tumors, a proper dosimetry system is needed to provide an accurate measurement of the dose delivered to the patients with minimum uncertainty. Thermoluminescence materials find its wide application in the dosimetry of ionizing radiation in personal, environmental, and clinical fields and thus we can use the thermoluminescence based dosimetry system in the field of radiotherapy with accurate precision in dose measurement. However, the use of TL materials in the dosimetry of

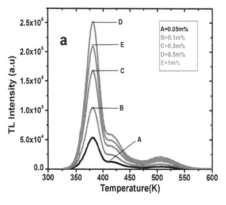 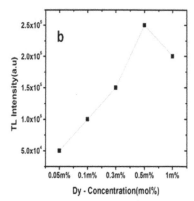

Fig. 11.8. TL glow curve of $Li_2BaP_2O_7$:Dy phosphor for different concentration of Dy, irradiated by C^{5+} ion beam. (Reprinted with permission from J.A. Wani, "Luminescence characteristics of C^{5+} ions and ^{60}Co irradiated $Li_2BaP_2O_7$:Dy^{3+} phosphor", *Nuclear Instruments and Methods in Physics Research B*, 349 (2015): 56–63)

HCP radiation is limited, which is mainly due to the saturation effect in most of the TL materials that arises even at very low doses. The reason behind such type of saturation effects in most of the TL materials has been already explained with the help of different theoretical models in this chapter. Until now only a few numbers of TL phosphors have been investigated which will be suitable for the dosimetry of carbon ion beam. However, the issue of the saturation effect can be resolved by using the TL phosphor in its nanocrystalline form and it necessitates investigating various types of TL materials which can be used in the dosimetry of carbon ion beam or other heavy charged particle beams in radiotherapy.

References

Andrievski, R.A. 2011. Behaviour of radiation defects in nanomaterials. Rev. Adv. Mater. Sci. 29, 54-67.

Attix, F.H. 1975. Further consideration of the track-interaction model for thermoluminescence in LiF(TLD–100). J. Appl. Phys. 46, 81-88.

Avila, O., Brandan, M.E. 2004. Low energy ion tracks in lithium fluoride. Nucl. Instr. and Meth. B. 218, 289-293.

Bagheri, S.C., Farhidvash, F., Perciaccante, J. 2004. Diagnosis and treatment of patients with trigeminal neuralgia. J. Am. Dent. Assoc. 135(12), 1713-1717.

Behbehani, R., Sergott, R.C., Savino, P.J. 2004. Orbital radiotherapy for thyroid-related orbitopathy. Curr. Opin. Ophthalmol. 15(6), 479-482.

Bert, C., Gemmel, A., Saito, N., Chaudhri, N., Schardt, D., Durant, M., Kraft, G., Rietzel, E. 2010. Dosimetric precision of an ion beam tracking system. Radiat. Oncol. 5, 61.

Berger, M.J., Seltzer, S.M. 1964. Studies in penetration of charged particles in matter. Natl. Acad. Sci. Publ. 1133, National Academy of Science, Washington, DC.

Bethe, H.A. 1930. Theory of the passage of fast corpuscular rays through matter. Ann. Physik. 5, 325.

Bhatt, A.D. 2013. Carbon ions – A new horizon in radiation oncology. J. Nucl. Med. Radiat. Ther. 4, 2.

Bichsel, H. 1972. Section 8d of American Institute of Physics Handbook. McGraw Hill.

Biersack, J.P., Haggmark, L.G. 1980. A Monte Carlo computer program for the transport of energetic ions in amorphous targets. Nucl. Instr. and Meth. 174, 257-269.

Brandan, M.E., Gamboa-deBuen, I., Rodriguez-Villafuerte, M. 2002. Thermoluminescence induced by heavy charged particles. Radiat. Prot. Dosim. 100(1-4), 39.

Boyle, R. 1664. Experiments and consideration touching colures. Royal Society. 413.

Cameron, J.R., Zimmerman, D., Kenney, G., Buch, R., Bland, R., Grant, R. 1964. Thermoluminescence radiation dosimeter utilizing LiF. Health Phys. 10(1), 25-29.

Castro, J.R., Linstadt, D.E., Bahary, J.P., Petti, P.L., Daftari, I., Collier, J.M., Gutin, P.H., Gauger, G., Phillips, T.L. 1994. Experience in charged particle irradiation of the skull base 1977-1992. Int. J. Radiat. Oncol. Biol. Phys. 29, 647-655.

Claffy, E.W., Klick, C.C., Attix, F.H. 1968. Proceedings of the Second international Conference Luminescence Dosimetry, Gatlinburg, AEC Conf. 680920, 302.

Daniels, F., Boyd, C.A. and Saunders, D.F. 1953. Thermoluminescence as a research tool. Science 117, 343-349.

Doan, N.V., Martin, G. 2003. Elimination of irradiation point defects in crystalline solids: Sink strengths. Phys. Rev. B. 67, 134107.

Eickhoff, H., Haberer, T., Kraft, G., Krause, U., Richter, M., Steiner, R., Debus, J. 1999. The GSI cancer therapy project. Strahlenther. Onkol. 175, 21.

Eric, H.J. 2000. Radiobiology for the radiologist. Philadelphia: Lippincott Williams Wilkins. p. 351.

Ertner, D.S., Jakel, O., Schlegal, W. 2006. Radiation therapy with charged particles. Semin. Radiat. Oncol. 16, 249.

Ezra, K.J., Horowitz, Y.S. 1982. Heavy charged particle thermoluminescence dosimetry: Track structure theory and experiments. Int. J. Appl. Radiat. Isot. 33, 1085-1100.

Fuks, E., Horowitz, Y.S., Horowitz, A., Oster, L., Marino, S., Rainer, M., Rosenfeld, A., Datz, H. 2011. Thermoluminescence solid state nanodosimtry – The peak 5A/5 dosemeter. Radiat. Prot. Dosim. 143(2-4), 416-426.

Garlick, G.F.J., Gibson, A.F. 1949. The luminescence of photo-conducting phosphors. J. Opt. Soc. Am. 39(11), 935-941.

Geiß, O.P., Kramer, M., Kraft, G. 1998. Efficiency of thermoluminescent detectors to heavy charged particles. Nucl. Instrum. Methods B 142, 592-598.

Geiss, O.B., Kramer, M., Kraft, G. 1998. Efficiency of thermoluminescent detectors to heavy charged particles. Nucl. Instr. and Meth. B. 142, 592-598.

Horowitz, Y.S., Moscovitch, M., Dubi, A. 1982. Response curves for the thermoluminescence induced by alpha particles: Interpretation using track structure theory. Phys. Med. Biol. 27, 1325-1338.

Horowitz, Y.S. 1984. Thermoluminescence and thermoluminescent dosimetry. Vol. II. CRC Press. Boca Raton.

Horowitch, Y.S., Rosenkrantz, M., Mahajna, S., Yossian, D. 1996. The track interaction model for alpha particle induced thermoluminescence supralinearity: Dependence of the supralinearity on the vector properties of the alpha particle radiation field. J. Phys. D. Appl. Phys. 29, 205-217.

Horowitz, Y.S., Satinger, D., Oster, L., Issa, N., Brandam, M.E., Avila, O., Rodriguez-Villafuerte, M., Gamboa-deBuen, I., Buenfil, A.E., Ruiz-Trejo, C. 2001. The extended track interaction model: Supralinearity and saturation He-ion TL fluence response in sensitized TLD-100. Radiat. Meas. 33, 459-473.

Horowitz, Y.S. 2001. Theory of thermoluminescence gamma dose response: The unified interaction model. Nucl. Instr. and Meth. B. 184, 68-84.

Horowitz, Y.S., Avila, O., Villafuerte. M.R. 2001. Theory of heavy charged particle response (efficiency and supralinearity) in TL materials. Nucl. Instr. and Meth. in Phys. Res. B 184, 85-112.

Horowitz, Y.S., Satinger, D., Avila, O. 2002. Theory of heavy charged particle thermoluminescence response: Extended track interaction model. Radiation Protection Dosimetry 100(1-4), 91-94.

Horowitz, Y.S., Oster, L., Biderman, S., Einav, Y. 2003. Localized transitions in the thermoluminescence of LiF:Mg,Ti: Potential for nanoscale dosimetry. J. Phys. D. Appl. Phys. 36, 446-459.

Horowitz, Y.S., Olko, P. 2004. The effects of ionization density on the thermoluminescence response (efficiency) of LiF:Mg,Ti and LiF:Mg,Cu,P. Radiation Protection Dosimetry 109(4), 331.

Horowitz, Y., Chen, R., Oster, L., Eliyahu, I. 2017. Thermoluminescence theory and analysis: Advances and impact on applications. Encyclopedia of Spectroscopy and Spectrometry, Third Edition 444-451.

Ishikawa, H., Tsuji, H., Murayama, S., Sugimoto, M., Shinohara, N., Maruyama, S., Murakami, M., Shirato, H., Sakurai H. 2019. Particle therapy for prostate cancer: The past, present and future. Int. J. Urol. 26, 971-979.

Kanematsu, N., Furukawa, T., Hara, Y., Inaniwa, T., Iwata, Y., Mizushima, K. 2019. New technologies for carbon-ion radiotherapy—Developments at the National Institute of Radiological Sciences, QST, Japan. Radiat. Phys. Chem. 162, 90-95.

Kanjilal, D., Chopra, S., Narayanan, M.M., Iyer, I.S., Vandana, J.J.R., Datta, S.K. 1993. Testing and operation of the 15UD Pelletron at NSC. Nucl. Instrum. Methods A. 328, 97-100.

Karger, C.P., Jakel, O., Palmans, H., Kanai, T. 2010. Dosimetry for ion beam radiotherapy. Phy. Med. Biol. 55, R193-R234.

Kellerer, A.M., Rossi, H.H. 1972. The theory of dual-radiation action. Cur. Top. Radiat. Res. Quart. 8, 85.

Kore, B.P., Dhoble, N.S., Lochab, S.P., Dhoble, S.J. 2014. A new highly sensitive phosphor for carbon ion dosimetry. RSC. Adv. 4, 49979-49986.

Kovacheva, D., Nikolov, V., Petrov, K., Rojas, R.M., Herrerob, P., Rojo, J.M. 2001. Synthesis and ionic conductivity of pure and Co-doped lithium barium diphosphates $Li_2BaP_2O_7$. J. Mater. Chem. 11, 444-448.

Kraft, G. 2000. Tumor therapy with heavy charged particles. Prog. Part. Nucl. Phys. 45, S473.

Krasheninnikov, A.V., Nordlund, K. 2010. Ion and electron irradiation-induced effects in nanostructured materials. J. Appl. Phys. 107, 071301.

Lindhard, J., Scharff, M. 1961. Energy dissipation by ions in the kev region. Phys. Rev. 124, 128-130.

Lyman, T. 1935. The transparency of the air between 1100 and 1300A. Phys. Rev. 48, 149-151.

Massillon-JL, G., Gamboa-deBuen, I., Brandan, M.E. 2006. Observation of enhanced efficiency in the excitation of ion-induced LiF:Mg,Ti thermoluminescent peaks. J. Appl. Phys. 100, 103521-103526.

McKeever, S.W.S., Horowitz, Y.S. 1990. Charge trapping mechanisms and microdosimetric processes in lithium fluoride. Int. J. Radiat. Appl. Instrum. Part C 36, 35-46.

Mische, E.F., McKeever, S.W.S. 1989. Mechanisms of supralinearity in lithium fluoride thermoluminescence dosimeters. Radiat. Prot. Dosim. 29, 159-175.

Moscovitch, M., Horowitch, Y.S. 1988. A microdosimetric track interaction model applied to alpha-particle-induced supralinearity and linearity in thermoluminescent LiF:Mg, Ti. J. Phys. D. Appl. Phys. 21, 804.

Nieder, C., Milas, L., Ang, K.K. 2000. Tissue tolerance to reirradiation. Semin. Radiat. Oncol. 10(3), 200.

Olko, P., Bilski, P., Budzanowski, M., Waligórski, M.P.R., Reitz, G. 2002. Modeling the response of thermoluminescence detectors exposed to low- and high-LET radiation fields. J. Radiat. Res. 43, S59-S62.

Overman, M.J., Kopetz, S., Lin, E., Abbruzzese, J.L., Wolef, R.A. 2010. Is there a role for adjuvant therapy in resected adenocarcinoma of the small intestine. Acta Oncologica 49, 474-479.

Oza, A.H., Dhoble, N.S., Dhoble, S.J. 2015. Luminescence study of Dy or Ce activated $LiCaBO_3$ phosphor for γ-ray and C^{5+} ion beam irradiation. Luminescence 30(7), 967.

Parker, R.G. 1985. Particle Radiation Therapy. Cancer 55, 2240.

Pour, N.H., Farajollahi, A., Jamali, M., Zeinali, A., Jangjou, A.G. 2018. Comparison of three and four-field radiotherapy technique and the effect of laryngeal shield on vocal and spinal cord radiation dose in radiotherapy of non-laryngeal head and neck tumors. Pol. J. Med. Phys. Eng. 24(1), 25-31.

Randall, J.T., Wilkins, M.H.F. 1945. Phosphorescence and electron traps. Proceedings of the Royal Society A184, 366.

Rosenkrantz, M., Horowitz, Y.S. 1993. Alpha particle induced TL supralinearity in TLD-100: Dependence on vector properties of the radiation field. Radiat. Prot. Dosim. 47, 27-30.

Salah, N., Sahare, P.D., Lochab, S.P., Kumar, P. 2006. TL and PL studies on $CaSO_4$:Dy nanoparticles. Radiat. Meas. 41, 40-47.

Salah, N. 2008. Carbon ions irradiation on nano- and microcrystalline $CaSO_4$:Dy. J. Phys. D: Appl. Phys. 41, 155302.

Salah, N. 2011. Nanocrystalline materials for the dosimetry of heavy charged particles. Radiat. Phys. Chem. 80, 1-10.

Samuel, E.J.J., Srinivasan, K., Poopathi, V. 2017. Radiological properties of plastics and TLD materials: its application in radiation dosimetry. IOP Conf. Series: Journal of Physics. Conf. Series 847, 012064.

Satinger, D. 2000. The extended track interaction model: Experimental and theoretical investigations of HCP induced supralinearity, sensitization and saturation. Ph.D. Thesis, Ben Gurion University, Negev, Israel.

Schmidt, P., Fellinger, J., Hubner, K. 1990. Experimental determination of the TL response for protons and deuterons in various detector materials. Radiat. Prot. Dosim. 33, 171-173.

Skowronek, J. 2017. Current status of brachytherapy in cancer treatment – Short overview. J. Contemp. Brachytherapy 9(6), 581-589.

Su, L.N., Wu, C.C., Hsu, P.C., Weng, P.S. 1985. Gonadal dose measurement in diagnostic nuclear medicine using thermoluminescent dosimeters. Radioisotopes 34, 137-143.

Tanaka, S., Furuta, Y. 1974. Estimation of gamma-ray exposure in mixed gamma-neutron fields by ^6LiF and ^7LiF thermoluminescence dosimeters in pair use. Nucl. Instr. and Meth. 117, 93-97.

Taylor, C.W., Nisbet, A., McGale, P., Darby, S.C. 1990. Cardiac exposures in breast cancer radiation therapy: 1950s-1990s. Int. J. Radiat. Oncol. Biol. Phys. 69(5), 1484-1495.

Tobias, C.A., Anger, H.O., Lawrence, J.H. 1952. Radiological use of high energy deuterons and alpha particles. Am. J. Roentgenol. 67, 1-27.

Torikoshi, M., Ogawa, H., Yamada, S., Kanazawa, M., Kohno, T., Noda, K., Sato, Y., Takada, E., Araki, N., Kawachi, K., Hirao, Y., Sudo, M., Takagi, H., Narita, K., Mizobata, M., Ueda, K. 1995. Performance of beam monitors used at a beam transport system of HIMAC. AIP Conference Proceedings 333, 412-418.

Tsuji, H., Kamada, T. 2012. A review of update clinical results of carbon ion radiotherapy. Jpn. J. Clin. Oncol. 42(8), 670.

Waligorski, M.P.R., Katz, R. 1980. Supralinearity of peak 5 and 6 in TLD – 700. Nucl. Instrum. Methods 175, 48–50.

Wani, J.A., Dhoble, N.S., Lochab, S.P., Dhoble, S.J. 2015. Luminescence characteristics of C^{5+} ions and ^{60}Co irradiated $Li_2BaP_2O_7:Dy^{3+}$ phosphor. Nucl. Instrum. and Methods in Phys. Resear. B, 349, 56-63.

Wiedman, E., Schmidt, G.C. 1895. Ann. Phys. Chem. Neut. Folge 54, 604.

Wilson, R.R. 1946. Radiological use of fast protons. Radiology. 47, 487-491.

Yamashita, T., Nada, N., Ohishi, H., Kitamura, S. 1971. Calcium sulfate activated by thulium or dysprosium for thermoluminescence dosimetry. Health Phys. 21, 295-300.

Yanwen, Z., Aidhy, S.D., Varga, T., Moll, S., Edmondson, D.P., Namavar, F., Jin, K., Ostrouchov, C.N., Weber W.J. 2014. The effect of electronic energy loss on irradiation-induced grain growth in nanocrystalline oxides. Phys. Chem. Chem. Phys. 16, 8051-8059.

Yu, H., Dong, Q., Yao, Z., Zhang, H.K., Kirk, M.A., Daymond, M.R. 2019. In-situ study of heavy ion irradiation induced lattice defects and phase instability in β-Zr of a Zr-Nb alloy. J. Nucl. Mater. 522, 192-199.

Zaider, M. 1990. Microdosimetry and Katz's track structure theory 1: One-hit detectors. Radiat. Res. 124, S16-S22.

Zhou, J., Yao, T., Cao, D., Lian, J., Lu, F. 2018. In-situ TEM study of radiation-induced amorphization and recrystallization of hydroxyl apatite. J. Nucl. Mater. 512, 307-313.

Upconversion Photoluminescence in the Rare Earth Doped Y_2O_3 Phosphor Materials

R.S. Yadav[1]*, Monika[2] and S.B. Rai[2]

[1] Department of Zoology, Institute of Science, Banaras Hindu University, Varanasi - 221005, India

[2] Laser & Spectroscopy Laboratory, Department of Physics, Institute of Science, Banaras Hindu University, Varanasi - 221005, India

Introduction

Luminescence is characterized by the emission of light from a phosphor material on excitation with a proper source of light. If the emission occurs due to interaction of the phosphor materials with the incident photon it is termed as photoluminescence. If the emitted light has a longer wavelength than the incident light it is called downconversion. However, the emission of light of lower wavelength than the incident light occurs in the upconversion process. When two or more low energy near-infrared (NIR) photons are converted in the high energy Ultra-Violet (UV) and Visible (Vis) photons the process is termed as upconversion. The rare earth ions are very rich to produce photoluminescence via upconversion and downconversion (Yadav et al. 2017; Muenchausen et al. 2007). In this chapter, the upconversion photoluminescence occurring from the rare earth ions is discussed.

The rare earth ions are also known as lanthanide ions and are the member of lanthanide series. They contain 17 elements from lanthanum with atomic number (57) to lutetium (Lu) with atomic number (71) alongwith two other chemically similar elements, such as Sc (21) and Y (39). The rare earth ions contain a large number of energy levels. The lifetime of these levels ranges from nano-second (ns) to milli-second (ms). These levels are responsible for the emission of light in phosphor materials. They are also

---

*Corresponding author: ramsagaryadav@gmail.com

responsible for an energy transfer from one rare earth ion to another. This increases the population of the ions in excited and long-lived levels of the rare earth ions. The energy transferred by an ion is called sensitizer/donor ion whereas the energy absorbed by another ion is known as activator/accepter ion. As a result, the emission intensity of one of the rare earth ion is enhanced significantly. This process has been observed not only in the case of upconversion but also in downconversion. However, the Yb^{3+} ion acts as a sensitizer in the upconversion process only (Yadav et al. 2015; Auzel 1996).

The host material is a very important part of a phosphor for initiating transition features in the rare earth ion. The Y_2O_3 has been extensively used as the phosphor host by various groups of researchers. It has a relatively low phonon energy (430-550 cm^{-1}), which is an essential parameter for the radiative transitions. It is also a structurally, chemically and thermally stable host. It has a wide band gap energy of the order of 5.6 eV (Muenchausen et al. 2007; Yadav et al. 2017a; Yadav et al. 2019b). The Y_2O_3 host has been utilized not only in upconversion but also in downconversion and quantum cutting processes (Auzel 1996; Yadav et al. 2019b; Jadhav et al. 2016; Wei et al. 2010). Muenchausen et al. have also studied the downconversion photoluminescence in Tb^{3+} doped Y_2O_3 nano-phosphor and observed green photoluminescence on excitation with 304 nm (Muenchausen et al. 2007). The downconversion photoluminescence of Sm^{3+} doped Y_2O_3 nano-crystalline phosphor was studied by our group and the effect of annealing and excitation wavelengths were considered. It was observed that the phosphor sample emits larger photoluminescence corresponding to the excitation band that appeared with larger intensity i.e. 407 nm (Yadav et al. 2019b). However, quantum cutting was observed by Jadhav et al. in the Tb^{3+}, Yb^{3+} co-doped Y_2O_3 nanophosphor in which the Tb^{3+} ions transferred their excitation energy to Yb^{3+} ions due to cooperative downconversion energy transfer process. They have calculated the energy transfer efficiency as 181.1% and suggested its application in photonics and solar cells (Jadhav et al. 2016). The quantum cutting was also observed in the Bi^{3+}, Yb^{3+} co-doped Y_2O_3 phosphor. In this case, the one Bi^{3+} ion transferred their excitation energy to the two Yb^{3+} ions cooperatively and the NIR emission occurs at 979 nm, which can be useful for solar cells (Wei et al. 2010).

On the other hand, the upconversion is very interesting phenomenon, which utilizes the absorption of low energy NIR photons by the materials. The upconversion process was initially led by Auzel in the Tm^{3+}, Yb^{3+} co-doped material using near-infrared (NIR) wavelength as an excitation source (Auzel 1996). He discussed the mechanisms of upconversion in the rare earth ions. However, we describe the upconversion emission arising in different combinations of rare earth doped Y_2O_3 phosphors in the present case. When the rare earth ions are co-doped in Y_2O_3 host it gives intense UV, visible and NIR emissions as it contains low phonon energy (Silver et al. 2001; Vetrone et al. 2003; Guo and Qiao et al. 2009; Hou et al. 2011; Zheng et al. 2010; Yadav et al. 2015). The upconversion emission in Er^{3+}, Yb^{3+} co-doped Y_2O_3 phosphor has been studied by Silver et al. and they discussed the effect of EDTA on the morphology, crystallinity and upconversion intensity of the phosphor (Silver et al. 2001). When the EDTA is present in the phosphor the particles size of the phosphor was affected. They have also observed that if the EDTA is in a smaller amount the particles size is larger; however, for larger EDTA amounts the particles size is found to be smaller. In this case, the Er^{3+} and Yb^{3+} co-doped Y_2O_3 phosphor gives Stokes and anti-Stokes emissions and the phosphor was excited with 632.5 nm wavelength of He/Ne laser. However, Vetrone et al. prepared the Er^{3+}, Yb^{3+} co-doped Y_2O_3 phosphor through solution combustion method (Vetrone et al. 2003). They used 978 nm radiation for exciting the Er^{3+}, Yb^{3+} co-doped Y_2O_3 phosphor. It has been observed that the phosphor emits green and red anti-Stokes emissions in which the intensity of the red band is larger than that of the green band. The sample prepared by the solution combustion method yields uniform emission intensity. When the phosphor was excited by 488 nm directly visible and NIR emissions were observed. The emission in the NIR region matches the absorption cross section area of Yb^{3+} ion. This confirms the energy transfer from Yb^{3+} to Er^{3+} ions, which enhances the emission intensity of the phosphor samples (Guo and Qiao et al. 2009).

The upconversion was also studied in other combinations of the rare earth ions. Hou et al prepared the RE/Yb co-doped Y_2O_3 transparent ceramics with RE ions (i.e. RE = Er, Ho, Pr, and Tm) through solid state reaction method and studied the upconversion luminescence on excitation with 980 nm (Hou et al. 2011). They observed green and red upconversion emissions in the Er, Yb co-doped sample while only green UC emission has been observed in the Ho, Yb and Pr, Yb co-doped samples. The Tm, Yb co-doped sample emits intense blue upconversion emission. They also monitored the transmittance spectra of the RE/Yb co-doped Y_2O_3 transparent ceramics and the spectra contain large transmittance in the NIR region for all the ceramics due to Yb ion. The ceramics also showed transparency in the visible region. The absorption spectra of the RE/Yb co-doped Y_2O_3 transparent ceramics reveal the highest absorption cross section for Yb at 980 nm. The Yb ion absorbs a large number of incident photons and transfers its excitation energy to Er, Ho, Pr, and Tm ions. As a result, the UC intensity of the co-doped ceramics enhances significantly. The pump power versus emission intensity plots have discussed the absorption of two, three and four photons for red, green, blue and violet light, respectively. In another work, Zheng et al. prepared the Pr, Yb co-doped Y_2O_3 nano-particle through laser ablation and sol-gel methods (Zheng et al. 2010). The sample gives intense green and weak blue and red UC emissions on excitation with 976 nm. The phosphor prepared using laser ablation method yields high UC intensity compared to the sol gel method. It also has a longer lifetime due to less optical quenching centers, such as OH^-, CO_3^{2-}, etc. The Pr^{3+}, Yb^{3+} co-doped Y_2O_3 nano-phosphor was prepared by our group using combustion method and found similar UC emissions (Yadav et al. 2015a). Thus, the rare earth doped Y_2O_3 phosphor materials show intense photoluminescence from UV to NIR regions.

Upconversion Mechanisms

It has been mentioned earlier that the UC mechanisms were demonstrated by F. Auzel in the Tm^{3+}, Yb^{3+} co-doped material upon NIR excitation (Auzel 1996). The upconversion contains various processes such as Ground State Absorption (GSA), Excited State Absorption (ESA), Energy Transfer Upconversion (ETU), Cooperative Energy Transfer (CET), etc. The GSA and ESA processes occur in the singly rare earth doped phosphor, while in the case of co-doped phosphor all these processes take place. Figure 12.1 shows the schematic UC processes, such as GSA, ESA, ETU, CET, etc. in the case of rare earth doped phosphor materials.

In Fig. 12.1(a), the activator ion absorbs the incident photon (hv) in the ground state (1) and promoted to the excited state (2) through GSA. The ion present in the state (2) further absorbs the (hv) energy and finally promoted to the excited state (3) through ESA. As a result, the transition from state (3) to state (1) gives an upconverted photon of energy (hv'), which has more energy than the incident photon i.e. (hv' > hv). This process generally occurs in the singly rare earth doped phosphor materials.

Figure 12.1(b) shows the schematic energy level diagram for the donor and activator ions. In this case, the GSA, ESA and ETU processes collectively take place. The donor ions absorb the (hv) energy, which promote them to its excited state and finally relaxed to the ground state. The donor ion on relaxation transfers its energy to the activator ion due to dipole-dipole interaction in the ground state (1). Due to this, the activator ion is promoted to the excited state (2) through GSA. The ion in the excited state (2) further absorbs the (hv) energy through ETU, which promotes the activator ion to the excited state (3). The transition from state (3) to state (1) gives an upconverted photon with larger energy (hv') i.e. (hv' > hv).

The upconversion via CET process is shown in Fig. 12.1(c) in which the donor and activator ions are excited with the same light source and the ions are excited to state (3) through GSA and ETU followed by state (2). In addition to this, the two excited donor ions combine to each other in the excited state (2) of the donor ion and form a virtual state. This virtual state

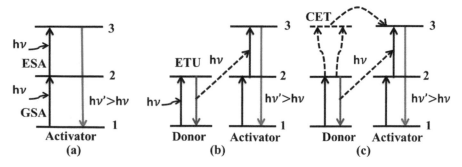

Fig. 12.1. Upconversion processes (a) GSA and ESA; (b) ETU and (c) CET in the rare earth ions.

is formed for a short time and an energy transfer takes place from this state to the excited state (3) of the activator ion. This is termed as the Cooperative Energy Transfer (CET) process. As a result, the emitted photon has larger energy than the incident photon i.e. ($hv' > hv$).

Synthesis: A Solution Combustion Technique

The rare earth doped Y_2O_3 phosphor materials have been prepared by various techniques such as high temperature solid state reaction method, sol-gel method, solution combustion method, laser ablation method, co-precipitation method, solvothermal method, etc. These methods yield the particles size ranging from sub-micrometer (μm) to nanometer (nm). Among these methods, the solution combustion method has been discussed for synthesizing the phosphor materials. It is a very simple method. The starting materials are dissolved for a short time. It needs very simple and low cost instruments like a glass beaker, hot plate magnetic stirrer, thermometer, magnetic bid, etc. The resultant phosphor contains the particles size in the nanometer (nm) range. The flow chart for the synthesis of phosphor materials using solution combustion technique is shown in Fig. 12.2.

Figure 12.2 illustrates the synthesis of phosphor materials in which the oxide forms of starting materials are dissolved in a small amount of HNO_3 and the urea solution is added to this drop wise. The final solution was stirred at a constant temperature maintained at 60 °C for 3-4 hours and the solution was turned to gel. This gel was placed in a closed furnace maintained at 600 °C for 10 minutes. The auto-ignition took place and the white powder was obtained. This is termed as the as-prepared phosphor (ASP). This phosphor was annealed at higher temperature to improve its structural and optical properties. The phosphor thus obtained is poly-crystalline and has the particles in the nanometer (nm) range. The phosphor emits uniform and intense photoluminescence throughout the sample.

Vetrone et al. have prepared the Er^{3+}, Yb^{3+} co-doped Y_2O_3 phosphor through solution combustion method and observed strong upconversion emission upon 978 nm excitation (Vetrone et al. 2003). The solution combustion method was also used by Pandey et al. to prepare the Tb^{3+}, Yb^{3+} co-doped Y_2O_3 phosphor and monitored the visible upconversion on excitation with 980 nm diode laser (Pandey et al. 2014). They also found an enhancement in the upconversion intensity in presence of Li^+ ion. We have also synthesized the Pr^{3+}, Yb^{3+} co-doped Y_2O_3 nano-phosphor using this method and observed large upconversion intensity in presence of Yb^{3+} ions due to energy transfer (Yadav et al. 2015a). Recently, Chen et al. have also prepared the Er^{3+}, Tm^{3+}, Yb^{3+} co-doped Y_2O_3 phosphor using this method and found an average particles size of 49 nm. The phosphor sample emits intense blue, green and red upconversion emissions upon 980 nm excitation (Chen et al. 2018). Figure 12.3 shows the Transmission Electron Microscope (TEM) image of the phosphor and the particles are in the range of 35-65 nm with an average grain size ~45 nm. The inset figure shows the high resolution

Solution Combustion Technique

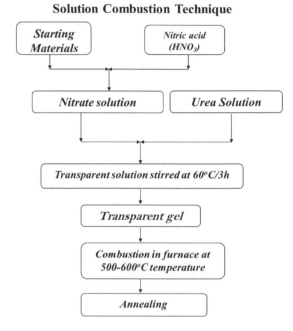

Fig. 12.2. Flow charts of solution combustion technique for the synthesis of phosphor materials.

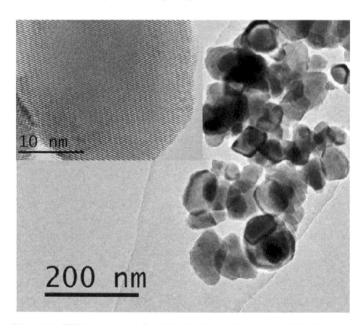

Fig. 12.3. TEM image of the Er^{3+}, Tm^{3+}, Yb^{3+} co-doped Y_2O_3 nano-phosphor and the inset shows HRTEM image of the sample. (Reproduced with copyright permission from Elsevier, 2018)

TEM image of the nanoparticles with d-spacing around 3.06Å corresponding to the lattice place (222) of Y$_2$O$_3$ phosphor. Thus, the solution combustion method yields the nanometer sized particles.

Application of Y$_2$O$_3$ Phosphors

As has been mentioned earlier, the Y$_2$O$_3$ is a low phonon frequency host and promotes the radiative transitions. When the rare earth ions are doped in this host the large photoluminescence intensity has been observed by various researchers (Jadhav et al. 2016; Wei et al. 2010; Silver et al. 2001; Vetrone et al. 2003; Guo and Qiao et al. 2009; Hou et al. 2011; Zheng et al. 2010; Yadav et al. 2015a; Pandey et al. 2014; Chen et al. 2018). The upconversion intensity of the Er^{3+} doped Y$_2$O$_3$ phosphor was enhanced significantly due to energy transfer from Yb^{3+} to Er^{3+} ions (Vetrone et al. 2003; Guo and Qiao et al. 2009). The Er^{3+}, Yb^{3+} co-doped Y$_2$O$_3$ phosphor emits intense upconverted green and red emissions on excitation with 978 nm diode laser. Pang et al. also found an enhancement in emission intensity due to energy transfer from Yb^{3+} to Ho^{3+} ions (Pang et al. 2011). Pandey et al. prepared the Tb^{3+}, Yb^{3+} co-doped Y$_2$O$_3$ phosphor and suggested the application of this phosphor in photonic applications (Pandey et al. 2014). They also prepared the Pr^{3+}, Yb^{3+} co-doped Y$_2$O$_3$ phosphor and pointed out its application in displays devices (Pandey et al. 2014). Diaz-Torres et al. also studied the optical properties in Er^{3+}, Yb^{3+} co-doped Y$_2$O$_3$ phosphor and observed high color purity and tunable emissions, which can be used to prepare the upconversion lamps (Diaz-Torres et al. 2017). In these systems, the Yb^{3+} ion transfers its energy to the rare earth ions and enhances their emission intensity.

The rare earth co-doped Y$_2$O$_3$ phosphors have been also used for the development of blue, green, red and white light materials (Yadav et al. 2013; Diaz-Torres et al. 2017; Pandey et al. 2012; Bai et al. 2009). Zheng et al. prepared the Tm^{3+}, Yb^{3+} co-doped Y$_2$O$_3$ phosphor and observed blue UC emission on excitation with 976 nm wavelength (Zheng et al. 2011). This phosphor acts as a blue emitting material. The green and red upconversion emissions were observed in the Er^{3+}, Yb^{3+} co-doped Y$_2$O$_3$ phosphor for higher concentrations of Yb^{3+} ions (Diaz-Torres et al. 2017). The color tunable emission was observed in the Tm^{3+}, Ho^{3+}, Yb^{3+} co-doped Y$_2$O$_3$ phosphor upon 980 nm excitation (Pandey et al. 2012). This phosphor was suggested to use in the latent finger printing, photodynamic therapy and fluorescent bio-labels. When the red, green and blue colors are mixed together it gives a white light. The white light emission was observed by different groups of researchers (Yadav et al. 2013; Bai et al. 2009; Lv et al. 2015). They have also observed that white light was achieved followed by color tunability. Actually, in tri-doped systems the white light emission greatly depends on the concentration of dopant ions. Bai et al. synthesized the Er/Tm/Yb/Li co-doped Y$_2$O$_3$ nanocrystals and observed blue emission from Tm/Yb and green and red emissions from Er/Yb ions. When these colors are mixed together white light emission takes place and its intensity is enhanced in presence of Li ions via structural modifications (Bai et al. 2009). We have also achieved white light emission in the $Tm^{3+}/Er^{3+}/Yb^{3+}$ co-doped Y_2O_3-ZnO nano-composite (Yadav et al. 2013). In this system, ZnO forms composite material with Y_2O_3 and improves the crystallinity and reduces the optical quenching centers, which collectively enhances the white light intensity.

Recently, the rare earth doped Y_2O_3 phosphor was also found suitable for other optical based applications such as optical thermometry. The Fluorescence Intensity Ratio (FIR) has been used to study the temperature dependent UC photoluminescence. The optical temperature sensing in the Er^{3+}, Tm^{3+}, Yb^{3+} co-doped Y_2O_3 phosphor was studied by Chen et al. The phosphor sample emits blue, green and red emissions from Tm^{3+} and Er^{3+} ions on excitation with 980 nm. The emission intensity versus pump power measurement shows the involvement of three and two photons for blue; green and red emissions, respectively. They have monitored not only the FIR of thermally coupled levels of Er^{3+} ions but also non-thermally coupled levels of Tm^{3+} and Er^{3+} ions, respectively at different temperatures i.e. from 298 K to 573 K. The results indicated that the temperature sensing ability was found higher in the case of the transitions due to the non-thermally coupled levels of Tm^{3+} and Er^{3+} ions. The temperature sensitivity was observed maximum as 1640×10^{-4} at 573K (Chen et al. 2018).

It is interesting to note that Lv et al. successfully prepared the Y_2O_3:Yb/Er-Cu_xS hollow spheres and studied their structural, optical and biological properties (Lv et al. 2015). The composite sample was excited by 980 nm and corresponding green and red UC emissions were observed. The result shows that the Y_2O_3:Yb/Er-Cu_xS hollow spheres can be used in bioimaging, chemotherapy and photothermal treatment (PTT). Figure 12.4 shows the

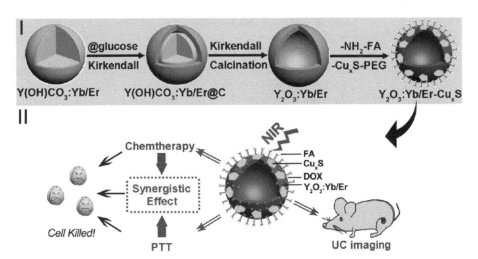

Fig. 12.4. Schematic diagram of the formation of Y_2O_3:Yb/Er-Cu_xS spheres and its bio-application. (Reproduced with copyright permission from American Chemical Society, 2015)

schematic diagram for the formation of Y_2O_3:Yb/Er-Cu$_x$S hollow spheres and its bio-applications.

It has also been observed that the rare earth doped other oxide phosphor materials are widely studied and contain low phonon frequency similar to the Y_2O_3 host. The rare earth ions were also doped in the La_2O_3 host material. The La_2O_3 host is hygroscopic in nature and is converted in the $La(OH)_3$ host due to absorption of moisture present in the atmosphere (Yadav et al. 2012; Yadav et al. 2014). It was noticed that there is an energy transfer from Yb^{3+} to Pr^{3+} ions in the phosphor and this leads to an enhancement in the green and NIR intensity of Pr^{3+} ions (Yadav et al. 2015a, b). The observation of NIR to NIR emission is remarkable and can be used in bio-thermal treatment. When Yb^{3+} and Er^{3+} ions are co-doped further in the La_2O_3 host there is an energy transfer from Yb^{3+} and Er^{3+} ions. The phosphor thus prepared has been suggested to be used in temperature sensors and display devices (Yadav et al. 2015b; Dey and Rai, 2014). The Sm^{3+} and Eu^{3+} co-doped Lu_2O_3 phosphor was also prepared and studied for energy transfer from Sm^{3+} to Eu^{3+} ions (An et al. 2008). It has been noted that the rare earth doped Gd_2O_3 phosphors were also studied by different researchers for upconversion and downconversion processes (Jeena et al. 2017; Qian et al. 2018; Guo et al. 2004). In a downconversion work, Qian et al have reported preparation, luminescence properties and growth mechanism in the Eu^{3+}/Tb^{3+} co-doped Gd_2O_3 phosphors (Qian et al. 2018). They observed the red color in Eu^{3+} doped Gd_2O_3 phosphor on excitation with 250 nm UV radiation. On the other hand, when the Tb^{3+} doped Gd_2O_3 phosphor is excited by 240 nm UV radiation it gives very intense green photoluminescence. They have suggested its applications in displays devices, optoelectronic devices and biosensors. The Er^{3+}, Tm^{3+}, Yb^{3+} co-doped Gd_2O_3 phosphor gives blue, green and red upconverted emissions excited by 980 nm (Guo et al. 2004). These emissions occur due to absorption of three and two photons. The presence of Yb^{3+} ion leads to a color tunability in the phosphor material.

Conclusions

This chapter describes the upconversion photoluminescence in the rare earth doped Y_2O_3 phosphor materials. The GSA, ESA, ETU and CET processes involved in the UC emissions have been discussed. The Yb^{3+} ion acts as a sensitizer since it transfers its excitation energy to the other rare earth doped phosphor materials. The absorption cross section of Yb^{3+} ion is larger for the NIR wavelength. The upconversion process in the different rare earth doped Y_2O_3 phosphor materials has been discussed in the presence of Yb^{3+} ion. The presence of Yb^{3+} ion enhances the emission intensity of the rare earth co-doped phosphor materials due to energy transfer. The applications of different rare earth doped Y_2O_3 phosphor materials have also been discussed in this chapter.

References

An, L., Zhang, J., Liu, M., Chen, S., Wang, S. 2008. Preparation and photoluminescence of Sm^{3+} and Eu^{3+} doped Lu_2O_3 phosphor. Opt. Mater. 30, 957-960.

Auzel, F. 1966. Computeur quantique par transfer d'energie de Yb^{3+}, Tm^{3+} dans un tungstate mixte et dans un verre gemanate. Compte Rendus de l'Acadmie des Sciences 263, 819-821.

Bai, Y., Wang, Y., Peng, G., Zhang, W., Wang, Y., Yang, K., Zhang, X., Song, Y. 2009. Enhanced white light emission in $Er/Tm/Yb/Li$ codoped Y_2O_3 nanocrystals. Opt. Comm. 282, 1922-1924.

Chen, G., Lei, R., Huang, F., Wang, H., Zhao, S., Xu, S. 2018. Optical temperature sensing behavior of $Er^{3+}/Yb^{3+}/Tm^{3+}:Y_2O_3$ nanoparticles based on thermally and non-thermally coupled levels. Opt. Comm. 407, 57-62.

Dey, R., Rai, V.K. 2014. Yb^{3+} sensitized Er^{3+} doped La_2O_3 phosphor in temperature sensors and display devices. Dalton Trans. 43, 111-118.

Diaz-Torres, L.A., Salas, P., Oliva, J., Resendiz-L, E., Rodriguez-Gonzalez, C., Meza, O. 2017. Tuning from green to red upconversion emission of $Y_2O_3:Er^{3+}-Yb^{3+}$ nanophosphors. Appl. Phys. A 123(25), 1-8.

Guo, H., Dong, N., Yin, M., Zhang, W., Lou, L., Xia, S. 2004. Visible upconversion in rare earth ion-doped Gd_2O_3 nanocrystals. J. Phys. Chem. B 108, 19205-19209.

Guo, H., Qiao, Y.M. 2009. Preparation, characterization, and strong upconversion of monodisperse $Y_2O_3:Er^{3+},Yb^{3+}$ microspheres. Opt. Mater. 31, 583-589.

Hou, X., Zhou, S.M., Jia, T., Lin, H., Teng, H. 2011. Investigation of up-conversion luminescence properties of RE/Yb co-doped Y_2O_3 transparent ceramic (RE = Er, Ho, Pr, and Tm). Physica B 406, 3931-3937.

Jadhav, A.P., Khan, S., Kim, S.J., Lee, S.Y., Park, J.-K., Cho, S.-H. 2016. Near-infrared quantum cutting in Tb^{3+} to Yb^{3+} in Y_2O_3 nanophosphors. Res. Chem. Intermed. 16, 2427-2429.

Jeena, T.R.M., Ezhil Raj, A., Bououdina, M. 2017. Synthesis and photoluminescent characteristics of Dy^{3+} doped Gd_2O_3 phosphors. Mater. Res. Express 4, 025019 pp. 1-10.

Lv, R., Yang, P., He, F., Gai, S., Yang, G., Lin, J. 2015. Hollow structured $Y_2O_3:Yb/Er-Cu_xS$ nanospheres with controllable size for simultaneous chemo/photothermal therapy and bioimaging. Chem. Mater. 27, 483-496.

Muenchausen, R.E., Jacobsohn, L.G., Bennett, B.L., McKigney, E.A., Smith, J.F., Valdez, J.A., Cooke, D.W. 2007. Effects of Tb doping on the photoluminescence of $Y_2O_3:Tb$ nanophosphors. J. Lumin. 126, 838-842.

Pandey, A., Rai, V.K. 2012. Colour emission tunability in $Ho^{3+}-Tm^{3+}-Yb^{3+}$ co-doped Y_2O_3 upconverted phosphor. Appl. Phys. B 109, 611-616.

Pandey, A., Rai, V.K. 2014. $Pr^{3+}-Yb^{3+}$ codoped Y_2O_3 phosphor for display devices. Mater. Res. Bull. 57, 156-161.

Pandey, A., Rai, V.K., Kumar, K. 2014. Influence of Li^+ codoping on visible emission of $Y_2O_3:Tb^{3+}$, Yb^{3+} phosphor. Spectrochim. Acta Part A 118, 619-623.

Pang, T., Cao, W.H., Xing, M.M., Luo, X.X., Xu, S.J. 2011. Preparation and luminescence properties of monodisperse silica/aminosilane-coated $Y_2O_3:Yb$, Ho upconversion nanoparticles. Chin. Sci. Bull. 56, 137-141.

Qian, B., Zou, H., Meng, D., Zhou, X., Song, Y., Zheng, K., Miao, C., Sheng, Y. 2018. Columnar $Gd_2O_3:Eu^{3+}/Tb^{3+}$ phosphors: Preparation, luminescence properties and growth mechanism. Cryst. Eng. Comm. 20, 7322-7328.

Silver, J., Martinez-Rubio, M.I., Ireland, T.G., Fern, G.R., Withnall, R. 2001. The effect of particle morphology and crystallite size on the upconversion luminescence properties of Erbium and Ytterbium co-doped yttrium oxide phosphors. J. Phys. Chem. B 105, 948-953.

Vetrone, F., Boyer, J.C., Capobianco, J.A., Speghini, A., Bettinelli, M. 2003. Effect of Yb^{3+} codoping on the upconversion emission in nanocrystalline Y_2O_3:Er^{3+}. J. Phys. Chem. B 107, 1107-1112.

Wei, X.T., Zhao, J.B., Chen, Y.H., Yin, M., Li, Y. 2010. Quantum cutting downconversion by cooperative energy transfer from Bi^{3+} to Yb^{3+} in Y_2O_3 phosphor. Chin. Phys. B 19, 077804, 1-5.

Yadav, R.S., Dwivedi, Y., Rai, S.B. 2012. Structural and optical characterization of nano-sized $La(OH)_3$:Sm^{3+} phosphor. Spectrochim. Acta Part A 96, 148-153.

Yadav, R.S., Verma, R.K., Rai, S.B. 2013. Intense white light emission in Tm^{3+}/Er^{3+}/Yb^{3+} co-doped Y_2O_3-ZnO nano-composite. J. Phys. D: Appl. Phys. 46, 275101, 1-8.

Yadav, R.S., Dwivedi, Y., Rai, S.B. 2014. Structural and optical properties of Eu^{3+}, Sm^{3+} co-doped $La(OH)_3$ nano-crystalline red emitting phosphor. Spectrochim. Acta Part A 132, 599-603.

Yadav, R.S., Verma, R.K., Bahadur, A., Rai, S.B. 2015a. Structural characterizations and intense green upconversion emission in Yb^{3+}, Pr^{3+} co-doped Y_2O_3 nano-phosphor. Spectrochim. Acta Part A 137, 357-362.

Yadav, R.S., Verma, R.K., Bahadur, A., Rai, S.B. 2015b. Infrared to infrared upconversion emission in Pr^{3+}/Yb^{3+} co-doped La_2O_3 and $La(OH)_3$ nano-phosphors: A comparative study. Spectrochim. Acta Part A 142, 324-330.

Yadav, R.S., Rai, S.B. 2017. Frequency upconversion and downshifting emissions from rare earth co-doped strontium aluminate nano-phosphor: A multi-modal phosphor. J. Lumin. 190, 171-178.

Yadav, R.S., Rai, S.B. 2017. Structural analysis and enhanced photoluminescence via host sensitization from a lanthanide doped $BiVO_4$ nano-phosphor. J. Phys. Chem. Solids 110, 211-217.

Yadav, R.S., Rai, S.B. 2017a. Surface analysis and enhanced photoluminescence via Bi^{3+} doping in a Tb^{3+} doped Y_2O_3 nano-phosphor under UV excitation. J. Alloys Compds. 700, 228-237.

Yadav, R.S., Rai, S.B. 2018. Effect of concentration and wavelength on frequency downshifting photoluminescence from a Tb^{3+} doped yttria nano-phosphor: A photochromic phosphor. J. Phys. Chem. Solids 114, 179-186.

Yadav, R.S., Rai, S.B. 2019b. Effect of annealing and excitation wavelength on the downconversion photoluminescence of Sm^{3+} doped Y_2O_3 nano-crystalline phosphor. Opt. Laser Technol. 11, 169-175.

Zheng, C.B., Xia, Y.Q., Qin, F., Yu, Y., Miao, J.P., Zhang, Z.G., Cao, W.W. 2010. Femtosecond pulsed laser induced synthesis of ultrafine Y_2O_3: Pr, Yb nanoparticles with improved upconversion efficiency. Chem. Phys. Lett. 496, 316-320.

Zheng, C.B., Xia, Y.Q., Qin, F., Yu, Y., Miao, J.P., Zhang, Z.G., Cao, W.W. 2011. Upconversion emission from amorphous Y_2O_3:Tm^{3+}, Yb^{3+} prepared by nanosecond pulsed laser irradiation. Chem. Phys. Lett. 509, 29-32.

Synthesis and Potential Application of Rare Earth Doped Fluoride Based Host Matrices

S.P. Tiwari[1]*, R.S. Yadav[2], S.K. Maurya[3], A. Kumar[3], Vinod Kumar[4] and H.C. Swart[1]

[1] Department of Physics, University of Free State, Bloemfontein – 9300, South Africa
[2] Department of Zoology, Institute of Science, Banaras Hindu University, Varanasi - 221005, India
[3] Department of Physics, IIT (ISM) Dhanbad - 826004, India
[4] Center for Energy, Indian Institute of Technology Delhi, New Delhi - 110016, India

Introduction

Rare earth doped fluoride nanoparticles are potential host matrices for intense upconversion (UC) emissions (Tan et al. 2009; Yi et al. 2006; Maurya et al. 2019; Kumar et al. 2007; Chen et al. 2015; Xu et al. 2015). Some research has been performed on different hosts such as GF_4, YF_4, LaF_4, $NaGdF_4$, $NaYF_4$, $NaLaF_4$, $BaGF_5$, $BaYF_5$, etc. (Li et al. 2018; Chen et al. 2012; Kumar et al. 2015; Qui et al. 2014; Maurya et al. 2018; Runowski et al. 2014; Cao et al. 2011; Ivanova et al. 2008; Shan et al. 2010). These research outputs confirm the ability of these particles for different innovative applications. The main applications viz., temperature sensing, latent fingermarks detection, solar energy harvesting, bioimaging, and photo-catalytic by using these particles have been reported (Cui et al. 2014; Kumar et al. 2009; Xing et al. 2012; Yang et al. 2013; Sun et al. 2011). The choice of fluoride hosts has been taken into account due to their high chemical stability and low cut off phonon frequency. Moreover, it can easily be dispersed in a colloidal form that shows high quantum efficiency (Yin et al. 2010; Liang et al. 2013). These are certain properties which ensure the suitability of these hosts in the above-mentioned applications.

All the arbitrary applications depend on typical synthesis techniques. Synthesis of rare earth doped fluoride phosphors using different methods

---
*Corresponding authors: sptiwari.ism@gmail.com

has been reported by many researchers (Lojpur et al. 2013; Yi et al. 2004; Liu et al. 2018; Lemyre et al. 2005; Pandey et al. 2018). The combustion, co-precipitation, hydrothermal, thermal decompositions etc. are the most common synthesis methods. These synthesis routes are rather significant for the development of particles like oxide, sulfide, phosphate, fluoride-based host materials. Depending on the temperature condition and the approach of the application of material the reaction time and use of organic solvents are mixed in a schematics ratio. Among these preparation methods, combustion synthesis requires very high temperature while other routes have rather low temperature for synthesis (Tiwari et al. 2015; Zeng et al. 2011; Niu et al. 2011). For the synthesis of fluorides, the temperature conditions should be low as at a higher temperature (>500 °C) the fluorides generally degrade to form oxides. Therefore, the structural and optical properties may change accordingly. The co-precipitation, hydrothermal and thermal decomposition methods are therefore better approaches for the preparation of the rare earth doped fluoride phosphor particles, which have been discussed in different parts of this chapter.

The fluoride phosphor particles may be stable in power as well as in the colloidal dispersed state in different solvents (Niu et al. 2011; Ye et al. 2010). For the colloidal stability, the particle size should be a minimum with a uniform shape (Gainer et al. 2014). There are several ways to check the structural properties of fluoride phosphor particles. The crystal structure and crystalline sizes can be confirmed through X-Ray Diffraction (XRD) characterization. The micro/nanoscopic structure can be observed through Field Emission Scanning/Transmission Electron Microscopy (FESEM/TEM) and the particle shape and sizes can be optimized. The use of different precursors with synthesis may cause impurities in the envelope fluoride particles. Generally, these impurities are organic and can be estimated through Fourier transform infrared (FTIR) spectroscopy. Further, the reaction of these impurities can be examined by the same characterization. Apart from these the Energy Dispersive X-Ray Analysis (EDS), X-ray Photoelectron Spectroscopy (XPS) etc. analysis are useful for elemental stabilizations/confirmations. Zeta potentials are useful for colloidal dispersion stability. In optical characterizations, the Uv-Vis analysis is useful for the estimation of the optical band gaps and different absorption bands in these materials. The Near-Infrared (NIR) to visible UC luminescence properties may be studied through the excitation of these materials via wavelengths of 976/980 nm from continuous wave laser sources. The emitted visible radiation could be detected using either a Photon Multiplier Tube (PMT) coupled monochromator or Charge-Coupled Device (CCD) camera detector. The detailed descriptions have been discussed in different sections of this chapter.

As it was described earlier fluoride phosphor nanoparticles are very significant in many applications. Meruga et al. (2012) demonstrated the security printing/writing applications of $NaYF_4:Er^{3+}/Yb^{3+}$. Wang et al. (2015) discussed the latent fingermarks detection application of fluoride phosphor particles. Ramasamya et al. (2014) successfully utilized the UCNPs in the achievement of the higher efficiency of dye synthesize solar cell. Chen et al. (2015) have demonstrated the powder $YF_3:Tm^{3+}/Yb^{3+}/Ca^{2+}$ in the application of non-contact type temperature sensors. In the bio-imaging fields, Wang et al. (2010) did significant research with $NaYF_4$ doped with different ions. In this present chapter, we have discussed the synthesis, characterization and UC emission in different fluoride hosts.

Synthesis of Fluoride Nanoparticles

The development of high-quality rare earths-doped phosphors with precise shape and size and crystalline phase is fundamental to tune their optical and chemical properties and investigate their applications in different fields. The biggest challenges for the preparation of phosphors are the reproducibility and mono-dispersion of the phosphors. The nucleation and growth process of luminescent phosphors are dependent on the initial crystallization. Victor LaMer had extensively studied the nucleation-growth mechanism and it is in general referred to as the LaMer mechanism (Pound et al. 1952; LaMer et al. 1950). A recent article has been published to explain the kinetic and thermodynamic processes in the growth of particles (Kumar et al. 2018). An overview of the different kind of UC luminescent phosphors are shown in Table 13.1. Many groups have reported the preparation of phosphor materials using organic solvents as well. The prepared crystalline particles need surface modification for mono-dispersion. For applications in aqueous environments, only colloidally stable phosphors can be used. The most common methods for the synthesis of fluoride phosphors are thermal decomposition, hydrothermal and co-precipitation. The other preparation strategies are described in Table 13.1 (DaCosta et al. 2014).

All the reported methods of synthesis of nanoparticles with different routes are described below.

Thermal Decomposition Method

This method is usually used for the preparation of fluoride-based phosphors and is almost solely applied in the creation of alkali rare earth tetrafluoride phosphor such as $LiYF_4$, $LiGdF_4$, $NaGdF_4$, $NaYF_4$, $NaLuF_4$, $KGdF_4$, etc. This method is also used to produce rare earth doped/co-doped nanoparticles based on further host lattices with $BaYF_5$, YF_3, CaF_2, etc. This technique gives excellent control on particle size and shape with comparatively short reaction times with respect to other methods and produces monodispersed phosphors Haase et al. 2011). Generally organometallic precursors such as rare earth-trifluoroacetates are used. The rare earth-trifluoroacetates precursors are prepared by reacting trifluoroacetic acid (TFA) with Ln-oxides in the laboratory and are also available commercially. A high boiling solvent 1-octadecene (ODE), Oleic Acid (OA) and olelyamine (OM) are used as surface ligands in the reactant dissolution. Typically, one of these surface ligands solution are heated to a temperature above the decomposition

Table 13.1. Some general synthesis route for rare earth doped UC phosphors

Synthetic strategy	Process	Advantages	Disadvantages	Materials prepared
Arrested precipitation	Poorly soluble product precipitated within a template or confined space	Simple and fast reaction, cost effective, does not require high temperatures or pressures	Little control over particle shape and size, aggregation is typical, high-temperature post reaction annealing/ calcination step required resulting in aggregation	$LuPO_4$ $YbPO_4$ $NaYF_4$ $BaYF_5$ $Y_3Al_5O_{12}$ Y_2O_3 $LaPO_4$ $NaGdF_4$
Combustion synthesis	Typically done to prepare metal oxides M_2O_3; nanoparticle formation occurs following rapid heat pulse from combustion of propellant	Very rapid synthesis, cost-effective, low energy cost	Very little control over particle shape, size, and purity, considerable aggregation of particles, observed	Y_2O_3 Lu_2O_3 $LaPO_4$ Gd_3GaO_{12} ZnO
Sol-gel method	Primarily implemented in the synthesis of metal oxides; prepared materials resemble those prepared by combustion synthesis	Metal acetate or metal alcoxide precursors are very inexpensive, simple reactions	Little control over particle shape and size, high-temperature post reaction annealing/ calcination step required resulting in aggregation	CaS ZrO_2 Gd_2O_3 $BaTiO_3$ TiO_2 LuF_3 $NaLuF_4$ $NH_4Lu_2F_7$

(Contd.)

Table 13.1. (*Contd.*)

Synthetic strategy	Process	Advantages	Disadvantages	Materials prepared
Microwave-assisted synthesis	Uses 0.3–300 GHz microwave irradiation to heat reaction mixtures	Increased reaction rates, milder reaction conditions decreased energy consumption, high reproducibility	Requires specialized microwave irradiators, limited solvent choice (must be effectively heated by microwaves)	Y_2O_3 InGaP InP M_2O_3 (M = Pr, Nd, Sm, Eu, Gd, Tb, Dy) $NaYF_4$ $LiYF_4$ GdF_3
Micro-emulsion or reverse Micelle method	Uses the interior aqueous environments of reverse micelles in organic solvents as nano-scale reactors	Very versatile, reproducible, produces homogenous monodisperse materials, control over size and morphology of produced materials	Organic solvents being used, very limited production capacity since relying on amount of aqueous phase that can be solubilized and precursor concentrations	$NaYF_4$ Y_2O_3 Gd_2O_3 YVO_4 XO_2 (X = Ce, Sn, Zr) $SrTiO_3$ Sr_2TiO_4 Ba_2TiO_4 $XZrO_3$ (X = Sr, Ba, Pb)
Flame synthesis	Gas-phase materials are passed through a flame source using a hydrocarbon fuel and O_2 as the oxidant. Heat energy permits formation of crystalline materials.	High-purity products obtained, narrow size distributions, time and cost effective, facile control over morphology, scalable for industrial applications	Considerable aggregation of particles is observed, problems synthesizing multicomponent materials	TiO_2 SiO_2 Y_2O_3 La_2O_3 Ga_2O_3

Reproduced with permission from Ref. (DaCosta et al. 2014) @ Copyright 2014 Elsevier.

temperature of the precursors (>300 °C). A required solution of Ln-trifluoroacetates is poured into a hot ligand solution. When the reactants are added, due to thermal decomposition, the particles are quickly released and rapid burst nucleation take place (Kumar et al. 2019). These ligands have polar head groups. During the reaction, they help to grow the particles by coordinating. These long hydrocarbon tails provide the bulk solution of the required solubility (Lin et al. 2012). These surface ligands control the particles growth by blocking the extension of the lattice and also stabilize the particle growth by stopping aggregation over the repulsive interactions in the solution. These metal precursors are expensive, air sensitive and toxic. That is why inert atmosphere is used for annealing purposes. The precipitated material is centrifuged and washed with organic polar solvents. For monodispersed particles, the organic solvents are extracted from the nanoparticles (NPs) and finally colloidally stable particles are collected for a long span (Boyer et al. 2006). The first thermal decomposition method was reported by Mai and his group for the synthesis of $NaREF_4$ (where RE stands for Pr to Lu, Y) co-doped Er^{3+}/Yb^{3+} and Tm^{3+}/Yb^{3+} nanoparticles in 2006 (Mai et al. 2006). They synthesized $NaREF_4$ UCNPs with a decomposition method with a cubic and hexagonal phase. Figure 13.1 shows the hexagonal-phase particles of synthesized UCNPs. The LaMer mechanism was followed for the growth of hexagonal-$NaYF_4$ and cubic-$NaYF_4$ (Boyer et al. 2006; Mai et al. 2006; Wang et al. 2009). Five nm monodispersed UCNPs were obtained by tuning and varying the reaction parameters such as concentrations, temperature and reaction time (Wang et al. 2009; Mai et al. 2007). The Ostwald ripening method can also be applied for obtaining monodisperse hexagonal-$NaYF_4$ using rare earth-doped cubic-$NaYF_4$ UCNPs (Wang et al. 2009). The hexagonal-$NaYF_4$ phase can be achieved by refining the reaction criteria via thermal decomposition. Hexagonal-$NaYF_4$ UCNPs doped with lanthanides with uniform shape and size of around 10 nm was obtained by reacting in pure oleylamine at 330 °C (Boyer et al. 2006).

Hydrothermal Method

Another synthesis method to generate monodispersed UCNPs of controlled phase, size and shape is the hydrothermal method. This method operates at a lower temperature as compared to thermal decomposition (Chen et al. 2012). This reaction takes place in a Teflon or titanium container and is protected by a thick steel wall. This synthesis technique employs temperature and pressure of the solvent to above the critical point (Chen et al. 2012). A hydrothermal process forms a homogeneous solution which involves a combination of rare earth species, a solvent such as water and a fluoride. Generally, lanthanide oxide, nitrate and chloride are used as the lanthanide precursors. For trifluoride NPs NH_4F or HF is used as the fluoride precursor and for alkali lanthanide fluoride KF or NaF is used. Li et al. have developed an easy and convenient method to prepare NPs in a series form which have been based on a liquid crystal-like structure and is shown in scheme 1 (Kumar et al. 2018; Wang et al. 2011; Wang et al. 2005).

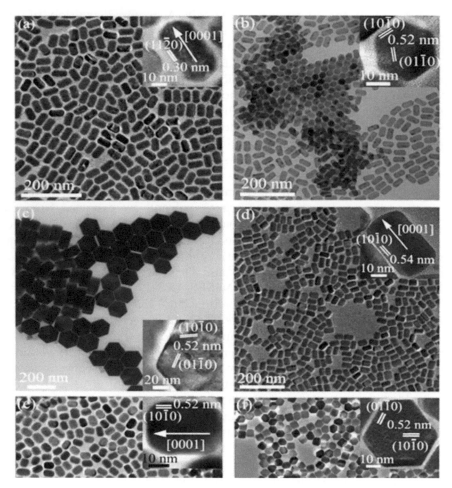

Fig. 13.1. TEM and HRTEM (inset) images of (a) hexagonal-NaYF$_4$ nanorods in 1:1 toluene/hexane and in (b) 1:1:0.48 toluene:hexane:ethanol, (c) hexagonal-NaYF$_4$ nanoplates, (d) hexagonal-NaNdF$_4$ nanorods, and (e) hexagonal-NaEuF$_4$ nanorods. (f) TEM and HRTEM (inset, upper: lying flat on the face; lower: standing on the side face from the highlighted square) images of hexagonal-NaHoF$_4$ hexagonal plates. (Reprinted with permission from Chen et al. 2012. Copyright 2006 American Chemical Society)

In this system, there are three phases: a liquid, a solid and a solution phase. In Scheme l, the oleic acid/ethanol, sodium inolate, and ethanol solution metal-containing ions are used as liquid, solid and solution, respectively. During the synthesis, a general separation mechanism with phase transfer occurs in the liquid, solid and solution phase. For example, the oleic acid is used as a stabilizing agent in the synthesis of NaYF$_4$ in the hydrothermal method (Wang et al. 2007). Using this method, Zhao and his group deposited a triangular array of nanoparticles (Wang et al. 2006). Lin and his group also

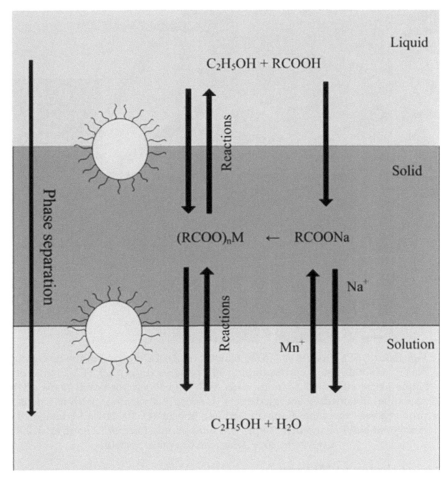

Scheme 13.1. Schematic description of synthetic strategy of phase transfer of the Liquid-Solid-Solution. (Reprinted with permission from Wang and Li 2007)

prepared a series of UC nanorods and nanowires using $(OH)_3$ of rare earths as precursors (Xu et al. 2009). These precursors retained the morphology of rare earth fluoride. In addition to this, cubic and hexagonal $NaYF_4$ nanoparticles have also been synthesized by a hydrothermal situ anion exchange reactions (Fig. 13.2) (Zhang et al. 2007; Zhang et al. 2008).

Different kind of nanoparticles have also been synthesized by a hydrothermal method based on rare earth fluorides hosts, for example $NaYF_4$, $NaLuF_4$, $NaYbF_4$, $NaGdF_4$, $BaYF_5$, LaF_3, YF_3, CaF_2, LuF_3, KMF_3 (M = Zn, Mn, Cd, or Mg), $KMnF_3$ (Fig. 13.2), SrF_2, and oxysalts or lanthanide oxides, such as Lu_2O_3, Y_2O_3, YVO_4, TiO_2, $LaPO_4$, $GdVO_4$ and $PbWO_4$. It was noted that morphology and crystalline phase of the obtained nanoparticles depend on parameters such as pH, solvent, reactant concentration, reaction time and hydrothermal temperature and dopant ratio (Wang et al. 2006).

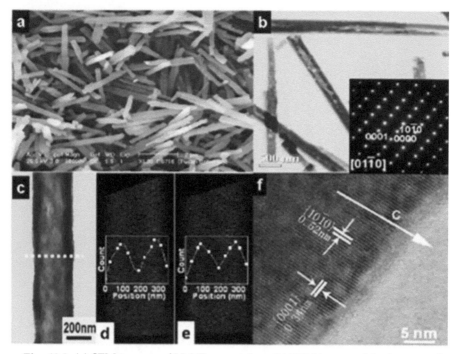

Fig. 13.2. (a) SEM images of $NaYF_4$ nanotubes. (b) TEM image of the hexagonal-$NaYF_4$ nanotubes and corresponding ED patterns (inset). (c) TEM image of a single hexagonal-$NaYF_4$ nanotube. (d, e) Y and F compositional maps of the nanotubes obtained by energy dispersive spectroscopy, along with its intensity profile across the tube along the dotted line in (c). (f) HRTEM image of a hexagonal-$NaYF_4$ nanotube. (Reprinted with permission from Zhang et al. 2008. Copyright 2009 American Chemical Society)

Co-precipitation Method

The most easy and convenient technique for UCNPs is the co-precipitation method. Using Ethylenediaminetetraacetic acid (EDTA), Yi et al. prepared the $NaYF_4$:Yb^{3+}/Er^{3+} by the co-precipitation route (Yang et al. 2012; Yi et al. 2004; He et al. 2013). By varying the molar ratio of the Ln^{3+} ions and EDTA, they controlled the size of the particles in the range of 35-165 nm. The pristine nanoparticles emitted weak UC emission. Though, when annealed between 400-600 °C the emission intensity enhanced by 40-folds. It was ascribed to the transition of the phase from cubic to hexagonal. Regrettably, the UCNPs aggregated and lost their shape after being annealed. Recently, He et al. synthesized hexagonal-$NaGdF_4$ co-doped Yb^{3+} and Er^{3+}/Tm^{3+}/Ho^{3+} microstructures using PVP, Na_2EDTA, sodium tartrate and sodium dodecyl sulfonate (Wang et al. 2011). In this method, the surfactants played a very significant role in the morphology, controlling the size, UC emission intensity and magnetic possessions of the microstructures. Although, the co-precipitation method is one of the easiest methods, the enhancement in the signal of the UC phosphors occur due to the larger size of the particles and due to the aggregation of tiny particles. After annealing the pristine samples, organic capping reagents such as PVP and EDTA are carbonized and reduced the hydrophilicity of the samples.

Applications of Rare Earth Doped Fluoride Nanoparticles

The rare earth doped phosphors are PL materials that are able to not only emit the primary colors such as blue, green and red but also emit complementary colors viz. cyan, yellow and magenta. These colors are observed when the rare earth doped phosphors are illuminated using a proper excitation source. The UC, downconversion (DC) and Quantum Cutting (QC) processes are responsible for these emissions in the rare earth doped phosphors. They also show high quantum yields. These materials give multipurpose emissions not only in the visible region but also in the NIR region. This is due to the low phonon energy of these phosphors. The low phonon energy host promotes the radiative transitions by reducing multiphonon relaxations of the ions in the excited energy levels (Yadav et al. 2017a; Yadav et al. 2016; Yadav et al. 2018b; Yadav et al. 2017a; Tiwari et al. 2016; Yadhav et al. 2018b; Yadhav et al. 2018; Yadav et al. 2016; Yadav et al. 2015; Yadav et al. 2013).

The rare earth doped fluoride phosphors have low phonon energy and high luminous efficacy (Tiwari et al. 2018; Van Der Ende et al. 2009; Kumar et al. 2018; Kumar et al. 2016a; Tiwari et al. 2017; Tiwari et al. 2016; Kumar et al. 2019; Tiwari et al. 2016; Tiwari et al. 2015). The fluoride-based phosphors have numerous applications extending from photonics, solar cells, cancer therapy, photo-thermal treatment to bio-imaging. They are termed as multifunctional materials.

The DC and QC emissions were obtained in the SrF_2:Pr^{3+}, Yb^{3+} phosphors in which there was an energy transfer from Pr^{3+} to Yb^{3+} ions (Van Der Ende et al. 2009). The authors have observed complete energy transfer from Pr^{3+} to Yb^{3+} ions at higher concentrations of Yb^{3+} ions. They have also suggested a decrease in the emission intensity at higher concentrations of Yb^{3+} ions, which was due to concentration quenching. Recently, Wang and Meijerink et al. (2018) have also observed DC and QC emissions from the same rare earth ions in the $NaYF_4$ host. They also found an energy transfer from Pr^{3+} to Yb^{3+} ions. The effective DC and QC emissions in the Pr^{3+}, Yb^{3+} co-doped $NaYF_4$ in absence and presence of Coumarin dye is shown in Fig. 13.3(a) and the energy transfers from Coumarin dye to Pr^{3+} and finally to Yb^{3+} ions are shown in Fig. 13.3(b).

In this case, the intensity of the NIR emission band initially increases with the increase in the concentrations of Coumarin dye up to 7.5 mg/mL. The Coumarin dye is easily excited by 397 nm wavelength and emits an intense blue emission at 462 nm. The light emitted by Coumarin dye is transferred to Pr^{3+} ions, which is ultimately transferred to Yb^{3+} ions. The Frequency Resonant Energy Transfer (FRET) takes place from Coumarin dye to Pr^{3+} ions. A concentration quenching takes place for higher concentrations

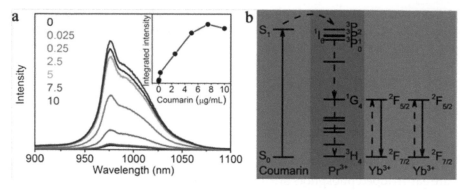

Fig. 13.3. (a) Emission spectra of 0.5 mg/mL NaYF$_4$:1%Pr^{3+}, 20%Yb^{3+} NCs without Coumarin (λ_{ex} = 443 nm) and with different Coumarin concentrations (0.025 to 10 µg/mL, λ_{ex} = 397 nm). The inset shows the integrated intensity of Yb^{3+} emission as a function of Coumarin concentration. (b) Energy level diagram of Coumarin-sensitized downconversion in Pr^{3+}, Yb^{3+} co-doped NCs. (Reproduced with permission from Wang and Meijerink 2018 copyright @ 2018 ACS)

of Coumarin dye i.e. 7.5 mg/mL and thereby decreases the intensity of NIR band. The inset in Fig. 13.3(a) shows the integrated emission intensity of Yb^{3+} emission as a function of Coumarin concentration and the optimum intensity was observed at 7.5 mg/mL concentrations of Coumarin dye. Thus, the dye-sensitized downconversion material can be used in photonics and photovoltaic applications (Wang and Meijerink et al. 2018).

The particles size of the fluoride-based phosphors can be confined to some nm. Some researchers have prepared the nano-phosphors in different host materials using various combinations of rare earth ions. They have observed a tunable UC photoluminescence with the increase in the concentration of rare earth ions (Zhao et al. 2016; Dou et al. 2011; Chen et al. 2010). The ultra-small particles have potential applications of these phosphors in biological studies. The controlled particles sizes of these phosphors shifted the applications towards not only in photonics and solar cells but also in cancer therapy, photothermal treatment, bio-imaging, etc. (Chen et al. 2010; Zeng et al. 2013). Zeng et al. prepared the multifunctional photosensitizer-conjugated core-shell Fe$_3$O$_4$@NaYF$_4$:Yb/Er nano-complexes and reported their applications in magnetic resonance (MR)/upconversion luminescence (UCL) imaging, photodynamic therapy (PDT), clinical diagnosis and treatment of cancers and therapy effect (Xing et al. 2012). Figure 13.4 shows (a) real photographs; (b) Uv-Vis absorption spectra; (c) UCL photographs; and (d) UCL spectra of (1) AlPcS$_4$, (2) NPs-AlPcS$_4$ and (3) NPs on excitation with 980 nm laser diode.

Figure 13.4(a) and (c) reveals the real and UCL photographs of (1) AlPcS$_4$, (2) NPs-AlPcS$_4$ and (3) NPs samples, respectively in the photosensitizer-conjugated core-shell Fe$_3$O$_4$@NaYF$_4$:Yb/Er nano-complexes.

In Fig. 13.4(a), the color of solutions is different from each other. It indicates that the AlPcS$_4$ photosensititzer were conjugated with the as-prepared NPs due to presence of free AlPcS$_4$ molecules by means of doing centrifugation and water washing. Figure 3.4(c) depicts UCL on excitation with 980 nm diode laser for the (1) $AlPcS_4$, (2) NPs-$AlPcS_4$ and (3) NPs samples. It is clear from the figure that the pure $AlPcS_4$ does not show any UCL. However, when the pure $AlPcS_4$ was treated with the NPs it gives intense visible upconversion fluorescence i.e. the sample (2). It is also clear from Fig. 13.4(c) that the pure Fe_3O_4@$NaYF_4$:Yb/Er NPs i.e. sample (3) also shows intense and distinct UCL.

Figure 13.4(b) shows three absorption peaks of $AlPcS_4$ photosensitizers at 351, 609 and 676 nm. It is also clear from the figure that the NPs-$AlPcS_4$ nano-complexes also contain the similar bands. However, no absorption band was observed in the absorption spectrum of the pure NPs. Thus, the $AlPcS_4$ was well conjugated with NPs. Figure 13.4(d) shows the UCL spectra of the NPs and NPs-$AlPcS_4$ nano-complexes, which contain UCL peaks centered at 524, 545 and 659 nm and they are attributed to the transitions from the $^2H_{11/2}$, $^4S_{3/2}$ and $^4F_{9/2}$ states to the ground state $^4I_{15/2}$ of Er^{3+} ions, respectively. There is no UCL peak observed in the UCL spectrum of the pure $AlPcS_4$ photosensitizer. The UCL peak of NPs-$AlPcS_4$ nano-complex at 659 nm matches with the excitation band of $AlPcS_4$ photosensitizer at 670 nm. Thus, the as-prepared NPs-$AlPcS_4$ nano-complexes may be used in PDT performance upon 980 nm excitation (Zeng et al. 2013).

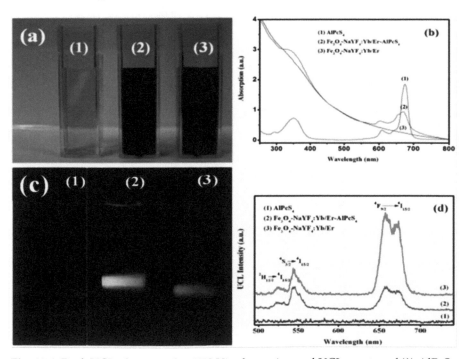

Fig. 13.4. Real, UCL photographs, UV-Vis absorption and UCL spectra of (1) $AlPcS_4$, (2) NPs-$AlPcS_4$, (3) NPs. (a) real photographs; (b) UV-Vis absorption spectra; (c) UCL photographs; (d) UCL spectra. (Reproduced with permission from Zeng et al. 2013 copyright @ 2013 RSC)

Figure 13.5 shows the photomicrographs of MCF-7 cells incubated with NPs-AlPcS$_4$ nano-complexes pre- and post-NIR irradiation at different time intervals (i.e. 0, 1, 3 and 5 minutes). The cytotoxicity of the NPs and NPs-AlPcS$_4$ nano-complexes has been studied by MTT (3-(4,5-dimethylthiazol-2-yl)-2,5-diphenyltetrazolium bromide, a tetrazole) assay method.

The MCF-7 cells were incubated with the as-prepared NPs and NPs-AlPcS$_4$ nano-complexes for 24 hours in the absence of a NIR source. The cell viability was observed as 93.2% using 75 mg/mL concentration of NPs. The cell viability of MCF-7 cells via incubation with NPs-AlPcS$_4$ nano-complexes was decreased by about 5%, which is due to the coupling of the AlPcS$_4$ photosensitizer. The as-prepared PEG-coated NPs and NPs-AlPcS$_4$ nano-complexes with low cytotoxicity are useful for *in vitro* and *in vivo* bio-imaging. Under 980 nm illuminations for 1, 3, and 5 minutes, the cell viability of the control cells was found to decrease less. When the MCF-7 cells were incubated with NPs the cell viability has decreased from 90 to 83%, whereas for AlPcS$_4$ it was from 97 to 85%, respectively. The cell viability of the control

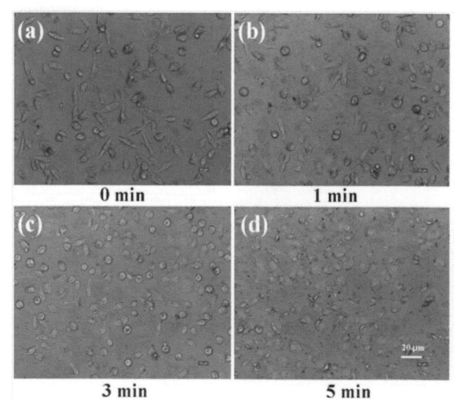

Fig. 13.5. Photomicrographs of MCF-7 cells incubated with NPs-AlPcS$_4$ nano-complexes pre- and post-NIR irradiation at different time intervals (i.e. 0, 1, 3 and 5 minutes). (Reproduced with permission from Zeng et al. 2013 copyright @ 2013 RSC)

MCF-7 cells was found to be 97% compared to the cells without irradiation. The cells incubated with NPs-AlPcS$_4$ nano-complexes have a selectively killing ability under 980 nm excitations. This material could be a potential PDT candidate for cancer therapy (Zeng et al. 2013).

The killed cells were characterized by a bio-microscope to see the changes in morphology of the cells. The images of MCF-7 cells incubated with NPs-AlPcS$_4$ nano-complexes under pre- and post-NIR light irradiation for different time intervals (i.e. 0, 1, 3 and 5 minutes) are shown in Fig. 13.5. It is evident from Fig. 13.5 (a) to Fig. 13.5(d) that a considerable change in the morphology of MCF-7 cells has been observed. Upon 980 nm laser illumination for 3 and 5 minutes, it has been observed that the shape of a large number of cells changes from spindle to circular and the boundary of cells was observed in fuzzy for 5 minutes irradiation. Thus, the multifunctional photosensitizer-conjugated core-shell Fe$_3$O$_4$@NaYF$_4$:Yb^{3+}/Er^{3+} nano-complexes have wide applications in various fields such as photonics, solar cells, MRI, UCL imaging, PDT, clinical diagnosis and treatment of cancers.

Conclusions

In the present chapter, the progress of novel fluoride based host has been extensively described. The controlled synthesis techniques and possible mechanisms of the particle growth have been discussed for both cubic and hexagonal phases of rare earth doped NaYF$_4$ UC phosphors. Among all the reported synthesis methods the thermal decomposition, hydrothermal and co-precipitation preparation techniques were explicitly illustrated. In terms of an *in vitro* imaging system, the surface modification of fluoride hosts with different organic moieties and core cells of Fe$_3$O$_4$ have been shown to detect the MCF-7 cells. Thus the selected host matrices are suitable agents for medical and clinical practitioners.

References

Boyer, J.C., Vetrone, F., Cuccia, L.A., Capobianco, J.A. 2006. J. Ameri. Chem. Soc. 128, 7444-7445.

Cao, T., Yang, Y., Gao, Y., Zhou, J., Li, Z., Li, F. 2011. Biomaterial. 32, 2959-2968.

Chen, G., Ågren, H., Ohulchanskyy, T.Y., Prasad, P.N. 2015. Chem. Soc. Rev. 44, 1680-1713.

Chen, G., Ohulchanskyy, T.Y., Kumar, R., Ågren, H., Prasad, P.N. 2010. ACS Nano. 4, 3163-3168.

Chen, G., Ohulchanskyy, T.Y., Liu, S., Law, W.C., Wu, F., Swihart, M.T., Prasad, P.N. 2012. ACS Nano. 6, 2969-2977.

Chen, J., Zhao, J.X. 2012. Sensor. 12, 2414-2435.

Cui, Y., Hegde, R.S., Phang, I.Y., Lee, H.K., Ling, X.Y. 2014. Nanoscal. 6, 282-288.

DaCosta, M.V., Doughan, S., Han, Y., Krull, U.J. 2014. Analyt. Chim. Acta. 832, 1-33.

Dou, Q., Zhang, Y. 2011. Langmuir. 27, 13236-13241.

Gainer, C.F., Romanowski, M. 2014. J. Innovat. Opt. Health Sci. 7, 1330007.
Haase, M., Schäfer, H. 2011. Angew. Chem. Int. Edi. 50, 5808-5829.
He, F., Niu, N., Wang, L., Xu, J., Wang, Y., Yang, G., Yang, P. 2013. Dalton Transaction 42, 10019-10028.
Ivanova, S., Pellé, F., Tkachuk, A., Joubert, M.F., Guyot, Y., Gapontzev, V.P. 2008. J. Lumines. 128, 914-917.
Kumar, A., Couto, M.H., Tiwari, S.P., Kumar, K., Esteves da Silva, J.C. 2018. Chemistry Select. 3, 10566-10573.
Kumar, A., da Silva, J.C.E., Kumar, K., Swart, H.C., Maurya, S.K., Kumar, P., Tiwari, S.P. 2019. Mater. Res. Bullet. 112, 28-37.
Kumar, A., Tiwari, S.P., da Silva, J.C.E., Kumar, K. 2018. Laser Phys. Lett. 15, 075901.
Kumar, A., Tiwari, S.P., Krishna, K.M., Kumar, K. 2016. In AIP Conference Proceedings 1731, 2016, May, 050135. AIP Publishing. doi.org/10.1063/1.4947789.
Kumar, A., Tiwari, S.P., Kumar, K. 2015. Adv. Sci. Lett. 21, 2632-2634.
Kumar, A., Tiwari, S.P., Kumar, K., da Silva, J.C.E. 2019. J. Alloy. Comp. 776, 207-214.
Kumar, A., Tiwari, S.P., Kumar, K., Rai, V.K. 2016. Spectrochimica Acta Part A: Molec. Biomolecular Spectr. 167, 134-141.
Kumar, A., Tiwari, S.P., Sardar, A., Kumar, K., da Silva, J.C.E. 2018. Sens. Actuator. A: Phys. 280, 179-187.
Kumar, A., Tiwari, S.P., Singh, A.K., Kumar, K. 2016. Appl. Phys. B 122, 190.
Kumar, G.A., Chen, C.W., Ballato, J., Riman, R.E. 2007. Chem. Mater. 19, 1523-1528.
Kumar, R., Nyk, M., Ohulchanskyy, T.Y., Flask, C.A., Prasad, P.N. 2009. Adv. Funct. Mater. 19, 853-859.
LaMer, V.K., Dinegar, R.H. 1950. J. Ameri. Chem. Soc. 72, 4847-4854.
Lemyre, J.L., Ritcey, A.M. 2005. Chem. Mater. 17, 3040-3043.
Li, T., Li, Y., Luo, R., Ning, Z., Zhao, Y., Liu, M., Bi, J. 2018. J. Alloy. Comp. 740, 1204-1214.
Liang, L., Liu, Y., Bu, C., Guo, K., Sun, W., Huang, N., Guo, S. 2013. Adv. Mater. 25, 2174-2180.
Lin, M., Zhao, Y., Wang, S., Liu, M., Duan, Z., Chen, Y., Lu, T. 2012. Biotech. Adv. 30, 1551-1561.
Liu, J., Kaczmarek, A.M., Van Deun, R. 2018. Chem. Soc. Rev. 47, 7225-7238.
Lojpur, V.M., Ahrenkiel, P.S., Dramićanin, M.D. 2013. Nanoscal. Res. Lett. 8, 131.
Mai, H.X., Zhang, Y.W., Si, R., Yan, Z.G., Sun, L.D., You, L.P., Yan, C.H. 2006. J. Ameri. Chem. Soc. 128, 6426-6436.
Mai, H.X., Zhang, Y.W., Sun, L.D., Yan, C.H. 2007. J. Phys. Chem. C 111, 13730-13739.
Maurya, S.K., Kushawaha, R., Tiwari, S.P., Kumar, A., Kumar, K., da Silva, J.C.E. 2019. Mater. Res. Exp. 6, 086211.
Maurya, S.K., Tiwari, S.P., Kumar, A., Kumar, K. 2018. J. Rare Earth 36, 903-910.
Meruga, J.M., Cross, W.M., May, P.S., Luu, Q., Crawford, G.A., Kellar, J.J. 2012. Nano Tech. 23, 395201.
Niu, W., Wu, S., Zhang, S. 2011. J. Mater. Chem. 21, 10894-10902.
Pandey, A., Tiwari, S.P., Dutta, V., Kumar, V. 2018. In Emerg. Synth. Tech. Lumin. Mater. 86-116. IGI Global, doi.org/10.4018/978-1-5225-5170-6.ch004.
Pound, G.M., Mer, V.K.L. 1952. J. Ameri. Chem. Soc. 74, 2323-2332.
Qiu, H., Chen, G., Fan, R., Yang, L., Liu, C., Hao, S., Prasad, P.N. 2014. Nanoscale 6, 753-757.
Ramasamy, P., Kim, J. 2014. Chem. Communicat. 50, 879-881.
Runowski, M., Ekner-Grzyb, A., Mrówczyńska, L., Balabhadra, S., Grzyb, T., Paczesny, J., Lis, S. 2014. Langmuir. 30, 9533-9543.
Shan, G.B., Demopoulos, G.P. 2010. Adv. Mater. 22, 4373-4377.

Sun, Y., Yu, M., Liang, S., Zhang, Y., Li, C., Mou, T., Li, F. 2011. Biomaterial. 32, 2999-3007.

Tan, M.C., Kumar, G.A., Riman, R.E., Brik, M.G., Brown, E., Hommerich, U. 2009. J. Appl. Phys. 106, 063118.

Tiwari, S.P., Kumar, A., Kumar, K. 2016. Research Frontiers in Sciences 23-42. ISBN: 978-81-931247-1-0.

Tiwari, S.P., Kumar, A., Singh, S., Kumar, K. 2017. Vacuum 146, 537-541.

Tiwari, S.P., Kumar, K. 2016. In Proceedings of International Conference on Perspectives in Vibrational Spectroscopy 48, ISBN: 978-93-5267-364-3.

Tiwari, S.P., Kumar, K., Rai, V.K. 2015. Appl. Phys. B 121, 221-228.

Tiwari, S.P., Kumar, K., Rai, V.K. 2015. J. Appl. Phys. 118, 183109.

Tiwari, S.P., Maurya, S.K., Yadav, R.S., Kumar, A., Kumar, V., Joubert, M.F., Swart, H.C. 2018. J. Vacu. Sci. Tech. B, Nanotech. Microelectronic.: Material., Proces. Measure. Pheno. 36, 060801. doi.org/10.1116/1.5044596.

Tiwari, S.P., Singh, S., Kumar, A., Kumar, K. 2016. In AIP Conference Proceedings. 1728, May. 020137. AIP Publishing. doi.org/10.1063/1.4946188.

Van Der Ende, B.M., Aarts, L., Meijerink, A. 2009. Adv. Mater. 21, 3073-3077.

Wang, F., Banerjee, D., Liu, Y., Chen, X., Liu, X. 2010. Analyst. 135, 1839-1854.

Wang, F., Liu, X. 2009. Chem. Soc. Rev. 38, 976-989.

Wang, G., Peng, Q., Li, Y. 2011. Account. Chem. Res. 44, 322-332.

Wang, L., Li, Y. 2006. Nano Letter 6, 1645-1649.

Wang, M., Zhu, Y., Mao, C. 2015. Langmuir. 31, 7084-7090.

Wang, X., Li, Y. 2007. Chem. Communicat. 28, 2901-2910.

Wang, X., Zhuang, J., Peng, Q., Li, Y. 2005. Nature 437, 121.

Wang, Z., Meijerink, A. 2018. J. Phys. Chem. Lett. 9, 1522-1526.

Xing, H., Bu, W., Zhang, S., Zheng, X., Li, M., Chen, F., Shi, J. 2012. Biomaterial 33, 1079-1089.

Xu, X., Wang, Z., Lei, P., Yu, Y., Yao, S., Song, S., Zhang, H. 2015. ACS Appl. Mater. Inter. 7, 20813-20819.

Xu, Z., Li, C., Yang, P., Zhang, C., Huang, S., Lin, J. 2009. Crystal Growth Design 9, 4752-4758.

Yadav, R.S., Dhoble, S.J., Rai, S.B. 2018. New J. Chem. 42, 7272-7282.

Yadav, R.S., Dhoble, S.J., Rai, S.B. 2018. Sensor. Act. B: Chem. 273, 1425-1434.

Yadav, R.S., Rai, S.B. 2017. J. Alloy. Comp. 700, 228-237.

Yadav, R.S., Rai, S.B. 2017. J. Phy. Chem. Solid. 110, 211-217.

Yadav, R.S., Rai, S.B. 2018. J. Phys. Chem. Solid. 114, 179-186.

Yadav, R.S., Verma, R.K., Bahadur, A., Rai, S.B. 2015. Spectrochimica Acta Part A: Molec. Biomolecular Spectr. 137, 357-362.

Yadav, R.S., Verma, R.K., Rai, S.B. 2013. J. Phys. D: Appl. Phys. 46, 275101.

Yadav, R.S., Yadav, R.V., Bahadur, A., Rai, S.B. 2016. RSC Adv. 6, 51768-51776.

Yadav, R.S., Yadav, R.V., Bahadur, A., Yadav, T.P., Rai, S.B. 2016. Mater. Res. Exp. 3, 036201.

Yang, L.W., Li, Y., Li, Y.C., Li, J.J., Hao, J.H., Zhong, J.X., Chu, P.K. 2012. J. Mater. Chem. 22, 2254-2262.

Yang, Y., Sun, Y., Cao, T., Peng, J., Liu, Y., Wu, Y., Li, F. 2013. Biomaterial 34, 774-783.

Ye, X., Collins, J.E., Kang, Y., Chen, J., Chen, D.T., Yodh, A.G., Murray, C.B. 2010. Proceedings of the National Academy of Sciences, 107, 22430-22435, doi. org/10.1073/pnas.1008958107.

Yi, G.S., Chow, G.M. 2006. Adv. Funct. Mater. 16, 2324-2329.

Yi, G., Lu, H., Zhao, S., Ge, Y., Yang, W., Chen, D., Guo, L.H. 2004. Nano Letter 4, 2191-2196.

Yin, A., Zhang, Y., Sun, L., Yan, C. 2010. Nanoscale 2, 953-959.

Zeng, L., Xiang, L., Ren, W., Zheng, J., Li, T., Chen, B., Wu, A. 2013. RSC Adv. 3, 13915-13925.

Zeng, S., Ren, G., Xu, C., Yang, Q. 2011. Cryst. Eng. Comm. 13, 1384-1390.

Zhang, F., Wan, Y., Yu, T., Zhang, F., Shi, Y., Xie, S., Zhao, D. 2007. Angewandte Chemie International Edition. 46, 7976-7979. doi.org/10.1002/anie.200702519

Zhang, F., Zhao, D. 2008. ACS Nano. 3, 159-164.

Zhao, S., Xia, D., Zhao, R., Zhu, H., Zhu, Y., Xiong, Y., Wang, Y. 2016. Nanotechnology 28, 015601.

Preparation of Thin Films Using Electron Beam Evaporation Deposition Method

D.S. Kshatri* and Shubhra Mishra

Department of Applied Physics, Shri Shankaracharya Institute of Professional Management and Technology, Raipur - 492015, India

Introduction

"A thin film is a layer of material ranging from fractions of a nanometer (monolayer) to several micrometers in thickness." Many of the electronic semiconductor devices are the main applications benefiting from thin film construction. The semiconducting materials, in thin film form, are of particular interest because they have a number of applications viz. transparent electrodes, photovoltaic devices, solar front panel display, surface acoustic wave devices, low emissivity coating for architectural glass, various gas sensors and heat reflectors for advanced gazing in solar cells. Due to surface and interface effects, properties of thin films differ considerably from those of bulk, and this dominates overall behavior of the thin films. Thin films play an important role in the nanotechnology and nanoscience development.

Thin Film Deposition Techniques

A solid material is said to be in thin film structure when it is developed as a thin layer on a solid substrate by controlled condensation of the individual atomic, molecular, or ionic species either by a physical process or ultra chemical reactions. There are dozens of deposition techniques for material formation (Maissel and Clang 1970; Vossen and Kern 1978; Bunshah 1982 and Ghandhi 1983).

Growth of thin films, as all phase transformation, involves the processes of nucleation and growth on the substrate or growth surfaces. The nucleation

---

*Corresponding author: dheer2713@gmail.com

process plays a very important role in determining the crystallinity and microstructure of the resultant films. For the deposition of thin films with thickness in the nanometer region, the initial nucleation process is even more important. In practice, the interaction between the film or nuclei and the substrate plays a very important role in determining the initial nucleation and the film growth. Many experimental observations revealed that there are three basic nucleation modes (Cao, 2004):

(1) Island or Volmer-Weber growth,
(2) Layer or Frank-van der Merwe growth and
(3) Island-layer or Stranski-Krastonov growth

Figure 14.1 illustrates three basic modes of initial nucleation in the film growth. Island growth occurs when the growth species are more strongly bonded to each other than to the substrate. Subsequent growth results in the island to coalesce to form a continuous film. The layer growth is the opposite of the island growth, where growth species are equally bound more strongly to the substrate than to each other. The most important examples of layer growth mode are the epitaxial growth of single crystal films. The island-layer growth is an intermediate combination of layer growth and island growth.

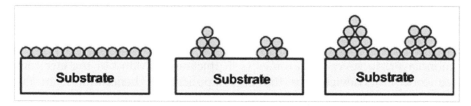

Fig. 14.1. Three basic modes of nucleation.

Practically, thin film deposition techniques are either purely physical or purely chemical. The two most important categories are (i) Physical Vapor Deposition (PVD), and (ii) Chemical Vapor Deposition (CVD). It is clear, that for each deposition technique, appropriate coating materials are required. The PVD process normally makes use of inorganic elements or compounds and gases, whereas the CVD process uses dip coating and spinning, liquid inorganic and organic compounds and gases. There are several techniques that are used to develop thin films of metals and semiconductors in both micro and nano-structures. Some of these techniques are: sol-gel process, chemical deposition method, hydrothermal method, pyrolysis method, chemical vapor deposition, and electro deposition method. Figure 14.2 shows various thin films deposition techniques.

Phosphor materials deposited in form of thin films provide important advantages over conventional bulk powders for display applications because of their higher resolution, more uniform density and increased thermal stability (Christoulakis et al. 2007). Strontium sulfide (SrS) semiconductor thin films have shown great potential for opto-electronic

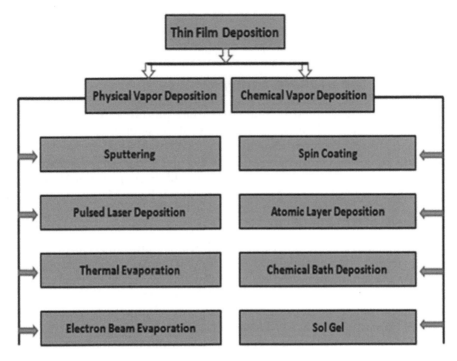

Fig. 14.2. Various thin film deposition techniques.

device fabrication. The rare earths doped SrS used for charge compensation in a blue luminescent phosphor with superior chromaticity as compared to other dopants like Sm, Tb, Pr and Eu, has attracted significant attention due to its potential applications in fabrication of full color electroluminescent (EL) displays (Summers et al. 2000). $SrS:Ce^{3+}$ thin films can also be used to develop opto-electronic devices (Park et al. 2003), solar cells (Fortunato et al. 2009) and civil and military applications (Fangli et al. 2010). $SrS:Ce^{3+}$ thin films have been grown using a variety of deposition techniques, such as, Pulsed Laser Deposition (PLD) (Choe et al. 2002), Radio Frequency (RF) sputtering (Warren et al. 1997), Atomic Layer Epitaxy (ALE) (Heikkinen et al. 1998), Metal Organic Chemical Vapor Deposition (MOCVD), etc. Most research on thin films has been devoted to improving the intensity, color purity and stability of blue component.

In the beginning of the last decade, many research groups including Warren et al. (1997), Barth et al. (2000) and Xu et al. (2000) synthesized SrS:Ce3+ thin films by popular methods like ALE and MOCVD, and reported their structural and optical properties in terms of Electron Paramagnetic Resonance (EPR), X-Ray diffraction, Atomic Force Microscopy (AFM), photoluminescence (PL) and electroluminescence (EL). Later in 2002, Fukada et al. (2002) studied stabilization of bluish-green luminescent Ce^{3+} centers by Rb^{+1} doping in $SrS:Ce^{3+}$ thin film EL devices.

Therefore the structural properties of SrS thin films deposited by Electron Beam Evaporation Deposition (EBED) method are considered here as a case study.

Outline of the Present Study

This chapter presents the structural and surface morphology of SrS thin film deposited by the EBED method at a substrate temperature of 250°C. The phase and structural properties of as-deposited SrS thin film are investigated by means of X-Ray Diffraction (XRD), while the surface morphologies are examined by Field Emission Scanning Electron Microscopy (FESEM), Energy Dispersive X- Ray Spectroscopy (EDX) and Atomic Force Microscopy (AFM).

Synthesis of SrS Thin Film

Phosphor materials, when deposited in form of thin films offer important advantages over conventional bulk powders for display applications because of higher resolution, more uniform density and increased thermal stability (Christoulakis et al. 2007). Ce^{3+} doped SrS is used for charge compensation in a blue luminescent phosphor with superior chromaticity as compared to other dopants like Sm, Tb, Pr and Eu, which have attracted significant attention recently due to their potential applications in the fabrication of full color EL displays (Summers et al. 2000). $SrS:Ce^{3+}$ thin films can be used to develop optoelectronic devices (Park et al. 2003; Yang et al. 2008), solar cells (Fortunato et al. 2009), and civil and military applications (Fangli et al. 2010). $SrS:Ce3^{+}$ thin films have been grown using a variety of deposition techniques, such as, Pulsed Laser Deposition (PLD) (Karner et al. 1997; Choe et al. 2002), Radio Frequency (RF) sputtering (Warren et al. 1997), Atomic Layer Epitaxy (ALE) (Heikkinen et al. 1998), etc. Most research on thin films has been devoted to improving their intensity, color purity and stability of the blue component.

Electron Beam Evaporation Deposition Method

Electron Beam Evaporation Deposition (EBED) is a method of using an electron beam generated from an electron source in a vacuum to irradiate an evaporant material, and heating and evaporating it so that the evaporated material forms a thin film on a substrate, generally glass.

A beam of electrons can be fixated on the sample, and by using voltage differences linking the filament emitting the electrons and the water-cooled hearth holding the sample, electrons can be passed through it causing heating and eventual vaporization as shown in Fig. 14.3. The main advantages of this method are that it can be scaled up willingly and, more importantly, high-boiling metals such as Pt, Rh, Mo, W, and U can be vaporized without materials problems, since the hearth can be kept cool.

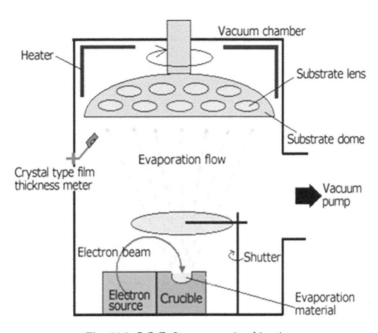

Fig. 14.3. SrS:Ce3⁺ evaporation kinetics.

The evaporation technique is compatible for non-refractory materials that vaporize at a temperature less than 1400 °C for thermal sources and less than 2200 °C for electron-beam sources. Typically, evaporation can generate good film stoichiometry for elements and simple compounds. Specifically, the evaporation technique has difficulty producing good films of intricate phosphors like rare earth oxysulfides because of the wide components of these compounds. On the other hand, the classical II-VI compounds form excellent films by evaporation (Ono 1995).

The kinetics of the evaporation, material conveys and film condensations are shown in Fig. 14.3. The individual atomic species are conveyed to the substrate in a line of sight trajectory provided that the pressure level is low enough to permit collisionless transport. The atoms then recombine on the substrate to form a II-VI compound. A feature of this recombination at the substrate is that it can be controlled to produce very stoichiometric films by adjusting the substrate temperature. The mechanism here is that the vapor pressure of the ingredient atoms, for example Sr and S, is largely sufficient for high substrate temperatures that neither Sr nor S atoms will hold to other similar atoms. Thus, film growth proceeds by arrangement of alternate layers of Sr and S atoms and stoichiometry is automatically achieved. The simple resistive heating of an evaporation source suffers from the disadvantages of possible contamination from the support material and the limitations of input potency, which make it difficult to evaporate high melting point materials (Anila 2008). These drawbacks are overcome by a capable source of heating by electron attack of the material. In principle, this type of source is capable of evaporating any material at rates ranging from fractions of an angstrom to microns per second. Thermal decomposition and structural changes of some chemical compounds may occur because of the intense heat and energetic electron bombardment.

Figure 14.4 shows a HINDHIVAC coating unit (model no.12A4D - T) for depositing SrS thin films. In this unit, the material to be deposited is kept in a boat shaped molybdenum sheet and tied to the two ends of copper electrodes. The electron gun (Fig. 14.5) fitted in the vacuum chamber consists of a filament, anode, permanent magnet, etc. It is operated by a high voltage power supply. During operation, the filament is heated by the low tension until it becomes incandescent and then emits electrons spontaneously and randomly. The anode plate then collects the electrons and forms them into a beam, which is accelerated through the high voltage potential.

Fig. 14.4. Electron beam evaporation deposition HINDHIVAC coating unit (Model no.12A4D - T).

Sample Preparation

The SrS nanophosphor powders for thin film deposition are synthesized by SSDM, as already discussed earlier (Mishra et al. 2015). The SrS nanophosphors so obtained are then deposited on ITO coated glass substrates using the EBED method to prepare thin films. Prior to deposition,

Fig. 14.5. Top view of an electron beam gun fixed in a vacuum coating unit.

the glass slides were cut into approximately 1.0 square inch pieces. First, the substrates are cleaned with a soap solution using a brush and then rinsed by deionized (DI) water. The substrates are then ultrasonicated in acetone for 15 minutes in an ultrasonic cleaner. Further, the ultrasonicated substrates are degreased in methanol in a degreasing unit. Then, the substrates are boiled in $DI:H_2O_2:NH_3$: 5:1:1 solution for 30 minutes and kept in DI water for 6 hours and dried in N_2 gas. The distance from the target to substrate is fixed at 6.0 cm. The deposition is done at two different pressures, 4.2×10^{-5} Torr and 6.4×10^{-6} Torr at substrate temperature of 250 °C. The substrate holder is rotated at a frequency ~11 Hz during the deposition to produce films of uniform thickness. All the samples are stored in complete darkness and in a clean and moisture-free [glove box (VAC, USA)] environment to prevent their possible oxidation and/or exposure. Figure 14.6 shows the thin film of nanophosphors.

Characterization

The pure SrS thin film sample prepared by the EBED method is characterized by both conventional and advanced characterization techniques such as XRD, FESEM, EDX and AFM.

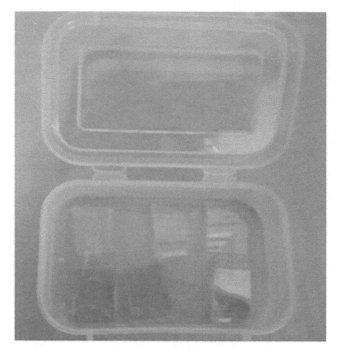

Fig. 14.6. Thin films.

X-ray Diffraction

The phase and structural properties of as-deposited films are confirmed by X-Ray Diffraction (XRD) measurement using PANalytical X-ray diffractometer. The XRD diffractogram corresponding to pure SrS thin film is shown in Fig. 14.7. The crystallographic phase and particle size of this film are examined using this technique in which $CuK_{\alpha 1}$ ($\lambda = 1.54$ Å) component of X-rays is applied within the 2θ range of 20°–70°. It is clearly observed that all the films are completely oriented along (111), (200), (220), (311) and (222) planes at 2θ values of 25.5°, 30.5°, 42°, 51° and 53°, respectively. Upon comparison with standard XRD pattern of SrS (PDF Card No: 03-065-4281), the rock salt cubic structure of SrS is confirmed. The average particle size calculated using Scherrer's formula is found to be nearly 27 nm (Mishra et al. 2016).

Field Emission Scanning Electron Microscopy

The surface morphology of pure SrS thin film deposited at 250 °C is presented in Fig. 14.8. From the FESEM image, it can be clearly seen that the pure SrS sample has nearly similar morphologies. Basically, grain growth in thin films is different from grain growth in bulk materials in several important ways (Moelans et al. 2007). The most important reason behind this is the fact that the interface of the film with the glass substrate and the top surface

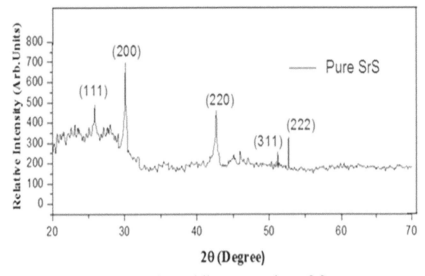

Fig. 14.7. X-ray diffractogram of pure SrS.

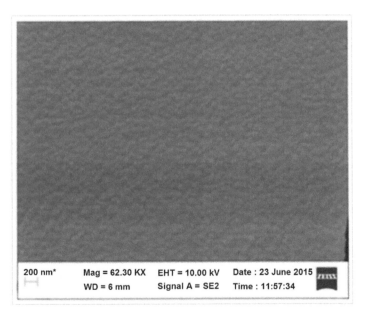

Fig. 14.8. FESEM image of pure SrS thin film.

of the film play an important role in suppressing normal growth of grains and encouraging secondary or abnormal grain growth; due to this reason anisotropic behavior of the energy of the surfaces or interfaces are exhibited. Secondary grains generally restrict crystallographic orientations that are affected by the atomic arrangement of the substrate interface as well as the environment of the top surface of the film (Thompson 1990).

It is also observed from figures that the granular surface of pure SrS thin film is smooth, uniformly dense and with tiny structures with average diameter ~30 nm is obtained with uniform distribution of grains sizing between 25 nm and 35 nm (Mishra et al. 2016).

Energy Dispersive X-ray Spectroscopy

The elemental compositions of pure SrS thin films determined through energy dispersive X-rays spectroscopic analysis are shown in Fig. 14.9. These results verify the presence of only strontium (Sr) and sulfur (S) for pure SrS. The corresponding atomic percentages are presented in Table 14.1.

Table 14.1. Atomic percent of elements in pure SrS thin film

Elements	Atomic % of pure SrS
Sr	45.47
S	54.53

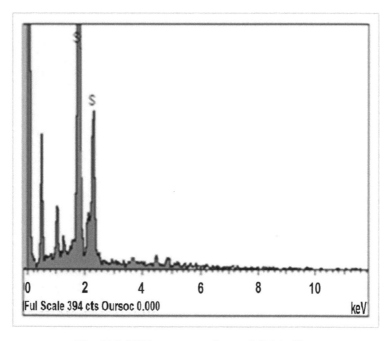

Fig. 14.9. EDX spectrum of pure SrS thin film.

Atomic Force Microscopy (AFM)

The two and three dimensional AFM microstructures and roughness profiles of pure SrS thin film are shown in Fig. 14.10. The continuous and smooth surfaces of as-deposited thin film are clearly observed, which facilitate to curtail the number of conducting paths between the absorption

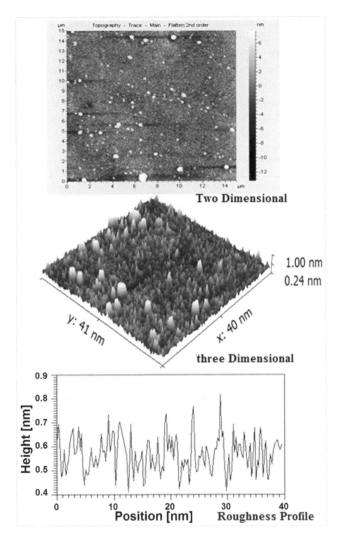

Fig. 14.10. Two and three dimensional AFM microstructures and roughness profiles of pure SrS thin film.

layer and electrode. The average particle size of this thin film estimated using Gwyddion AFM software is found to be ~16 nm, which is smaller in comparison to the size estimated by Scherrer's formula (~27 nm) and FESEM (~30 nm). The larger size of particles in case of XRD analysis may be due to strain or presence of some artefacts, which can be observed as white spots (Bushroa et al. 2012) in the two dimensional figure of thin film. Two types of topographical information are witnessed by the observation of two and three dimensional AFM images shown in Fig. 14.10. First, the film exhibits regular island like structures, which are uninfluenced by the increasing Ce^{3+} ions concentration. This observation is firmly supported by roughness profile of the corresponding film, also shown in Fig. 14.10. Second, the low value average roughness, the Root Mean Square (RMS) roughness, maximum height of roughness, maximum roughness valley depth and maximum roughness peak height show how smooth film is prepared and can be observed by the data presented in Table 14.2. To reduce the reflection loss due to roughness induced surface scattering, the surface smoothness is a highly desired parameter in opto-electronic applications (Faraj et al. 2010).

Table 14.2. Parameters obtained from AFM microstructure of pure SrS thin film

Parameters	SrS thin film
Crystallite size (nm)	~16
Average roughness (nm)	0.041
RMS roughness (nm)	0.032
Maximum height of roughness (nm)	0.401
Maximum roughness valley depth (nm)	0.92
Maximum roughness peak height (nm)	0.101

Conclusions

Pure SrS thin films are successfully deposited on glass substrates by the EBED method. The structural and optical properties of as-deposited thin film are inter-related to each other. XRD analysis reveals the cubic structure of deposited films with average grain size ~27 nm. The FESEM images illustrate smooth, uniformly dense and tiny structures of deposited films. The EDX profiles of different phosphor samples confirm the presence of main constituents, strontium, sulfur and cerium, in the thin film. The average particle size of thin films estimated using Gwyddion AFM software is found to be ~16 nm.

The SrS thin films doped with different rare earths ions can be selected for further research in alternating current thin film electroluminescent (ACTFEL) device fabrication and analysis.

References

Anila, E.I., Arun, M.K., Jayaraj, A. 2008. The photoluminescence of SrS: Cu nanophosphor. Nanotechnology 19, 145604.
Barth, K.W., Lau, J.E., Peterson, G.G., Endisch, D., Kaloyeros, A.E., Tuenge, R.T., King, C.N. 2000. Metallorganic chemical vapor deposition of SrS:Ce for thin film electroluminescent device applications. J. Electrochem. Soc. 147, 2174.
Bunshah, R.F. 1982. Deposition Technologies for Films and Coatings: Developments and Applications. Noyes Publications, Park Ridge, NJ.
Bushroa, A.R., Rahbari, R.G., Masjuki, H.H., Muhamad, M.R. 2012. Approximation of crystallite size and microstrain via XRD line broadening analysis in TiSiN thin films. Vaccume 86, 1107.

Cao, G. 2004. Nanostructures and Nanomaterials: Synthesis, Properties and Applications. Imperial College Press, London.

Choe, J.Y., Blomquist, S.M., Morton, D.C. 2002. Blue- and green-emitting SrS:Cu electroluminescent devices deposited by the atomic layer deposition technique. Appl. Phys. Lett. 80, 4124.

Christoulakis, S., Suchea, M., Katsarakis, N., Koudoumas, E. 2007. Substrate temperature effect on the structural and photoluminescence properties of $(Y-Gd)_3Al_5O_{12}:Ce^{3+}$ thin films prepared by pulsed laser deposition. Appl. Surf. Sci. 253, 8169.

Fangli, D., Ning, W., Dongmei, Z., Yingzhong, S. 2010. Europium and samarium doped calcium sulfide thin films grown by PLD. J. Rare Earth 28, 391.

Faraj, M.G., Ibrahim, K., Eisa, M.H., Pakhuruddin, M.K.M., Pakhuruddin, M.Z. 2010. Comparison of zinc oxide thin films deposited on the glass and polyethylene terephthalate substrates by thermal evaporation technique for applications in solar cells. Optoelectron. Adv. Mater. 4, 1587.

Fortunato, E., Goncalves, A., Pimentel, A., Barquinha, P., Goncalves, G., Pereira, L., Ferreira, I., Martins, R. 2009. Zinc oxide, a multifunctional material: From material to device applications. Appl. Phys. A Mater. 96, 197.

Fukada, H., Sasakura, A., Sugio, Y., Kimura, T., Ohmi, K., Tanaka, S., Kobayashi, H. 2002. Stabilization of bluish-green luminescent Ce^{3+} centers by Rb doping in SrS: Ce thin film electroluminescent devices. Jpn. J. Appl. Phys. 41, L941.

Ghandhi, S.K. 1983. VLSI Fabrication Principles. John Wiley & Sons, New York.

Heikkinen, H., Johansson, L.S., Nykänen, E., Niinistö, L. 1998. An XPS study of SrS:Ce thin films for electroluminescent devices. Appl. Surf. Sci. 133, 205.

Karner, C., Maguire, P., Mc Laughlin, J., Laverty, S. 1997. Electron microscopic studies of internal gettering of nickel in silicon. Philos. Mag. Lett. 76, 111.

Maissel, L.I., Clang, R. 1970. Handbook of Thin Film Technology. McGraw-Hill, New York.

Mishra, S., Khare, A., Kshatri, D.S., Tiwari, S. 2015. Structural and optical properties of SrS nanophosphors influenced by Ce^{3+} ions concentrations and particle size reduction. Superlattices Microstruct. 86, 73.

Mishra, S., Kshatri, D.S., Khare, A. Tiwari, S., Dwivedi, P.K. 2016. SrS:Ce^{3+} thin films for electroluminescence device applications deposited by electron-beam evaporation deposition method. Mat. Lett. 183, 191.

Moelans, N., Blanpain, B., Wollants, P. 2007. Pinning effect of second-phase particles on grain growth in polycrystalline films studied by 3-D phase field simulations. Acta. Mat. 55, 2173.

Park, W.I., Yi, G.C., Kim, J.W., Park, S.M. 2003. Schottky nanocontacts on ZnO nanorod arrays. Appl. Phys. Lett. 82, 4358.

Ono, Y.A. 1995. Electroluminescent Displays. Singapore: World Scientific Publishing Co.

Summers, C.J., Wagner, B.K., Tong, W., Park, W., Chaichimansour, M., Xin, Y.B. 2000. Recent progress in the development of full color SrS-based electroluminescent phosphors. J. Cryst. Grow 214, 918.

Thompson, C.V. 1990. Grain growth in thin films. Annu. Rev. Mater. Sci. 20, 245.

Vossen, J.L., Kern, W. 1978. Thin Film Processes. Academic Press, New York.

Warren, W.L., Vanheusden, K., Tallant, D.R., Seager, C.H., Sun, S.-S., Evans, D.R., Dennis, W.M., Soininen, E., Bullington, J.A. 1997. Atomic level stress and light emission of Ce activated SrS thin films. J. Appl. Phys. 82, 1812.

Warren, W.L., Seager, C.H., Sun, S.S., Naman, A., Holloway, P.H., Jones, K.S., Soininen, E. 1997. Microstructure and atomic effects on the electroluminescent efficiency of thin film devices. J. Appl. Phys. 82, 5138.

Xu, C., Cui, Y., Xu, X. 2000. Electronic transport in thin film electroluminescence of SrS:Ce. J. Appl. Phys. 88, 4623.

Yang, J., Gao, M., Yang, L., Zhang, Y., Lang, J., Wang, D., Wang, Y., Liu, H., Fan, H. 2008. Low-temperature growth and optical properties of Ce-doped ZnO nanorods. Appl. Sur. Sci. 255, 2646.

Index

About the Editors

Dr. Vikas Dubey obtained his Masters degree in Physics from the Govt. V.Y.T.PG. Auto. College Durg, India.. He completed his PhD from National Institute of Technology Raipur. He is currently working as Assistant Professor in the department of Physics Bhilai Institute of Technology Raipur, India. During the last ten years of his research career, he has published more than 100 research papers in various reputed international journals, with more than 1500 citations and h-index of 28 and i10 index 48. He has published 6 authored books and 4 edited books and several book chapters in several excellent books. He is editor of various reputed journals and edited some special issue in various journals. Completed two minor research project and two projects are under progress.

Dr. Sudipta Som is working as a Postdoctoral Researcher in the Department of Chemical Engineering at National Taiwan University, Taiwan. During the last ten years of his research career, he has published more than 90 research papers in various reputed international journals, with more than 2600 citations and h-index of 27. He has published one book and several book chapters in several excellent books. He has edited Virtual Special Issues of VACUUM and Materials Today: Proceedings as a guest editor.

As an independent researcher, he has skilled in chemical lab work and different characterization techniques, especially for defects in metal oxides. He has admirable skills in the synthesis of materials in 0D, 1D, 2D, and 3D forms with various shapes and sizes via different physical and chemical synthesis routes. He has in-depth knowledge of materials physics. His speciality is photoluminescence spectroscopy and its applicability towards display device application, LED fabrication, fingerprint detection, barcode detection, and security ink applications. Presently he is involved with the fabrication of flexible white light-emitting diodes based on inorganic materials and development of Mini/ Micro- LEDs with the help of perovskite quantum dots.

Dr. Vijay Kumar is presently working as Assistant Professor, Department of Physics, National Institute of Technology Srinagar (J&K), India. He received his Ph.D. in Physics at the beginning of 2013 from the SLIET Longowal. From

2013 to 2015, he was a postdoctoral research fellow in the Phosphor Research Group, Department of Physics, University of the Free State, South Africa. He has received the Young Scientist Award under the fast track scheme of the Department of Science and Technology (Ministry of Science and Technology, Government of India), New Delhi. He has been a nominated member of the Scientific Advisory Committee for Initiative for Research and Innovation in Science (IRIS) by DST. He is currently engaged in the research of functional materials, solid-state luminescent materials, nanophosphors, ion beam analysis, smart polymers, biodegradable composites, and biomedical applications. He has more than 70 research papers in international peer-reviewed journals, 10 peer-reviewed conference proceedings, 4 book chapters and edited 3 books. He has more than 2750 citations with a h-index of 30.

T - #0399 - 071024 - C282 - 234/156/13 - PB - 9780367541170 - Gloss Lamination